Construction
Supply Chain
Management
HANDBOOK

Construction
Supply Chain
Management
HANDBOOK

Edited by
William J. O'Brien
Carlos T. Formoso
Ruben Vrijhoef
Kerry A. London

CRC Press
Taylor & Francis Group
Boca Raton London New York

CRC Press is an imprint of the
Taylor & Francis Group, an **informa** business

CRC Press
Taylor & Francis Group
6000 Broken Sound Parkway NW, Suite 300
Boca Raton, FL 33487-2742

© 2009 by Taylor & Francis Group, LLC
CRC Press is an imprint of Taylor & Francis Group, an Informa business

No claim to original U.S. Government works
Printed in the United States of America on acid-free paper
10 9 8 7 6 5 4 3 2 1

International Standard Book Number-13: 978-1-4200-4745-5 (Hardcover)

**Visit the Taylor & Francis Web site at
http://www.taylorandfrancis.com**

**and the CRC Press Web site at
http://www.crcpress.com**

Contents

SECTION II   Organizational Perspectives

SECTION III   Information Technology

Authors

Dr. Carlos Formoso has a PhD awarded by the University of Salford, UK. He is an associate professor at the Department of Civil Engineering of the Federal University of Rio Grande do Sul (UFRGS), where he has been teaching both undergraduate and graduate courses, and conducting research in the area of construction management since 1991. He is the head of the Construction Management Research Group at UFRGS, leading a team of around 25 researchers. His research is concerned with the design and management of production systems in the construction industry, including themes such as production control, organizational learning, safety management, performance measurement, and product development. He has led several projects in partnership with the industry in the last 15 years, most of them funded both by the Brazilian Government and private companies. He has been a very active member of the International Group for Lean Construction since 1996. He was a visiting scholar at the University of California, Berkeley, USA, between October 1999 and September 2000.

Dr. Kerry London is an associate professor in the School of Architecture and Built Environment in the Faculty of Engineering and Built Environment at the University of Newcastle, Newcastle, Australia and received her PhD from the University of Melbourne. Dr. London joined Newcastle in 2001 and teaches undergraduate and postgraduate students in both architecture and construction management programs. She is assistant dean for Community Engagement. She is director of the Centre for Interdisciplinary Built Environment Research [CIBER] and has published widely on the construction supply chain concept. She was a visiting scholar at Lund University, Sweden in 2000 and at the University of Reading, UK in 2006. Dr. London is also the author of *Construction Supply Chain Economics*, and her construction supply research areas include chain economics, capital infrastructure policy and management analysis, and information flow and supply chain sustainability.

William J. O'Brien, PE, PhD, focuses his professional efforts on improving collaboration and coordination among firms in the design and construction industry. Dr. O'Brien is an expert on construction supply chain management and electronic collaboration. He conducts research, teaches, and consults on both systems design and implementation issues. Dr. O'Brien is currently an assistant professor in the Department of Civil,

Architectural, and Environmental Engineering in the Cockrell School of Engineering at the University of Texas at Austin. From 1999 to 2004, he was a faculty member at the University of Florida (Gainesville). Prior to returning to academia, Dr. O'Brien led product development and planning efforts at Collaborative Structures, a Boston-based Internet start-up focused on serving the construction industry. Dr. O'Brien holds a PhD and an MS degree in Civil Engineering and an MS degree in Engineering-Economic Systems from Stanford University. He also holds a BS degree in Civil Engineering from Columbia University.

Ruben Vrijhoef graduated from the Faculty of Civil Engineering of the Delft University of Technology in the Netherlands in 1998. From 1997 to 1998, he worked at VTT Technical Research Centre of Finland as part of his graduate studies. From 1998 to 2007, he worked as a research consultant in the area of construction process innovations at TNO Built Environment and Geosciences in The Netherlands. In addition, from 1998 to 2000, he worked on a large innovation program with the Dutch leading construction firm HBG (currently Royal BAM). In 2007, he took a position in the Design and Construction Processes research group at the Faculty of Civil Engineering of the Delft University of Technology in the Netherlands. He has been an active contributor to the PSIBouw national innovation program for the Dutch construction industry, and an editor for the Dutch construction journal "Building Innovation." His research interests include construction management, supply chain management and integration, lean construction, and comparative studies with other industries.

Contributors

Burcu Akinci
Department of Civil
  and Environmental
  Engineering
Carnegie Mellon University
Pittsburgh, Pennsylvania

Luis F. Alarcón
Department of Construction
  Engineering and
  Management
Pontificia Universidad
  Católica de Chile
Santiago, Chile

C.J. Anumba
Department of
  Architectural
  Engineering
The Pennsylvania State
  University
University Park, Pennsylvania

Roberto J. Arbulú
Strategic Project Solutions
San Francisco, California

Marcelo Azambuja
Department of Civil,
  Architectural and
  Environmental
  Engineering
The University of Texas at
  Austin
Austin, Texas

Glenn Ballard
Department of Civil
  and Environmental
  Engineering
University of California
Berkeley, California

Bart A.G. Bossink
VU University Amsterdam
Amsterdam, The Netherlands

J. Stal-Le Cardinal
Industrial Engineering
  Research Laboratory
Ecole Centrale de Paris
Paris, France

**Anders Kirk
Christoffersen**
NIRAS Consulting
  Engineers
Allerød, Denmark

Rachel Cooper
Lancaster Institute for the
  Contemporary Arts
Lancaster University
Lancaster, England,
  United Kingdom

Andrew Cox
Birmingham Business School
University of Birmingham
Birmingham, England,
  United Kingdom

A.F. Cutting-Decelle
Industrial Engineering
  Research Laboratory
Ecole Centrale de Lille
Lille, France

B.P. Das
Department of Mechanical
  and Manufacturing
  Engineering
Loughborough University
Leicestershire, England,
  United Kingdom

Nashwan Dawood
School of Science and
  Technology
University of Teesside
Middlesbrough, England,
  United Kingdom

Stephen M. Disney
Cardiff Business School
Cardiff University
Cardiff, Wales,
  United Kingdom

Erica Dyson
Trafford General Hospital
Trafford National Health
  Service Trust
Greater Manchester,
  England,
  United Kingdom

Stephen Emmitt
Department of Civil and
   Building Engineering
Loughborough University
Leicestershire, England,
   United Kingdom

Esin Ergen
Department of Civil
   Engineering
Istanbul Technical
   University
Istanbul, Turkey

Carlos T. Formoso
NORIE—Building
   Innovation Reasearch Unit
Federal University of Rio
   Grande do Sul
Porto Alegre, Brazil

**Séverine Strohhecker
(née Hong-Minh)**
Boehringer Ingelheim
   Pharma GmbH & Co.KG
Ingelheim, Germany

Will Hughes
School of Construction
   Management and
   Engineering
University of Reading
Reading, England,
   United Kingdom

Eduardo L. Isatto
NORIE—Building
   Innovation Reasearch Unit
Federal University of Rio
   Grande do Sul
Porto Alegre, Brazil

Richard H.F. Jackson
FIATECH
Austin, Texas

Mike Kagioglou
School of the Built
   Environment
University of Salford
Greater Manchester,
   England, United Kingdom

Philip Kaminsky
Department of Industrial
   Engineering and
   Operations Research
University of California
Berkeley, California

Semiha Kiziltas
Department of Civil
   and Environmental
   Engineering
Carnegie Mellon University
Pittsburgh, Pennsylvania

Kerry London
The University of Newcastle
Callaghan, Australia

Sergio Maturana
Department of Industrial
   and Systems Engineering
Pontificia Universidad
   Católica de Chile
Santiago, Chile

Mohamed M. Naim
Cardiff Business School
Cardiff University
Cardiff, Wales,
   United Kingdom

William J. O'Brien
Department of Civil,
   Architectural and
   Environmental
   Engineering
The University of Texas at
   Austin
Austin, Texas

Anu Pradhan
Department of Civil
   and Environmental
   Engineering
Carnegie Mellon University
Pittsburgh, Pennsylvania

Rafael Sacks
Technion—Israel Institute of
   Technology
Haifa, Israel

Ignacio Schonherr
Pontificia Universidad
   Católica de Chile
Santiago, Chile

Pingbo Tang
Department of Civil
   and Environmental
   Engineering
Carnegie Mellon University
Pittsburgh, Pennsylvania

Iris D. Tommelein
Department of Civil
   and Environmental
   Engineering
University of California
Berkeley, California

Denis R. Towill
Cardiff Business School
Cardiff University
Cardiff, Wales,
   United Kingdom

**Patricia
Tzortzopoulos**
School of the Built
   Environment
University of Salford
Greater Manchester,
   England, United Kingdom

**Kalyanaraman
Vaidyanathan**
i2 Technologies
Cambridge, Massachusetts

Ruben Vrijhoef
Delft University of
   Technology
Delft, The Netherlands

R.I. Young
Department of Mechanical
   and Manufacturing
   Engineering
Loughborough University
Leicestershire, England,
   United Kingdom

1

William J. O'Brien
The University of
Texas at Austin

Carlos T. Formoso
Federal University of
Rio Grande do Sul

Kerry A. London
The University of Newcastle

Ruben Vrijhoef
Delft University
of Technology

Introduction

Construction supply chain management (CSCM) is an emerging area of practice. It is inspired by but differs substantially from manufacturing supply chain management, where the emphasis is on modeling volume production. CSCM is more concerned with the coordination of discrete quantities of materials (and associated specialty engineering services) delivered to specific construction projects. The organization and sourcing of materials is becoming increasingly complex across the global construction industry. Mounting emphasis on CSCM is due to both global sourcing of materials and assemblies provided by advances in transportation technologies as well as a shortage of craft labor (reaching crisis proportions in many parts of the world) that force increasing amounts of value-added work to be conducted off-site deep in the supply chain. At the same time, construction clients are demanding faster, more responsive construction processes and higher-quality facilities. These demands generally involve both more responsive production chains and closer coordination between the owner and the construction team. For all these reasons, effective construction project execution will mean effective CSCM.

Despite increasing interest in CSCM, it remains difficult for both practitioners and academics to educate themselves on current practice and advances. Just as the global construction industry is distributed and diverse, the literature on CSCM is spread across numerous journals, conference proceedings, trade magazines, and specialty books. A range of perspectives on supply chain management exist as research and execution draw from diverse areas such as production and operations management, organizational arrangements, and information technology. As such, there is not a single definition of CSCM, nor is there a single dominant paradigm. Nonetheless, there is mounting evidence of improvements in project performance through taking a supply chain perspective.

The purpose of this volume is to compile in one place an overview of the diverse research and examples of construction supply chain practice around the globe. We hope this book can serve as a useful reference for academics and practitioners who seek to understand and advance project supply chain management. This book takes an interdisciplinary perspective with contributions from leading authors in three major areas: production and operations analysis, organizational perspectives, and information technology.

We are aware that the emerging field of CSCM has aspects that are not covered within this volume, but believe that the main thrusts of research and practice have been surveyed. We have sought to keep a global perspective, inviting contributions from around the world. Each section has both a lead chapter intended to provide a broad overview of the area, and concludes with a commentary on the set of chapters in each section.

Section 1: Production and Operations Analysis

This section surveys the current literature on modeling construction supply chain production and describes a set of approaches and methods for designing and operating project supply chains, with references to design as well as materials production. It provides the basic framework for understanding the challenges and approaches to representing and improving supply chain performance.

Chapter 2, by Marcelo Azambuja and William J. O'Brien, provides a broad review of existing approaches to modeling production in construction supply chains and a characterization of construction supply chains. Based on the gaps identified in the literature, a structured approach for modeling construction supply chains is proposed, which consists of a sequence of steps suggesting metrics, level of detail, elements, and attributes that need to be considered in the modeling process.

Chapter 3, by Carlos T. Formoso and Eduardo Isatto, presents a framework for planning and controlling construction projects, emphasizing those subprocesses that are concerned with the coordination of project supply chains. The main roles of production planning and control systems in the coordination of project supply chains are discussed, based on the theory of coordination and the Language–Action Perspective.

Chapter 4, by Patricia Tzortzopoulos, Mike Kagioglou, Rachel Cooper, and Erica Dyson, discusses the dynamics of supply chains involved in the product development process of construction projects, focusing on its influence on the overall process and more specifically on requirements management. A case study on the delivery of primary healthcare facilities in the UK is presented, emphasizing the need to integrate supply chain members within the overall product development process in order to allow the generation of appropriate value and the reduction of waste.

Chapter 5, by Stephen Emmitt and Anders Kirk Christoffersen, describes a structured approach to establish client values in construction projects. This approach is based on a series of facilitated workshops, which bring key supply chain members together to discuss and agree values. These workshops also promote interpersonal communication and a team ethos among supply chain members throughout the project.

Chapter 6, by Iris D. Tommelein, Glenn Ballard, and Philip Kaminsky, presents a thorough review of supply chain management concepts, adopting a lean construction perspective. According to that perspective, a project becomes a part of a supply chain by design, acquiring goods and services from a combination of preexistent and custom-made supply chains, each providing goods and services to the project customer.

Chapter 7, by Roberto Arbulú, is concerned with development of integrated materials management strategies for different types of materials. Two case studies are presented to illustrate the application of integrated approaches to materials management, the use of *kanban* techniques for managing non-task-specific materials, and the use of real-time materials management and production control systems for managing task-specific materials.

Chapter 8, by Rafael Sacks, discusses the motivations and interests of subcontractors in allocating resources. It presents an economic model that helps to understand the interdependent behavior of project managers and subcontractors using game theory. This model can be used for developing strategies for reducing the undesirable impacts of subcontracting.

Finally, Chapter 9, by Severine Hong-Minh, Stephen M. Disney, and Mohamed M. Naim, discusses different types of simulation that can be used for understanding and improving the dynamic behavior of a supply chain, emphasizing one particular type of simulation, namely systems dynamics. Some guidelines for developing a simulation study are presented, based on a case study in a low-value fit-out construction supply chain.

Section 2: Organizational Perspectives

A salient characteristic of construction projects is the large number of firms involved in designing, procuring, and assembling construction. It is difficult to address supply chain production improvements without also considering arrangements between organizations. This section reviews various perspectives on understanding and improving organizational issues in the supply chain.

In Chapter 10, Ruben Vrijhoef and Kerry London present a concise review of organizational approaches to the construction supply chain. The approaches are observed on three main levels of increasing complexity of interfirm relationships: (1) intraorganizational: within organizations within an individual supply chain; (2) interorganizational: relationships between organizations within projects; and (3) cross-organizational: relationships between many organizations in clusters of the construction industry. The chapter discusses dominant views on the construction supply chain existing on those three levels.

In Chapter 11, Dennis Towill tests the premise that construction projects and the sector as such are often viewed as being unique. On the other hand, because of similarities to other sectors, the application of concepts and techniques from sectors other than construction would be feasible and beneficial. This is demonstrated based on the application of the concept of business process re-engineering, aimed at an effort to compress total cycle time of the construction process. The example presented demonstrates the benefits of such an approach for construction projects and its stakeholders, and in fact for the construction industry as a whole if carefully managed.

In Chapter 12, Andrew Cox elaborates further on the path of stakeholder benefits, but from the viewpoint of procurement and thus the client. Construction clients must have a good understanding of the market structure to be good procurers. In addition, capital (CAPEX) and operational (OPEX) expenditure decisions influence the relationships between procurers and construction partners. The project-based nature of construction projects supports CAPEX decisions primarily, which is a totally different approach to procurement based on OPEX decisions in repetitive relationships. Therefore, in construction win–win situations are hard to achieve. Construction clients need to consider different relationships for different types of construction procurement depending on whether it is a CAPEX or OPEX situation.

In Chapter 13, Kerry London presents a model that focuses on supplier procurement based on industrial organization economic theory and object-oriented modeling. The discussion of the model provides a detailed investigation into the nature of the procurement relationships that are formed in the case of a façade supply chain. The façade supply chain has been categorized as a complex core commodity supply chain. A complex commodity chain is one where the nexus of contracts to the project contract is complex in either technology or managerial complexity. This requires unique, specialist, and innovative design and construction solutions and a high level of integrative managerial capacity. These types of supply chains characterized by innovative design, new materials, and numerous different types of suppliers, require sourcing and integration of suppliers not typically managed previously.

In Chapter 14, Bart A.G. Bossink and Ruben Vrijhoef consider the extent to which innovation can be managed in the context of the complex structure of the construction industry. In this chapter the application of a series of innovation management tools in the construction supply chain is observed on three levels: (1) transfirm level (for all firms in the construction supply chain); (2) interfirm level (for groups of firms cooperating in projects); and (3) intrafirm level (for individual firms). An interview survey indicated that the innovation management tools found are used in various manners to stimulate and facilitate innovation on all three levels, and improve the quality of cooperative ties with other organizations and firms in the construction supply chain.

Section 3: Information Technology

Just as there are a large number of organizations involved in a construction supply chain, there is a need for large flows of information within and across firms. Information technologies promise support for information sharing and analysis, being an essential enabler of improved performance. This section provides an overview of a range of information technologies that can contribute to supply chain performance, as well as examples of effective use.

Chapter 15, by Kalyanaraman Vaidyanathan, provides an overview of several technologies that are relevant to supply chain performance. The author brings a perspective of deployment of advanced supply chain applications within the manufacturing sector, and highlights lessons learned from that sector with respect to opportunities and challenges for information technology deployment in construction.

Chapter 16, by Semiha Kiziltas, Burcu Akinci, Esin Ergen, Pingbo Tang, and Anu Pradhan, reviews field information technology applications and tools. While these field applications are often seen as supporting traditional construction productivity, the close coordination of field production needs and off-site supply production is a necessary component of supply chain performance.

Chapter 17, by Ignacio Schonherr, Luis F. Alarcón, and Sergio Maturana, relates experience with deployment of an e-marketplace for bidding on projects in Chile. The chapter provides a practice example of using information technology to aid supply chain coordination early in a project, before final contracts are set.

Chapter 18, by A. F. Cutting-Decelle, R. I. Young, B. P. Das, and Chimay Anumba, addresses the role of data standards in supporting supply chain communication. The firms in a construction supply chain have diverse information technology systems that generally are not interoperable. Data standards are essential components to overcome communication challenges.

Chapter 19, by Nashwan Dawood, reviews the application of 3D CAD and constraint reasoning systems to improve production coordination among members of the construction supply chain. By example, the chapter holds lessons for deployment of related applications.

Acknowledgments

This book would not have been possible without the hard work of the authors, editors, and the many reviewers who helped make this a high quality, peer reviewed publication. As experts in their fields, many of the contributors also acted as reviewers. Many others also served as reviewers and we wish to acknowledge them here: Dr. Thaís Alves, Federal University of Ceará; Dr. Reza Beheshti, Delft University of Technology; Dr. Bart. A.G. Bossink, VU University Amsterdam; Dr. Carlos Caldas, The University of Texas at Austin; Dr. Diego Echeverry, Universidad de los Andes; Dr. Nuno Gil, University of Manchester; Prof. Lauri Koskela, University of Salford; Prof. Hennes A.J. de Ridder, Delft University of Technology; Dr. Kenneth Walsh, San Diego State University; and Dr. Jennifer Whyte, University of Reading.

I

Production and Operations Analysis

2

Construction Supply Chain Modeling: Issues and Perspectives

Marcelo Azambuja
*The University of
Texas at Austin*

William J. O'Brien
*The University of
Texas at Austin*

2.1 Introduction

Over the last two decades, most manufacturing firms have recognized supply chain management (SCM) as a new way of doing business. The implementation of this new approach was a consequence of various changes in manufacturing environments, such as development of information technology (Internet), globalization, and sophisticated customers who demand increasing product variety, lower cost, better quality, and faster response. Competition is shifting from firm versus firm to supply chain (SC) versus SC [Vonderembse et al. 2006; Min and Zhou 2002]. Successful firms such as Wal-Mart and Dell Computer have survived and achieved a high level of performance through organizing, planning, and controlling a SC as a whole.

SCs are very complex systems for which final performance depends upon a combination of hundreds of decisions made by multiple independent firms. Because there are so many decision variables to assess, models and tools have been developed to support decision making and help practitioners realize the effect of their local decisions on the whole SC performance. Different modeling approaches used in the manufacturing context include, for example, spreadsheet-based inventory and mathematical programming models (e.g. linear and integer programming), discrete event simulation models [Kleijnen 2005], game theory, and decision support systems that utilize the Internet, data mining, and geographical information systems [see Min and Zhou 2002 for a complete review of SC modeling perspectives]. An early case application of a modeling tool for SC decision support is described by Davis (1993). In his paper, Davis shows a successful implementation of SC modeling at Hewlett-Packard, where risk and variability modeling help to optimize inventory levels and locations in a global SC.

In recent years, especially during the last decade, the construction industry has also recognized the importance of SCM to improve the performance of projects [O'Brien 1998; Vrijhoef and Koskela 2000]. As in the manufacturing context, construction companies are facing increasing competition and customers are requiring lower costs, higher quality, shorter execution durations and more reliable schedules.

Initial research has shown how complex and ineffective construction SCs are. To reach these conclusions, researchers had to develop SC models using tools developed for the manufacturing context to guide their assessment. These models, usually studied in an ad hoc manner, generated useful knowledge about modeling production in construction. However, this knowledge is not yet structured in a manner that helps researchers and practitioners interpret and utilize a wide body of existing tools and techniques.

This chapter aims to provide a structured approach to SC modeling in construction. The authors review key concepts and perspectives in modeling SC production, including the definition of manufacturing SC capabilities, decisions, and technologies in Section 2.2. Section 2.3 focuses on the characterization of construction SCs and provides an overview of existing tools and techniques for construction SC modeling. A detailed description and analysis of extant models enable us to identify modeling types, variables, and metrics that support the identification of gaps. Section 2.4 presents a structured approach to modeling construction SCs based on the gaps identified in Section 2.3. First, a conceptual framework is presented to link SC models' purposes with models' attributes. Building from the conceptual framework, a structured approach to defining construction SC models is detailed.

2.2   Perspectives and Concepts of SC Production Modeling

2.2.1   SC Capabilities and Decisions

The literature on SCM presents many similar definitions for SCs [see Bechtel and Jayaram 1997]. In this chapter we use the definition provided by the Committee on Supply Chain

Integration [National Research Council 2000], which describes a SC as "an association of customers and suppliers who, working together yet in their own best interests, buy, convert, distribute, and sell goods and services among themselves resulting in the creation of a specific end product."

A typical SC may involve a variety of stages including raw material (RM) and component suppliers, manufacturers, distributors, and customers [Chopra and Meindl 2004]. One or more companies, geographically dispersed, may be involved at each stage, for example, a manufacturer—in general the focal company—may receive material from several suppliers and then supply several distributors. Figure 2.1 shows the structure of a typical manufacturing SC.

A SC is complex, dynamic and involves the constant flow of information (forecast, orders, schedules, etc.), material (components, end products, etc.), and funds between different and independent stages. The appropriate management of these flows is required in order to respond to customers' expectations and keep SC costs at an adequate level.

Understanding of customers' expectations and SC uncertainty (demand and supply) that a company faces is essential for developing the right capabilities or abilities to serve its markets. A SC may need to emphasize either its responsiveness or efficiency capabilities, depending on a set of final product characteristics and expected performance.

A responsive SC is able to deal with a wide range of quantities demanded, meet short lead times, handle a large variety of products, meet a very high service level and handle supply uncertainty. However, there are many costs associated with responsiveness. This

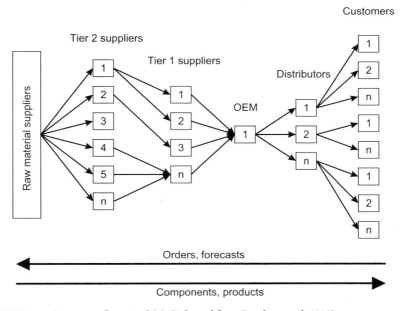

FIGURE 2.1   Structure of a typical SC. (Adapted from Lambert et al. 1998)

increase in cost leads to the second capability, which is efficiency. SC efficiency is the cost of making and delivering a product to the customer. The higher the costs, the lower the SC efficiency. Chopra and Meindl (2004) and Hugos (2006) presented a list of questions that support the identification of SC characteristics and consequently its required capabilities:

- Do customers want small or large quantities of products?
- Do customers expect quick service or is a longer lead time (time from order to the delivery of the product) acceptable?
- Do customers look for a narrow and well-defined bundle of products or a wide selection of different products?
- Do customers expect all products to be available for immediate delivery or are partial deliveries acceptable?
- Do customers pay more for convenience or do they buy based on the lowest price?

SC capabilities of responsiveness and efficiency are results of decisions made about five SC drivers [Hugos 2006; Chopra and Meindl 2004]. The right combination of these decisions determines the capabilities of a SC:

- Production: what products does the market want? How much of which products should be produced and by when?
- Inventory: what inventory should be stocked at each stage in a SC? How much inventory should be held as RMs, work in process, or finished goods (FG)?
- Location: where should facilities for production and inventory storage be located? Where are the most cost efficient locations for production and for storage of inventory? Should existing facilities be used or new ones built?
- Transportation: how should inventory be moved from one SC location to another? When is it better to use which mode of transportation?
- Information: how much data should be collected and how much information should be shared? Timely and accurate information holds the promise of better coordination and decision making.

Table 2.1 summarizes the characteristics of responsive and efficient SCs based upon these decision drivers.

Such decisions are made at different time periods and frequencies. The next section discusses a SC decision framework that provides a perspective on how these decisions can be categorized.

2.2.2 Categories of SC Decisions: A Hierarchical View Decision Framework

Chopra and Meindl (2004) presented three SC decision categories based on the frequency with which they are made and the time frame over which a decision has an impact. The categories are the following: SC strategy, SC tactical planning, and SC operation. Strategic decisions are typically made for the long term, such as SC configuration, location

TABLE 2.1 Characteristics of Responsiveness and Efficient Supply Chains (from Hugos 2006)

Decision Drivers	Responsive SCs	Efficient SCs
Production	Excess capacity	Little excess capacity
	Flexible manufacturing	Narrow focus
	Many smaller factories	Few central plants
Inventory	High inventory levels	Low inventory levels
	Wide range of items	Fewer items
Location	Many locations close to customers	A few central locations serve wide areas
Transportation	Frequent shipments	Few, large shipments
	Fast and flexible mode	Slow, cheaper modes
Information	Collect and share timely, accurate data	Cost of information drops while other costs rise

and capacities of production and warehousing facilities, and modes of transportation to be made available along different shipping legs.

The SC tactical planning involves a set of operating policies within which a SC will function over a period of time. Tactical decisions include inventory control, production and distribution coordination, material handling, and order and freight consolidation.

At the operational level, the SC configuration is considered fixed and planning policies are already defined. The time horizon is weekly or daily and decisions deal with operational routines such as workforce scheduling, vehicle routing and scheduling, material replenishment, and packaging. The goal of SCM at the operational level is to reduce uncertainty and optimize SC performance given the constraints established by the configuration and planning policies.

The following scenario provides an example of these three categories [adapted from Mathur and Solow 1994]. A steel manufacturer produces two types of steel (high- and low-grade) at its two plants in the United States. One plant can process up to 1200 tons per year and the other, more modern and with lower processing costs, can produce at most 600 tons. These plants receive orders from the United States, Mexico, Korea, and Brazil. The RM (iron ores) are supplied by two American mines. The managers have collected data on purchase costs, ore shipping costs, processing costs at plants, demand per customer, and steel shipping costs. They face strategic, tactical, and operational decisions.

In the above scenario, these are examples of strategic decisions: Is it necessary to build another plant to fulfill the global demand? Should that plant be located in Asia or South America? How much high-grade and low-grade steel should be produced in each plant per year? Will it be necessary to build any warehouses in Brazil? Should steel be transported overseas by air or by ship?

Managers of the steel manufacturer are also supposed to make tactical decisions such as: deciding how much ore to order from the mines each quarter, adjusting the manufacturing capacity of each plant based on short-term demands from each market, finding alternatives to consolidating orders from Mexico and Brazil to reduce shipping costs, and coordinating shipping schedules with customers' needs.

Finally, these are examples of operational decisions: Determining how many days or shifts per week each plant will produce, determining which routes are faster to ship to each customer, and determining how many tons of steel can be shipped in each delivery.

Mathematical optimization models based on linear and integer programming have been widely used to solve such complex decisions. However, while mathematical modeling is important, a qualitative understanding of SC configuration and concepts is needed because the complexity of these decisions cannot be fully represented mathematically. Knowledge of issues regarding SC configuration can facilitate, for example, the identification of potential supply risks that optimization programs might not handle, and also support the decisions to achieve the required SC capabilities. In the following section, some of these basic concepts and issues are described.

2.2.3   Issues in SC Configuration

Knowledge of a SC configuration—each individual firm, processes and products, as well as interfaces among different firms—can provide preliminary qualitative insights about overall SC performance and risks.

Manufacturing environments vary greatly with respect to their process structure, that is, the manner in which material moves through the plant [Hopp and Spearman 2000]. Hayes and Wheelwright (1979) classified manufacturing environments by process structure into four categories: job shops, batch, assembly line, and continuous flow. Also, each environment is suited to different types of products.

A job shop environment typically produces low-volume and highly customized products. Each product is usually of an individual nature and requires interpretation of the customer's design and specifications. In a job shop, small batches are produced with a high variety of routings through the plant. The batch size refers to the number of units of a particular product type that will be produced before beginning production of another product type. Thus, the outputs differ significantly in form, structure, materials, and processing required.

In the batch environment, firms provide similar items on a repeated basis, usually in low volumes. For this reason, many processes are repeated, creating a smoother flow of work-in-process (WIP) throughout the shop, although many of the characteristics of job shop production are retained.

Assembly line is the appropriate process for products with high demand. The same operations are executed for each production run in a standard and usually automated flow and all products are very similar. Automobile production is a classic example of line flow processes.

Finally, the continuous flow process encompasses a continuous product flow automatically down a fixed routing. This process is used to produce highly standardized products in extremely large volumes (e.g., refinery products). They are often characterized as a commodity.

A product–process matrix summarizes this description, presenting the associations between manufacturing processes and product types (Figure 2.2). SCs are usually

Process structure vs. product type	Low volume (one of a kind)	Low volume (multiple products)	Higher volume (standardized product)	Very high volume (commodity)
Job shop	Job shop			
Batch		Batch		
Assembly line			Assembly line	
Continuous flow				Continuous

FIGURE 2.2   Product–process matrix. (After Hayes and Wheelwright 1979)

comprised of firms located in all different quadrants of this matrix. Therefore, this matrix can be used as a tool to provide an initial evaluation of SC capability.

In our example of the steel manufacturer, the RM suppliers provide iron ores (commodity), which will be transformed into steel by two plants (assembly-line process structure). Once the two types of steel are produced, they are shipped to their final destination. The capacity of each plant is 1200 and 600 tons of steel per year. Capacity or maximum throughput rate is defined as the maximum quantity of output (tons) that can be processed per unit of time (year). These plants produce the two types of steel in long production runs, and one of them has capacity limitations, thus any rush order has a high probability of delivery delay. This is due to the lack of flexibility of these types of plants to include an additional order in the long production run cycles and also the lack of capacity. Such brief descriptions of products and process characteristics—based on the configuration of the steel SC and the matrix evaluation—has provided an initial qualitative insight into SC risks.

Other SC risks are the consequence of the occurrence of variations that appear to be out of one's control—variability. According to Davis (1993), the three sources of variability that may affect SCs are supplier performance, the manufacturing process itself (e.g., process type as well as machines reliability), and customer demand. In the steel manufacturer scenario, customer demand is fairly stable, the older plant presents higher variability due to old machinery, and delivery performance to Brazil and Korea are much more variable than deliveries to the United States and Mexico due to overseas shipments.

To mitigate these risks, derived from the firms' characteristics and SC overall variability, managers can locate buffers at strategic locations of the chain. Buffers are resource cushions used to protect the production system against variability or resource starvation. There are three different types of buffers: inventory, capacity and time [Hopp and Spearman 2000]. Inventory buffers are material stockpiles of RM, WIP, or FG. Capacity buffers are extra available capacity that may be used in case production falls behind schedule. Time buffers are extra time embedded in schedules to deal with potential variations. In our scenario, customers in Brazil and Korea need to store much more steel (inventory buffer) than customers in the United States and Mexico. Otherwise, these customers are highly exposed to the risks of delay caused by overseas transportation. Another strategy

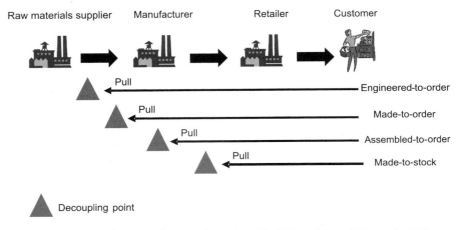

FIGURE 2.3   Decoupling point locations in various SCs. (Adapted from Naim et al. 1999)

that could be adopted by these customers is the inclusion of time buffers in the procurement schedule—ordering the steel well in advance. As for the capacity buffer, the plant that has the higher capacity to produce steel can be utilized to fulfill unexpected demands of the plant that has limited capacity in certain months of the year.

Another concept related to buffering is the decoupling point, which is a strategic stock that buffers the SC from changes in customer demand, in terms of both volume and variety [Naim, Naylor, and Barlow 1999]. The decoupling point differs between product groups (Figure 2.3). Starting with the most upstream location of a decoupling point we have engineered-to-order (ETO), followed by made-to-order (MTO), assembled-to-order (ATO), and then made-to-stock (MTS) products. In the steel SC, the steel plants can mass produce the two types of steel and keep them stored to better respond to variations in local and global demand. However, inventory is a source of cost and the manufacturer needs to judge how much to hold. Another example of such a concept is presented by Walsh et al. (2004). These authors discuss the strategic positioning of inventory in the SC for stainless steel pipes.

This section of the chapter has covered fundamental concepts, different decisions, and phases involved in the configuration of manufacturing SCs. The modeling of production in this context is supported by a structured framework that comprises a set of SC capabilities, categories of decisions to achieve them, and the different phases during which these decisions should be made. The following section addresses these issues in the construction industry scenario.

2.3   Review of Modeling Approaches for Construction SCs

The modeling of construction SCs is a subject that has been mainly investigated since the early 1990s. The goal is to explore how manufacturing concepts can be transferred to the construction context in order to improve production efficiency and reduce project costs. Initial case study descriptions and partial implementation have been reported by many researchers, especially those associated with the International Group for Lean Construction (IGLC). Their findings have shown insightful solutions or suggestions for improvement which demonstrate the usefulness of modeling production beyond the boundaries of construction sites.

However, construction industry characteristics differ substantially from the manufacturing SCs presented in the previous section of this chapter. Recent publications [see Vaidyanathan and O'Brien 2003; Green, Fernie, and Weller 2005; London and Kenley 2001] have investigated the application of manufacturing SC concepts in construction and have highlighted differences and opportunities. We summarize some of the key differences in Table 2.2. These are useful for understanding the difficulties in applying SCM concepts in construction as well as illustrating how some SC practices, such as buffering and capacity planning, are distinct between these contexts.

TABLE 2.2   Manufacturing vs. Construction SCs

Characteristics	Manufacturing SCs	Construction SCs
Structure	Highly consolidated High barriers to entry Fixed locations High interdependency Predominantly global markets	Highly fragmented Low barriers to entry Transient locations Low interdependency Predominantly local markets
Information flow	Highly integrated Highly shared Fast SCM tools (factory planning and scheduling, procurement, SC planning)	Recreated several times between trades Lack of sharing across firms Slow Lack of IT tools to support SC (no real data and workflow integration)
Collaboration	Long-term relationships Shared benefits, incentives	Adversarial practices
Product demand	Very uncertain (seasonality, competition, innovation, etc.) Advanced forecasting methods	Less uncertain (the amount of material is known somewhat in advance)
Production variability	Highly automated environment (machines, robots), standardization, production routes are defined—lower variability	Labor availability and productivity, tools, open environment (weather), lack of standardization and tolerance management, space availability, material and trade flows are complex—higher variability
Buffering	Inventory models (EOQ, safety inventory, etc.)	No models Inventory on site to reduce risks Use of floats (scheduling)
Capacity planning	Aggregate planning Optimization models	Independent planning Infinite capacity assumptions Reactive approach (respond to unexpected situations, for example, overtime)

Modern construction production modeling is somewhat broader than traditional site operations analysis. Terms like buffer, variability, and uncertainty are not yet common among experienced construction managers. However, on-site production inefficiency is often caused by poor production planning (which includes decisions on buffers) and limited planning concerning the impact of off-site production and delivery variability. It is a common practice to keep large amounts of inventory on job sites to reduce risk of delays on site production (decoupling strategy). However this material requires site space, resources to manage it, and represents, in most cases, an unnecessary investment. Another traditional practice is the development of optimistic schedules that do not include time buffers to protect against uncertainties (e.g. manufacturing and delivery of materials). Moreover, scheduling tools consider infinite capacity assumptions that are not real and need further investigation. We believe that SC modeling can guide managers to better allocate different types of buffers and include uncertainty issues in their production planning.

The need for change in construction practices is clear. First, owners and general contractors (GCs) have to understand the increasing importance of suppliers for achieving project goals. The identification of key suppliers can lead to the adoption of partnerships or long-term relationships that may facilitate the implementation of modern SC practices. Second, there is still much room for the development and implementation of SC modeling tools in construction. For example, a more systematic and formal method to model SCs for construction planning or even during the preproject planning to support strategic analysis and identification of major risks seems necessary.

The following sections advance the discussion on construction projects' characteristics, key production decisions that should be taken into account along any project life cycle, and current efforts on production modeling in construction.

2.3.1   Projects as Composed of Multiple SCs

O'Brien, London, and Vrijhoef (2002) presented a conceptual view of the construction project SC, which is shown in Figure 2.4. Figure 2.4 gives an indication of the complexity of SC production operations, depicting part of the large number of firms that compose a construction project SC. Owners, designers, GCs, subcontractors, and multiple types of suppliers that supply distinct technologies at different project phases are organized to form a unique SC configuration. Thus, the scope of analysis includes production both on-site and off-site. When extended to multiple subcontractors, each with their own supply structure, Figure 2.4 indicates that the scale of a construction SC is extensive even for small projects. As such, we take the perspective that construction projects are composed of multiple SCs, each with specific behaviors. Their behavior is strongly determined by the type of product that is delivered to a project site.

Elfving (2003) and Arbulu, Koerckel, and Espana (2005) described the characteristics of four general types of construction products: ETO, MTO, ATO, and MTS. ETO products are specially made for the customer following detailed specifications (e.g., power distribution equipment, preassembled rebar components), commonly characterized by long lead times and complex engineering processes for product specifications. MTO products are usually products manufactured once customer orders have been placed (e.g., cast-in-place concrete, prefabricated panels). Usually, MTO manufacturers don't hold stock and lead times can be either long or short, depending on the manufacturing

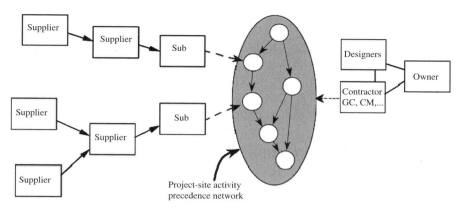

FIGURE 2.4   Conceptual view of the project SC. (O'Brien et al. 2002)

complexity. ATO products are also assembled (manufactured) after customer orders, however these products are usually standard or made of standard components (e.g., doors, windows). Lead times are usually short (or shorter than MTO lead times) and some stock is held by manufacturers, who need to manage an uncertain mix of orders. Finally, MTS products are commodities (e.g. consumables such as bricks, bolts) characterized by short lead times. MTS manufacturers usually hold stock, however managing the physical distribution of such products may be complex.

In construction, all material flows converge to on-site production. However, the focal point (job site) usually has no power to coordinate the SC in the same way as large manufacturing companies. Job sites constitute the demand that needs to be fulfilled by all SCs. This demand—typically when something is needed as opposed to how much—is often unstable due to the lack of reliability of site production systems. In cases of changes in demand, information should flow quickly to suppliers so that they can respond. However, access to demand information (which includes material orders, construction schedules, and site conditions) is somewhat limited to a few suppliers and subcontractors. As a consequence, the coordination of material flows is not efficient and much waste is spread through SCs [see Naim and Barlow 2003 for a case study on housing SC].

Another source of complexity in construction is the involvement of subcontractors with multiple projects at the same time [O'Brien and Fischer 2000]. Therefore, subcontractor resource availability has become a critical performance factor for any project. Also, some subcontractors are responsible for coordinating upstream material flows with their own suppliers. Then, the subcontractor's capability to coordinate this flow effectively is fundamental to reducing risks of material delays on construction job sites.

As for the suppliers, many variables influence their performances. Most risk resides either in those suppliers that provide long lead time products or those that have limited capacity to handle market demand. In general, these types of suppliers prioritize orders based on preferred customers or to gain internal efficiency. Thus, strategies to mitigate risks associated with the supply of materials provided by them should always be addressed by construction managers.

Finally, owners and designers may also influence SC performance. When owners are directly involved in the construction phase, they may delay approval processes, require design changes, and so on. The designers, on the other hand, may delay the procurement process of certain products (delay of detailing design conclusion) and also may affect on-site production, either in the form of change orders or lack of details interrupting production (bad quality design). The involvement of suppliers in the design process for ETO products also complicates SC coordination and increases risk.

2.3.2   Issues in Construction Project Configuration

The way construction projects' SCs are configured may determine their final success or failure. The configuration of any construction project describes how materials and information flow between companies. Usually, this configuration is detailed over the project life cycle, as a result of various decisions made from preproject planning to the end of construction phases.

Major decisions regarding projects' SC configuration are made in early phases, even before the start of conceptual design. Major decisions such as site location and selection of complex and long lead time technologies are mostly made during the preproject planning phase. For example, will the structure be cast-in-place or prefabricated? What are the dominant process technologies (for industrial plants)? In this phase, it is expected that most client requirements will be captured and further guide the design process. The output of the preproject planning phase provides a good idea of the final product complexity and the design effort coordination needed (e.g., number of different system designers), and a fair idea which subsystems are going to be outsourced or going to be executed by the GC (make or buy decision).

The design phase is fundamental for project SC performance. During this phase designers recommend systems' specifications, develop the detailed drawings that indicate the bill of materials, and assess constructability issues that might cause problems throughout project construction. Decisions regarding standardization and modularization of systems' components may make a substantial difference in the project's total cost and schedule. These differences may be reflected in the increased installation productivity of modular components, economy of scale during the manufacturing of modules, or possible price discounts when negotiating with manufacturers. These design decisions are increasingly influenced by input from GCs and key subcontractors and suppliers (particularly those supplying ETO systems). When designers make independent decisions without considering SC capabilities, costs can be higher and performance reduced. As an example of potential SC collaboration during design, Azambuja and Formoso (2003) provide an example of how different elevator shaft sizes may affect the prices among different elevator suppliers and then affect the project's total cost.

The procurement phase (which increasingly overlaps with the design phase) greatly defines the final configuration of a construction project SC. In this phase, most of the suppliers and subcontractors are selected. Usual criteria to select those companies are price, safety, quality, and schedule performance. Once the negotiation and contracts with the selected suppliers are concluded, a good number of the project SCs' constraints can be specifically recognized. The location of suppliers is already known, their capabilities (e.g. manufacturing and transportation, quality, safety, schedule reliability) were evaluated based on the project-specific criteria, the lead times for delivery of products are established with critical suppliers, and they have declared resource availability (labor or equipment) for the project (although that availability may change due to commitments to other projects). At this moment, any construction project manager can potentially understand how efficient or responsive to possible changes their project SC is. However, they should plan and protect their projects from the negative impacts that variability and uncertainty may cause. Common causes of problems affecting construction schedule and cost performance are related to suppliers; some examples include: delay of fabrication or deliveries by material suppliers, damaged or wrong products arriving at the job site, subcontractors having low productivity, and so on.

The construction stage is when the detailed SC operation and coordination take place. It is also the stage where construction planning has to assure that all uncertainties are considered and actions are taken in order to protect the job site production from the off-site environment. In this phase, mitigation strategies such as the use of time, inventory, or capacity buffers are used to shield construction operations from the off-site uncertainties. Another general concern is job site space utilization and work packages sequencing. These decisions directly affect the manner in which the materials are delivered and stored on site. As a result, important questions that need to be constantly considered during construction are: (a) how often should different types of materials be ordered and in what batch size? (b) how much inventory should be kept? (c) where should the inventory be located? Another important issue is the job site capacity to accomplish the proposed schedule. Having excess labor, equipment, and overtime are common strategies used to mitigate uncertainty problems. However those strategies usually have a negative impact on construction costs and should be used only when there is no other alternative.

Figure 2.5 illustrates a spectrum of SC decisions that should be made in each project phase and relates them to the strategic, tactical, or operational purposes set in the manufacturing context. The figure shows these stages in the classic, sequential order of project phases; increasingly, however, project phases overlap, collapsing the time in which decisions must be made.

Unlike manufacturing, research in construction has not shown any structured framework to categorize SC decisions and suggest when they should be made along the project life cycle. In this section, we have suggested such categories in order to provide an initial point for further discussion on the subject.

2.3.3 Construction SC Modeling: Capability Review and Needs

The spectrum of SC decisions from strategic to operational listed in Figure 2.5 have been studied in part by construction researchers. Studies have focussed on specific decisions, mostly at the tactical level. In addition, most studies have also focussed on descriptive, rather than prescriptive, analysis. This leads to several gaps in current approaches to construction SC modeling. This section reviews the current studies to identify these gaps in the context of the modeling discussion above and in support of the framework presented in the next section.

Beyond the classification of SC questions described above, the further separation of models as descriptive and prescriptive is useful for examining the current state of construction SC modeling. Descriptive models are usually deterministic, meaning that

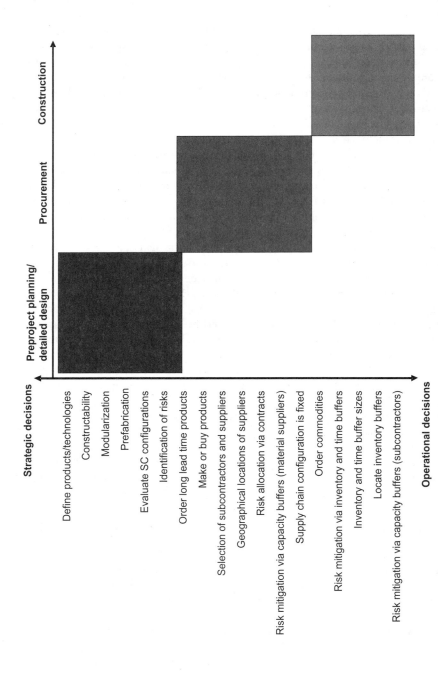

FIGURE 2.5   Construction project phases and associated SC decisions.

all model parameters are assumed to be known and certain. Descriptive models are also often static, illustrating a snapshot of an SC current state. Prescriptive models, by contrast, generally take into account uncertain and random model parameters. They are dynamic and aim to mimic SC behavior and performance, allowing prediction and discovery of emergent behavior in the SC for optimization.

Perhaps due to the complexity of the construction SC, most studies have focused on case descriptions of SCs for specific products (for example, rebar [Polat and Ballard 2003] and heating, ventilation, and air conditioning (HVAC) ductwork [Alves and Tommelein 2003]). These descriptions usually deploy visual process modeling tools to illustrate or "map" the respective SCs. Among the available mapping tools, construction researchers have predominantly adopted variants of Value Stream Mapping (VSM), a tool developed by the Lean Enterprise Institute. VSM is a process of representing the flows of information and materials, and other parameters such as inventory size and cycle time as they occur, summarizing them visually, and envisioning a future state with better performance [Jones and Womack 2003]. The objective is to identify inefficiencies or waste in the SC and remove them. This is usually measured comparing SC lead time and throughput time. VSM can be used to model current states of the SC as well as represent future states. In this sense the tool crosses between descriptive and prescriptive process analysis, although the tool lacks support for dynamic modeling or optimization. As such, it is employed primarily for descriptive analysis in construction.

Tommelein and Li (1999) and Tommelein and Weissenberger (1999) used the VSM to analyze the possibility of adopting a just-in-time production system through the strategic location of buffers (inventory and time) in the SC. Subsequent applications of VSM have mainly been used to support evaluation of different SC configurations [Arbulu and Tommelein 2002; Elfving, Tommelein, and Ballard 2002; Polat and Ballard 2003; Azambuja and Formoso 2003]. These models are simple diagrams or maps that typically include a partial sketch of engineering, procurement, fabrication, and installation processes specific to the SC problem addressed. Those authors used these visual maps to enrich their case descriptions and to provide insights for supporting strategic SC decisions, such as improvement of coordination and communication among companies, location of buffers to mitigate risks, and elimination of processes to reduce lead times.

VSM models have been used to guide tactical and operational SC decisions. Akel, Tommelein, and Boyers (2004), and Fontanini and Picchi (2004) presented more detailed models which included not only processes, but also material and information flow data. These models also dealt with lead time reduction, providing additional insights into transportation, batching, and material ordering issues. Supporting these studies, simulation-based models have also supported analysis of tactical and operational decisions. In particular, simulation models have predicted lead times and/or throughput performance of construction SCs. Several studies have assessed the effects of buffer and batch sizes [Al-Sudairi et al. 1999; Arbulu et al. 2002; Alves and Tommelein 2004; Walsh et al. 2004; Jeong, Hastak, and Syal 2006], variability in process durations [Alves, Tommelein, and Ballard 2006], and product standardization [Tommelein 2006] on the above-mentioned performance measures. While generally based on a descriptive study, the simulation models allow exploration of variation in parameters and hence support either optimization or broader generalization of, for example, the value of buffers or deleterious effects of variability.

Perhaps due to the nature of the modeling tools, VSM and simulation-based models tend to focus on SC configurations and processes that are well described in terms of production units (plants or work centers depending on the level of analysis), buffers, and materials flows. Measures such as throughput and time are standard in these models. As such, these models tend to focus on a subset of the SC decisions—in particular, strategic and tactical ones about specific and reasonably detailed SC configurations. The tools and models as deployed do not easily scale to earlier strategic decisions for a given project where details about specific process and materials flows have not been firmly identified. At the same time, the models have not supported wide-scale analysis of operational decisions, perhaps due to a lack of data and the challenges incumbent in modeling and calibrating such detailed models.

Models that address different parameters are relatively rare. O'Brien (1998) proposed a qualitative model to predict SCs' performance behavior. The model included a set of firms that were classified according to a typology of suppliers based on the suppliers' production technology, as well as the impact of site demand (uncertainty) on their performance. The combination of different classes of suppliers resulted in a high-level view of SC performance capabilities and intrinsic risks. These high-level models support rapid analysis without the need for detailed models, and could be used to support a variety of strategic configuration decisions, but their use has not been further explored.

Not until recently have SC models reported results which include cost performance. For example, Vidalakis and Tookey (2006) simulated the flow of materials from building merchants to construction sites and assessed inventory and transportation costs. Additionally, Polat, Arditi, and Mungen (2007) built a model to assist contractors in selecting the most economical rebar management system prior to the start of construction by recommending batch sizes, scheduling strategy, and buffer sizes. The inclusion of cost analysis offers a promising next step in construction SC modeling. For example, many existing models address lead time reduction, but the costs associated with the reduction of time are not clear. Time-cost tradeoff type analyses can help managers make more informed decisions.

The early, existing construction SC models support a variety of decisions for specific decisions, but do not address the full scope of decisions as described in the preceding section. Construction lacks models that enable professionals to draw inferences about construction SC strategic decisions. Examples of such decisions include: (a) determining buffer sizes and locations based on companies' geographic locations; (b) establishing the locations of projects to reduce overall logistics costs (especially transportation costs); and (c) identifying the demand per region to reduce the risk of delay of material delivery. These decisions are important as global materials sourcing is becoming common practice on a range of projects.

A broader view of various independent SCs converging to "multiple" construction projects is another strategic perspective that deserves further modeling effort in the construction industry. Multiple construction projects means either different projects executed by one GC or projects executed by different GCs. Current models have largely focused on the analysis of individual SCs that convert to a single or multiple construction project executed by one GC. This focus of analysis is useful to identify problems in the interface between one supplier (or subcontractor) and one GC. However, it does not support the understanding of more complex issues, such as the capacity of suppliers to react to multiple projects' demand spikes or change orders. In order to carry out the broader level of analyses, researchers and practitioners need to understand the typology of suppliers involved in their projects [e.g., O'Brien 1998].

Beyond increased scope for strategic decisions, SC modeling of tactical and operational decisions also needs to become more flexible and expressive to better capture project complexities and constraints. Existing models effectively support the identification of companies, processes, or deterministic lead times of SCs. However, new data have to be included or modified; for example, clarifying the location and type of buffers; including maximum, minimum, and most likely durations of SC processes (if possible, showing the probability distribution based on real data). Current models address lead time reduction. However, it is not clear how much time one can reduce in each process (e.g., is there a minimum duration?). Making models more capable, perhaps with some standard prespecified boundary conditions for classes of suppliers, would allow more rapid as well as prescriptive modeling for a range of tactical and operational decisions. The extension of these models to include cost will also spread the range of model applications.

Moreover, as noted by Azambuja and O'Brien (2007), a defined set of performance measures for the evaluation of construction models is needed to complement model development and evaluation. Ideally, modelers would select and monitor performance based on a proposed set of metrics that depend on specific model purpose and level of detail. As a result, different scenarios could be evaluated and compared, helping managers select the right SC configuration to achieve project goals.

2.4 Framework for Developing Construction SC Models

As noted above, construction modeling efforts are capable but need to be extended to better include a range of strategic as well as tactical and operational decisions. It is unclear to what extent the basic modeling constructs for SCs need fundamental extension to address various decisions or to what extent existing tools can address the decisions. A broader need, however, is discipline in applying the various approaches. In this section, we propose and describe a conceptual framework to model construction SCs. The framework, shown in Figure 2.6, consists of five sequential steps which support the modeling process. The steps are as follows: (a) define SC model purpose; (b) establish SC performance measures; (c) determine product type; (d) define SC configuration; and (e) characterize SC elements. It is expected that this framework will provide the necessary support for modelers to develop a comprehensive SC model that includes the model goal and metrics, as well as adequate boundaries, elements, and attributes.

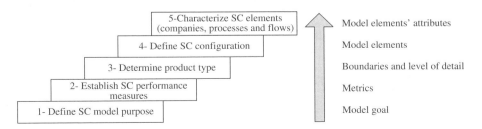

FIGURE 2.6   Framework for supply chain modeling.

The first step of the framework is to define the SC model purpose or goal. SC models can serve various purposes, which vary from choosing the best SC configuration to reducing material supply risks, to those which assess how companies' internal processes affect SC lead time and throughput performance. Table 2.3 presents a comprehensive list of purposes that have been modeled in construction and manufacturing over the last decade.

Once the modeler defines the purpose of the modeling exercise, a set of performance measures need to be associated to the model goal (step 2). For example, if the model goal is to support lead time reduction, metrics such as order processing time, engineering time, manufacturing/assembly time, and delivery time are fundamental. If a manager is interested in reducing the inventory buffer in a construction job site, the SC performance metrics to be monitored could be the number of items in stock, average waiting time in stock, average inventory turnover, installation demand rate, and

TABLE 2.3   SC model purposes

Evaluate the supply chain configuration

    Evaluate the best SC configuration to fulfill the demand (improve response time)

    Assess SC complexity (# of members, information and material flows, coordination effort)

    Assess location of facilities

    Show which company is responsible for each process (ownership decisions)

Reduce product lead time (eliminating or combining activities)

    Identify the number of processes performed to deliver the product

    Identify the time spent in each process (conversion and flow)

    Classify each process performed (value-added or non-value-added)

    Simplify the SC (eliminating non-value added activities, relocating inventories, consolidating points for distribution)

Evaluate buffering decisions

    Locate inventory buffers in the SC (decoupling points)

    Identify the type and size of buffers

    Influence of buffer location and size on the product lead time

Evaluate production decisions

    Assess batch decisions and their influence in the final product throughput

    Influence of set-up time on companies' delivery performance

    Evaluate the effect of capacity decisions on the SC (inventory behavior, lead time, throughput)

Evaluate transportation decisions

    Assess how transportation (type and frequency) affects SC (lead time, delivery performance, costs)

Assess SC costs

    Inventory costs, process costs, transportation costs, ordering costs, cost of resources

    Understand the risk of materials delay (on the construction schedule)

Illustrate the SC information coordination (IT application)

    Identify the frequency, content, and type of information exchanged (and how it is transferred)

    Evaluate the risk of communication errors and delays on material flows and inventory buffers (Bullwhip Effect)

Assess the impact of product complexity (standardization, # of parts) on the SC response time supplier delivery time and frequency. Table 2.4 provides a list of metrics that may be associated with other SC model purposes.

The third step in the process is to determine which type of product is going to be modeled. The type of product can provide valuable insights about the model boundaries and level of detail. So far, researchers have not considered how different products might affect SC model boundaries. A model of an ETO product needs to include various processes executed by different actors such as designers, engineering firms, GCs, and suppliers. Several information flows need to be taken into account when modeling SC processes for this kind of product since these flows directly affect the manufacturing process and product delivery. By contrast, a model of an MTS product involves fewer actors, usually the GC or subcontractor, and the supplier. The flow of information is usually restricted to the transaction process, although processes directly or indirectly related to material flows often need to be modeled.

The understanding of general types of products may also support a qualitative assessment of SC capabilities and potential risks. For example, a supplier of an ETO product is certainly a job shop producer that manufactures very low volumes and has difficulties reacting to short-term changes in production. Table 2.5 shows potential manufacturing environments, actors, and processes that should be taken into account when modeling different products. The outcome of this step is an initial indication of the SC model boundaries and level of detail (i.e., which SC processes may be included in the model).

The fourth modeling step is the definition of SC configuration. In this step, the model elements—SC actors, processes, activities, material flow, information flow, inventory buffers, and resources—are arranged to build the SC model. The knowledge of previous framework steps is important for refining the model level of detail. The model

TABLE 2.4   SC Performance Measures Associated with Model Purposes

SC Model Purposes	Metrics
Identify material supply risks	SC average throughput; SC lead time variability; percent on-time deliveries; lateness of deliveries
Decrease transportation costs	Delivery frequency; minimum batch size; distance; cost per unit; handling cost
Decrease manufacturing costs	Labor and machine costs; labor and machine utilization; process cycle time; capacity utilization; total inventory costs
Increase SC throughput	SC throughput; buffer sizes; batch sizes; number of processes; process cycle time; manufacturing lead time; delivery lead time
Measure SC reliability	SC lead time variability; percent on-time deliveries; lateness of deliveries; supply quality (shipping errors, customer complaints); stockout probability
Evaluate supply flexibility	Production volume (capacity); production mix (variety of products); or delivery dates (change planned delivery dates)
Evaluate SC complexity	Number of processes; number of different companies; geographical locations; flows of materials (number of stages); flow of information (centralized vs. decentralized)

TABLE 2.5   Suggestion of Model Boundaries and Processes Based on Different Technologies

Product Type	Manufacturing Environment	SC Boundaries	SC Processes
Made-to-stock	Assembly line or continuous flow	MTS supplier; warehouses; subcontractor or GC; job site	Subcontractor or GC places order; supplier checks product availability, picks and delivers; unloads on job site; product installation
Assembled-to-order	Assembly line	ATO supplier; upstream supplier of critical RM component; subcontractor or GC; job site	Subcontractor or GC places order; supplier checks RM availability; either starts assembling the product if RM available or needs to wait for RM; temporary FG inventory; delivery; unload on job site; product installation
Made-to-order	Job shop or batch	MTO supplier; upstream suppliers of critical RM components; GC; designer; owner	GC receives detailed design and places order; supplier checks design and order; if information is complete and RM are available it starts manufacturing, otherwise waits for RM or GC, designers and owners to review and approve changes; product manufacturing; temporary FG inventory or immediate delivery, unload on construction site, installation
Engineered-to-order	Job shop	ETO supplier; upstream suppliers of critical RM components; architect; owners; engineering company	Supplier either fully designs or only details the design received from an engineering company; the owner checks the detailed design; if design is accepted, the supplier can start manufacturing the product, otherwise it waits for owner response; product manufacturing; temporary FG inventory or immediate delivery; unload on construction site and installation

boundaries and processes identified in step 3 support the SC configuration. In order to configure the SC, those processes are assigned to the actors who are responsible for their execution. Then, the relationships between actors and processes can be visualized, as well as their logical sequence in the process.

Detailed SC models are generally complex and confusing; therefore a hierarchical approach is proposed to deal with this issue. The first level of the hierarchy depicts the SC actors, their respective product types, geographical location, and transportation flows (Figure 2.7).

FIGURE 2.7 SC configuration: Level 1.

Once the relevant SC actors are identified, classified and located, the model can be further detailed in order to describe each actor's processes and boundaries (level 2 of the hierarchy). At this level, the main processes are listed and their interfaces can be linked. Figure 2.8 through Figure 2.13 depict SC actors' common internal processes as well as those that define their external interfaces (arrows crossing the rectangle borders). The actors' names (top left of boxes) are followed by a number, which identifies their horizontal tier in relation to the job site. This identification serves to locate each actor participating in the material flow. For example, an MTS supplier 1 sends material directly to the job site while an MTS supplier 2 sends its product to the other location (e.g. warehouse) before delivering to the job site. The MTS supplier shown in Figure 2.7 is an instance of MTS 2. As for subcontractors, they can always be identified as number 1 as long as they are responsible for buying and storing material in their own location before the final installation on the job site. If they are only responsible for the installation, the install process (Figure 2.8) characterization is enough to describe the subcontractor's role.

Owners, designers, engineering firms, and GCs often are not part of off-site material physical flows. Instead, they are responsible for coordinating information flows and usually are the ones who exchange information with suppliers (e.g., orders, design approvals, detailing) to support the material flows. Thus, these types of actors do not require an identification number (SC horizontal tier). The responsibility for information coordination varies depending upon each project's organizational structure. For example, who will order materials: owner, GC, or both? Who will provide a detail design: engineering firm or supplier? For this reason we decided not to represent those actors' generic process models. However, they are part of the modeling effort. Their interfaces with product suppliers are represented by the arrows pointing in and out of processes, such as receiving orders, design detailing, and approvals (Figure 2.10 through 2.13). Consideration

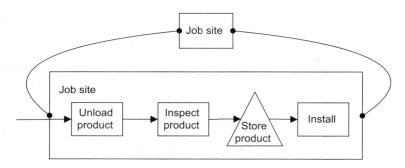

FIGURE 2.8   SC configuration level 2: Job site processes.

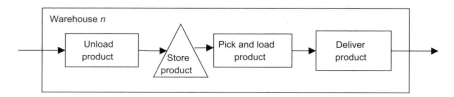

FIGURE 2.9   SC configuration level 2: Warehouse processes.

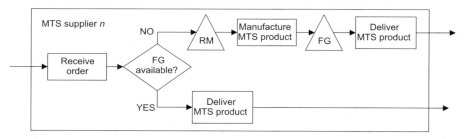

FIGURE 2.10   SC configuration level 2: MTS supplier processes.

and inclusion of designers or engineering firms is important in construction SC modeling because the design is required to execute tasks on-site. Often the design is pulled due
to site and material needs, thus the modeling of designers/engineering activities may
provide some indication of risks. This is particularly common in fast-track projects.

The outcome of this hierarchy level (level 2)—after linking all actors' processes—
is a representation very similar to the visual models presented by Arbulu and Tommelein (2002) and Polat and Ballard (2003), which were used to assess different SC
configurations.

The third and last level of the hierarchy is depicted in Figure 2.14. This level details the
processes presented in level 2. Level 3 serves to illustrate and describe which activities,
buffers, and resources may be included in the model. Figure 2.14 shows all the activities
required to manufacture an MTO product.

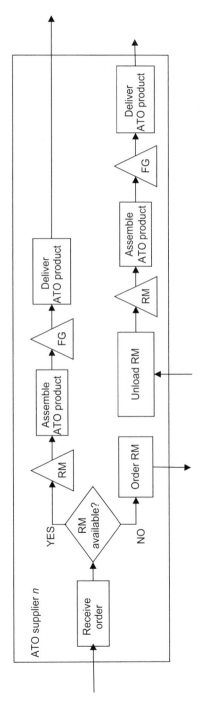

FIGURE 2.11 SC configuration level 2: ATO supplier processes.

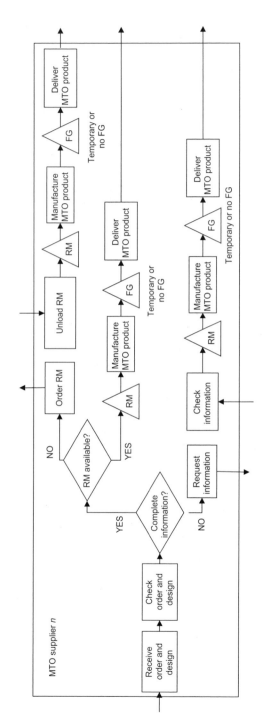

FIGURE 2.12   SC configuration level 2: MTO supplier processes.

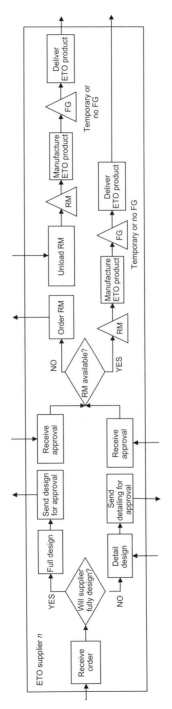

FIGURE 2.13 SC configuration level 2: ETO supplier processes.

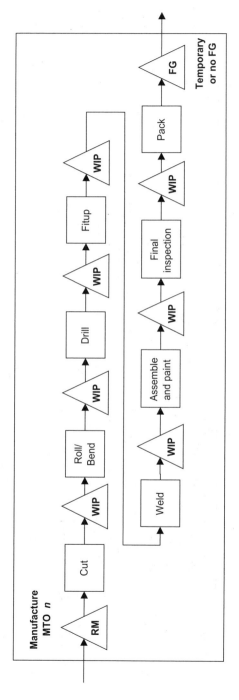

FIGURE 2.14   SC configuration level 3: Manufacturing process of MTO supplier. (Adapted from Arbulu and Tommelein 2002)

Note that the name of the process still includes the actor's identification. Hence, each actor's properties previously identified in level 2 can be transferred to this level. The hierarchical modeling approach used to configure SCs is useful to define and visualize any model elements. The selection of which level of detail (hierarchy levels) one needs to model depends on the model's purpose, boundaries, and processes required by each type of product.

The fifth and last step of the modeling approach is to characterize the SC model elements. This step encompasses the description of each model element in order to provide complete information on the current SC model and helps users understand the model behavior. Each model element will then have its own set of attributes. These attributes are necessary for the creation of prescriptive models such as simulation tools and for improving the descriptive power of construction SC models. Table 2.6 presents a list of SC elements and attributes.

Figure 2.15 provides one example of the attribute list applied for the pack activity. The standard list of attributes provides valuable information about the pack activity that is part of the manufacturing process of an MTO product located at the horizontal level *n* of a SC.

TABLE 2.6 Attributes of SC Elements

SC Elements	Attributes
SC actors	Name, location, processes; interfaces (boundaries), product type (ETO, MTO, ATO, MTS), final product
Processes and/or activities	Name (verb+noun), type (material or information conversion), cycle time (deterministic or probability distribution), resources, production batch size, cost
Material flow (delivery)	Product (size, weight), batch size, time, frequency, transportation mode, cost
Information flow	Information type (order, design drawing, schedule), time, frequency
Inventory buffers	Type (FIFO, LIFO), product in stock, maximum # of parts, holding costs
Resources	Name, type (labor, equipment), quantity, capacity, costs, failure or maintenance (downtimes), schedule (fixed or variable capacity)

Activity	Attributes
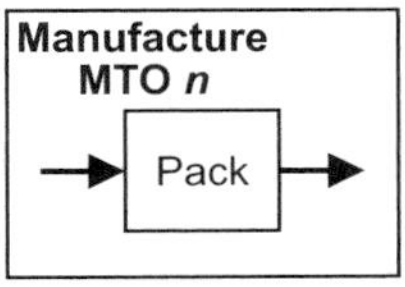 **Manufacture MTO *n*** → Pack →	Name: Pack MTO product
	Location: MTO *n* supplier
	Activity type: material conversion
	Cycle time: 2–4 hours
	Batch size: 100
	Resource type: labor
	Resource quantity: 2
	Schedule: 1 shift
	Total cost: 100–200 USD

FIGURE 2.15 Attributes of model elements: Pack activity.

2.5   Prospects for Construction SC Modeling

This chapter provides an overview of various concepts and issues regarding SC production modeling in manufacturing and construction contexts. We reviewed the literature on construction SC modeling and briefly described the extant modeling tools and their initial applications in case studies. The review allowed us to identify some modeling gaps that need to be addressed to make construction SC models more capable with respect to supporting a range of decisions. Finally, we proposed a conceptual modeling approach that consists of sequential steps and emphasizes issues that need to be considered when modeling construction SCs, such as a structured set of metrics, product characteristics and level of detail, and assignment of attributes to model elements.

Construction SC modeling is, in the authors' belief, very much the next generation of productivity modeling for construction. Much as construction benefited from traditional productivity studies (e.g., the application of time and motion to the job site) and their extension to constructability analysis (i.e., the application of construction knowledge in design to improve productivity), construction projects can benefit from SC models to improve tactical and operational decisions and apply that knowledge to generate wisdom for strategic decisions. At the same time, the history of productivity modeling poses challenges for the development of SC models. Productivity modeling suffered from a long period of limited development and a lack of widespread deployment for many years; only the relatively recent introduction of lean concepts has invigorated productivity modeling. SC models could suffer the same fate; the current models that are predominantly descriptive and specific to individual projects (and hence do not easily generate broad findings) do not yet represent a suite of analysis tools and associated understandings that will actively let managers make effective and informed SC decisions for their projects. It is the hope of the authors that the discussion and framework above will stimulate development of a wider and more capable range of construction SC models that can effectively enhance practice.

References

Akel, N. G., Tommelein, I. D., and Boyers, J. C. 2004. Application of lean supply chain concepts to a vertically-integrated company: A case study. *Proceedings of the 12th Annual Conference of the International Group for Lean Construction*, IGLC 12, Copenhagen, Denmark, 560–74.

Al-Sudairi, A. A., Diekman, J. E., Songer, A. D., and Brown, H. M. 1999. Simulation of construction processes: Traditional practices versus lean construction. *Proceedings of the 7th Annual Conference of the International Group for Lean Construction*, IGLC 7, Berkeley, CA, 39–50.

Alves, T. C. L., and Tommelein, I. D. 2003. Buffering and batching practices in the HVAC industry. *Proceedings of the 11th Annual Conference of the International Group for Lean Construction*, IGLC 11, Blacksburg, VA, 13.

_____. 2004. Simulation of buffering and batching practices in the interface detailing-fabrication-installation of HVAC ductwork. *Proceedings of the 12th Annual Conference of the International Group for Lean Construction*, IGLC 12, Copenhagen, Denmark, 277–89.

Alves, T. C. L., Tommelein, I. D., and Ballard, G. 2006. Simulation as a tool for production system design in construction. *Proceedings of the 14th Annual Conference of the International Group for Lean Construction*, IGLC 14, Santiago, Chile, 341–53.

Arbulu, R., Koerckel, A., and Espana, F. 2005. Linking production-level workflow with materials supply. *Proceedings of the 13th Annual Conference of the International Group for Lean Construction*, IGLC 13, Sydney, Australia, 199–206.

Arbulu, R. J., and Tommelein, I. D. 2002. Alternative supply-chain configurations for engineered or catalogued made-to-order components: Case study on pipe supports used in power plants. *Proceedings of the 10th Annual Conference of the International Group for Lean Construction*, IGLC 10, Gramado, Brazil, 197–209.

Arbulu, R. J., Tommelein, I. D., Walsh, K. D., and Hershauer, J. C. 2002. Contributors to lead time in construction supply chains: Case of pipe supports used in power plants. *Proceedings of the Winter Simulation Conference 2002 (WSC02)*, San Diego, CA, 1745–51.

Azambuja, M., and Formoso, C. T. 2003. Guidelines for the improvement of design, procurement and installation of elevators using supply chain management concepts. *Proceedings of the 11th Annual Conference of the International Group for Lean Construction*, IGLC 11, Blacksburg, VA, 306–18.

Azambuja, M., and O'Brien, W. J. 2007. A qualitative evaluation of construction supply chain visual process modeling tools. *Proceedings of the Construction Research Congress*, Grand Bahamas, 9.

Bechtel, C., and Jayaram, J. 1997. Supply chain management: A strategic perspective. *The International Journal of Logistics Management*, 8, 15–35.

Chopra, S., and Meindl, P. 2004. *Supply Chain Management: Strategy, Planning and Operations*. Pearson Education, Upper Saddle River, NJ.

Davis, T. 1993. Effective supply chain management. *Sloan Management Review*, 34 (4), 35–46.

Elfving, J. A. 2003. Exploration of opportunities to reduce lead times for engineered-to-order products. PhD dissertation, University of California, Berkeley, CA.

Elfving, J. A., Tommelein, I. D., and Ballard, G. 2002. Reducing lead time for electrical switchgear. *Proceedings of the 10th Annual Conference of the International Group for Lean Construction*, IGLC 10, Gramado, Brazil, 237–49.

Fontanini, P. S. P., and Picchi, F. A. 2004. Value stream macro mapping: A case study of aluminum windows for construction supply chain. *Proceedings of the 12th Annual Conference of the International Group for Lean Construction*, IGLC 12, Copenhagen, Denmark, 576–90.

Green, S. D., Fernie, S., and Weller, S. 2005. Making sense of supply chain management: A comparative study of aerospace and construction. *Construction Management and Economics*, 23, 579–93.

Hayes, R. H., and Wheelwright, S. C. 1979. Link manufacturing process and product life cycles. *Harvard Business Review*, 57, 133–40.

Hopp, W. J., and Spearman, M. L. 2000. *Factory Physics*. 2nd ed. McGraw-Hill International Editions, Boston.

Hugos, M. 2006. *Essentials of Supply Chain Management*. John Wiley & Sons, Hoboken.

Jeong, J. G., Hastak, M., and Syal, M. 2006. Supply chain simulation modeling for the manufactured housing industry. *Journal of Urban Planning and Development*, 132, 217–25.

Jones, D., and Womack, J. 2003. *Seeing the Whole: Mapping the Extended Value Stream*. Lean Enterprise Institute, Brookline, MA.

Kleijnen, J. P. 2005. Supply chain simulation tools and techniques: A survey. *International Journal of Simulation & Process Modeling*, 1, 82–89.

Lambert, D. M., Pugh, M., and Cooper, J. 1998. Supply chain management. *The International Journal of Logistics Management*, 9 (2), 1–19.

London, K., and Kenley, R. 2001. An industrial organization economic supply chain approach for the construction industry: A review. *Journal of Construction Management and Economics*, 19 (8), 777–88.

Mathur, K., and Solow, D. 1994. American steel company case. In *Management Science; The Art of Decision Making*. Prentice Hall, 86–88.

Min, H., and Zhou, G. 2002. Supply chain modeling: Past, present and future. *Computers and Industrial Engineering*, 43, 231–49.

Naim, M., and Barlow, J. 2003. An innovative supply chain strategy for customized housing. *Construction Management and Economics*, 21, 593–602.

Naim, M., Naylor, J., and Barlow, J. 1999. Developing lean and agile supply chains in the UK housebuilding industry. *Proceedings of the 7th Annual Conference of the International Group for Lean Construction*, IGLC 7, Berkeley, CA, 159–70.

National Research Council. 2000. *Surviving Supply Chain Integration: Strategies for Small Manufacturers*. National Academy Press, Washington, DC.

O'Brien, W. J. 1998. Capacity costing approaches for construction supply chain management. PhD dissertation, Stanford University, Stanford, CA.

O'Brien, W. J., and Fischer, M. A. 2000. Importance of capacity constraints to construction cost and schedule. *ASCE Journal of Construction Engineering and Management*, 125 (6), 366–73.

O'Brien, W. J., London, K., and Vrijhoef, R. 2002. Construction supply chain modeling: A research review and interdisciplinary research agenda. *Proceedings of the 10th Annual Conference of the International Group for Lean Construction*, IGLC 10, Gramado, Brazil, 129–47.

Polat, G., Arditi, D., and Mungen, U. 2007. Simulation-based decision support system for economical supply chain management of rebar. *Journal of Construction Engineering and Management*, 133, 29–39.

Polat, G., and Ballard, G. 2003. Construction supply chains: Turkish supply chain configurations for cut and bent rebar. *Proceedings of the 11th Annual Conference of the International Group for Lean Construction*, IGLC 11, Blacksburg, VA, 14.

Tommelein, I. D. 2006. Process benefits from use of standard products: Simulation experiments using the pipe spool model. *Proceedings of the 14th Annual Conference of the International Group for Lean Construction*, IGLC 14, Santiago, Chile, 177–89.

Tommelein, I. D., and Li, A. E. 1999. Just-in-time concrete delivery: Mapping alternatives for vertical supply chain integration. *Proceedings of the 7th Annual Conference of the International Group for Lean Construction*, IGLC 7, Berkeley, CA, 97–108.

Tommelein, I. D., and Weissenberger, M. 1999. More just-in-time: Location of buffers in structural steel supply and construction processes. *Proceedings of the 7th Annual Conference of the International Group for Lean Construction*, IGLC 7, Berkeley, CA, 109–20.

Vaidyanathan, K., and O'Brien, W. J. 2003. Opportunities for IT to support the construction supply chain. *Proceedings of the 4th Joint International Symposium on Information Technology in Civil Engineering*, Nashville, TN, 19.

Vidalakis, C., and Tookey, J. E. 2006. Conceptual functions of a simulation model for construction logistics. *Proceedings of the Joint International Conference on Computing and Decision Making in Civil and Building Engineering*, Montreal, Canada, 906–15.

Vonderembse, M. A., Uppal, M., Huang, S. H., and Dismukes, J. P. 2006. Designing supply chains: Towards theory development. *International Journal of Production Economics*, 100, 223–38.

Vrijhoef, R., and Koskela, L. 2000. The four roles of supply chain management in construction. *European Journal of Purchasing & Supply Management*, 6, 169–78.

Walsh, K. D., Hershauer, J. C., Tommelein, I. D., and Walsh, T. A. 2004. Strategic positioning of inventory to match demand in a capital projects supply chain. *Journal of Construction Engineering and Management*, 130, 818–26.

3

Production Planning and Control and the Coordination of Project Supply Chains

Carlos T. Formoso
Federal University of
Rio Grande do Sul

Eduardo L. Isatto
Federal University of
Rio Grande do Sul

3.1   Introduction

In spite of the attention that supply chain management (SCM) has attracted among practitioners and researchers, the translation of its concepts, practices, and techniques into the construction industry is still a challenging issue, mostly due to differences that exist between this sector and other industries [O'Brien 2002]. SCM was originated in industries where demand is predictable, the requirements for variety are low, and volume is high [Christopher 2000]. Most contemporary academic work on SCM is concerned with high-volume industries, where a large-scale (hence economically powerful) manufacturer is supported by smaller (economically weaker) suppliers or subcontractors [Bresnen 1996]. Many studies adopt the assumption of long-term relationships and the active control—and even the design—of the supply chain by a single company.

By contrast, construction project supply chains are essentially temporary multiorganizations, arising at the start of the project, developing, and finally disbanding at the end of it [Cherns and Bryant 1984]. The management of such project supply chains involves the coordination of a fairly large number of stakeholders, and often no single firm has the power or the ability to individually coordinate the whole supply chain, due to the large number of work specialties involved and the fragmentation of the overall process among supply chain members. In fact, the poor coordination of stakeholders has been pointed out as a major cause of conflicts and controversies in construction projects [Olander and Landin 2005].

A construction project supply chain can be defined as a particular human system that is set up with the purpose of delivering a construction project, organized as a network of multiple firms bound together by economic linkages. Although emergent properties and layers of hierarchy will usually arise as a natural consequence of the economic links that connect each firm to the construction project, the same does not necessarily happen with the alignment of interests (sharing a common purpose with the others) and the effective coordination of actions among its members (communication and control) [Isatto 2005]. Indeed, these two characteristics are major concerns in supply chain coordination [Ballou, Gilbert, and Mukherjee 2000].

Stadtler (2005) points out that SCM is essentially the integration of organizational units along the supply chain and the coordination of materials, information, and financial flows in order to fulfill customer demands. Integration is concerned with the supply chain leadership, the establishment of a network of organizations, and the choice of partners, while coordination involves production planning and control, process orientation, and information systems [Stadtler 2005]. Integration relies mostly on the existence of adequate conditions in terms of knowledge sharing and cooperation among supply chain members, which may not always exist. By contrast, coordination does not necessarily assume the presence of cooperation and comprehensive knowledge about the supply chain processes, playing a very important role when some supply chain members enjoy high levels of technical or economic autonomy.

This chapter is focused on the production planning and control process, considering the need of coordinating construction project supply chains. Although the effectiveness of this process has been pointed out as a key factor for improving project performance, there is much dissatisfaction with the application and results of planning and control in the construction industry. In fact, this theme has been the focus of many construction management researchers since the early 1960s when the Critical Path Method (CPM) started to be disseminated [Laufer and Tucker 1987].

In the SCM literature there are also indications of dissatisfaction with the performance of production planning and control, despite important advances in theory and practice in the last 30 years [Wiendahl, von Cieminski, and Wiendahl 2005]. Traditional planning systems, such as enterprise resource planning (ERP), deal mostly with standard business work flows, while the biggest impact on business performance is created by exceptions and variability [Stadtler 2000].

Two theoretical approaches, the theory of coordination [Crowston 1991] and the Language–Action Perspective [Flores 1982], are used for discussing two important roles of production planning and control in the coordination of project supply chains: the definition of goals for guiding actions of individual supply chain members, and the management of commitments, respectively.

From a practical perspective, this chapter provides some guidelines for devising and implementing production planning and control for coordinating project supply chains. It describes a framework for planning and controlling construction projects, which adopts some core lean production[1] concepts and principles, being partly based on the Last Planner™ system [Ballard 2000]. The framework contains a description of the main planning and control activities, which are organized in three main hierarchical levels, emphasizing those that are concerned with the coordination of project supply chains.

The development of this framework was based on multiple case studies, carried out by the Building Innovation Research Unit (NORIE) of the Federal University of Rio Grande do Sul (UFSGS) between 1996 and 2004. Those studies were undertaken in small- and medium-sized construction companies from the south of Brazil, mostly in residential, industrial, commercial, and hospital projects. Some details on how the framework has been developed and the assessment of its impacts are reported elsewhere [Alves and Formoso 2000; Bernardes and Formoso 2002; Soares, Bernardes, and Formoso 2002; Schramm, Costa, and Formoso 2004; Schramm, Rodrigues, and Formoso 2006; Bortolazza and Formoso 2006].

3.2 Supply Chain Coordination

3.2.1 To Make or To Buy

Companies need to manage suppliers because it has been decided at some point in the past that buying a resource would be more advantageous for them than making it themselves. There are several possible reasons for such a decision. Some of them are of a strategic nature, such as, for example, to simplify the overall management effort by transferring part of it to suppliers, usually focusing on the core competences of the company and outsourcing the rest [Venkatesan 1992]. Others are more closely associated to economic factors. Resources that are rarely needed by the company will not offer enough scale economies to justify in-house production, it being more advantageous to buy them in the market. By contrast, the more specific a resource is for the company (i.e., there are few other companies that need the same resource), the more difficult it will be to find a supplier that is willing to produce that resource [Williamson 1985].

In construction projects, the end product usually has a high level of complexity and demands a large number of made-to-order or engineered-to-order components. Consequently, a wide range of specialties are necessary, making outsourcing a preferred strategy in many situations.

As far as production planning and control is concerned, outsourcing brings some disadvantages. The buyer company has less power to influence production pace and

---

[1] The lean production philosophy is discussed in Chapter 6.

methods than it would have if making them itself. Also, much of the information regarding production is kept by the supplier who is responsible for the job. Thus, the company's power to react to undesired situations is reduced, as is its ability to identify the imminence of such situations by monitoring current production status. Moreover, the complexity of managing such situations is not limited to the buyer–supplier linkages but also encompasses the management of the linkages among distinct suppliers.

In SCM, not all suppliers should be considered equally important: a selection of key suppliers must be made in order to make the supply chain more manageable [Lambert and Cooper 2000]. Several criteria may be used for choosing key suppliers. An initial selection is often based on the critical processes or subsystems, such as engineered-to-order or made-to-order suppliers. However, willingness to cooperate and the awareness of a common purpose are also crucial factors to consider [Cox and Ireland 2002].

3.2.2   Theory of Coordination

According to Malone and Crowston (1994), coordination is the act of managing dependences among activities. By contrast, March and Simon (1958) and Thompson (1967) regard coordination from the perspective of dependences among actors. These approaches are closely related, since dependences among actors arise from the dependences among the activities that they are assigned to. Nonetheless, those dependences demand some kind of management, thus making it necessary to design and implement specific coordination tasks aimed at this purpose [Crowston 1991]. From this perspective, production planning and control is essentially a coordination process (a sum of coordination tasks) since its primary aim is to manage the dependences among production activities or among the people to whom such activities are assigned.

Building on March and Simon (1958), Thompson (1967) identified three types of dependences among actors, and suggested a number of coordination mechanisms to manage each one of them:

(a) Pooled dependences: each actor can act independently. Coordination among actors is obtained by standardizing their actions; that is, using rules to constrain their actions to be consistent with others.

(b) Sequential dependences: each actor must readjust if the previous one does not do what is expected. Such dependences among actors can be managed by using schedules of their actions.

(c) Reciprocal dependences: all must readjust whenever one changes. Each actor must provide new information during their actions, allowing mutual adjustment.

Mutual adjustment, also known as coordination by feedback, is typical of organizations that face a high degree of uncertainty, while standardization is more appropriate to situations in which variability is low and predictability is high [March and Simon 1958].

Crowston (1991) defined coordination problems as tasks creating and using resources, identifying three possible types of task–resource dependences: (a) a task produces the resource that is used by another task (flow dependences); (b) a set of tasks uses or creates common resources; and (c) dependences among tasks and subtasks.

Flow dependences typically arise between suppliers that are responsible for sequential activities in the same process. In case of flow dependences, three coordination problems must be addressed simultaneously: (a) the resource must be made available by one task at the time it is needed by the other (prerequisite condition); (b) the quality of resources must be adequate for its use (usability condition); and (c) the resources must be available in the right place (availability) [Crowston 1991].

In construction projects, there are several resources that need to be shared by different tasks, such as suppliers competing for the same dock for delivering materials on site, or different designers that work individually to produce a joint drawing. In situations that involve the use or generation of a common resource, Crowston (1991) pointed out that the nature of the common resource will be critical to determining the best coordination method to be adopted. If the common resource is shareable, no conflict will arise from its simultaneous use, but if two tasks create the same resource then it will be necessary to eliminate duplicate effort or to negotiate the object, choosing the best of them. By contrast, if the resource is nonshareable but reusable then the coordination will demand programming the use of the resource (as in flow dependences), or combining the outcome of tasks if the resource is created by two or more. However, if the common resource is nonshareable and consumable, then the coordination task will need to involve favoring one activity over another (in the case of the use of common resources) or seeking economies of scale (in the case of the creation of common resources) [Crowston 1991]. Obviously, the management of such dependences may also involve some kind of change in the nature of the resource, such as, for instance, replacing paper copies of a drawing (a nonshareable but reusable resource) by electronic copies (a shareable resource).

The decomposition of a task into subtasks is necessary for two main reasons. One reason is the need to manage the overall process by breaking down a task into manageable subtasks and assigning them to different actors. The second reason is to subsequently transfer some subtasks to an actor through task assignment or outsourcing. The original task should be considered as a goal to be achieved by those actors since the accomplishment of the main task only happens when all sub-tasks have been finished. Also, subtasks should usually be considered as intermediary or secondary goals [Crowston 1991]. Nonetheless, the decomposition process must be conducted with care because it generates dependences among subtasks, introducing the need for subsequent coordination of those dependences. The application of a work breakdown structure (WBS) for producing construction plans is closely related to this kind of decomposition purpose.

The way tasks are assigned to different actors will greatly influence the need for subsequent coordination. For example, if a task is decomposed into two sequential tasks, flow dependences arise between them. In the case that each of these subtasks is transferred to a distinct actor, those responsible for the original task will need to coordinate the flow of resources (in terms of prerequisite, availability, and usability) from one of the actors to the other. However, if both subtasks are assigned to the same actor, the coordination task itself is usually transferred to that actor too.

An adequate choice of strategy to be used to decompose the original task is very important, since it will influence the kind of coordination problem that will arise among actors. March and Simon (1958) suggested two forms of specialization to guide the process of task decomposition and assignment: (a) specialization by process, grouping people that know only a particular process (e.g., engineering, design, procurement); or (b) specialization by purpose, grouping people that work only in a particular part of the product (e.g., subcontractors of the same specialty for different parts of the project). The choice for one or another will greatly influence the need for further coordination. It is expected that specialization by process will generate flow dependences among actors because each actor is responsible for a single part of the whole process, and that specialization by purpose will demand a greater use or creation of common resources.

In practical terms, outsourcing strategies usually apply a combination of these approaches. For example, a company may choose to outsource the subsystems of a project from different suppliers, thus generating the need to coordinate the use and generation of common resources. However, the company may also decide to outsource from each supplier the design, production, and installation of the same subsystem, thus transferring to them the coordination of flow dependences between those sequential tasks.

3.2.3   Language–Action Perspective

The Language–Action Perspective, proposed by Flores (1982), sustains that the effective coordination of actions can be considered the same as the management of commitments, and the progress of work can be traced by watching speech acts in the communications of those coordinating. Moreover, Winograd and Flores (1986) pointed out that the breakdowns experienced by organizations are a direct consequence of the failures that occur in those speech acts.

Based on Flores (1982) and on Winograd and Flores (1986), Denning and Medina-Mora (1995) suggested that commitments must be regarded as workflow loops. A workflow loop always arises as a response to—or an anticipation of—a breakdown, and involves the execution of some kind of action in the real world in order to deliver something according some agreed schedule.

According to Denning and Medina-Mora (1995), a workflow loop (commitment) is a business transaction that involves two parties, one of them (the performer) promising to satisfy a request from the other (the customer). They also pointed out that four phases must be present in every workflow loop, and the completion of each phase occurs by the utterance of a specific kind of sentence by the performer or the customer. As shown in Figure 3.1, each one of the four stages is separated from the other by a specific kind of speech act [Denning and Medina-Mora 1995]:

(a) Request: the customer makes a request to the performer (or accepts an offer made by the performer), by uttering the speech act "I request."

(b) Negotiation: they negotiate the conditions that will satisfy the customer, culminating in the performer's promise (implied contract) to fulfil those conditions by declaring "I promise."

(c) Performance: the performer does the work and ends by stating that it is done ("I am done").

(d) Satisfaction: the customer accepts the work and declares satisfaction ("I am satisfied"), indicating that the implied contract has been fulfilled.

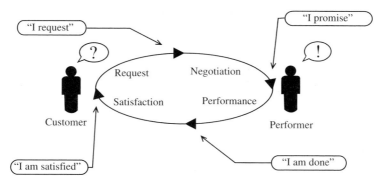

FIGURE 3.1 The commitment loop. (Based on Denning, P.J. and Medina-Mora, R., *Interfaces*, 25, 42–57, 1995. With permission.)

Denning and Medina-Mora (1995) also argued that every organization is a network of commitment loops, thus suggesting that a map of those interconnected commitment loops could be used as a guide to design and to manage work processes and their supporting information technologies. A construction project supply chain involved in the production planning and control process can be perceived as such a network, in which a number of commitment loops are intentionally interconnected to achieve some kind of change in the real world—in this particular case, the effective and efficient delivery of a construction project. From this perspective, production planning and control can be regarded as a mechanism to guide the process of progressively breaking down a project into a number of subtasks, which are assigned to different people as commitments. The associated control mechanism relies on the monitoring of the accomplishment of these commitments instead of the direct observation of action, which is impossible in many situations, as mentioned above.

A study carried out by Azambuja et al. (2006) in two made-to-order construction supply chains indicated that most of the breakdowns are caused by incompleteness of workflow loops or failures in connection among them. Therefore, both workflow completeness and connectivity must be addressed with caution when implementing a planning and control system:

(a) Every workflow loop must be complete, meaning that all four phases and associated speech acts must be undertaken in order to obtain effective commitments. In some cases this may imply the need to devise foolproof devices to avoid breakdowns due to the incompleteness of the workflow, such as, for example, the need to explicitly mention the date a task completion is expected to occur, the formal agreement by signing production plans, or the use of printed forms that force a clear statement of completed tasks.

(b) The planning and control process and the supporting information system must be designed in order that workflow loops are well connected to provide a network of commitments. In some cases, databases may be necessary to assure the consistency of information used by distinct work flows.

3.3   Planning and Control in Project Supply Chain Management

3.3.1   What is Planning?

Planning is a cognitive activity familiar to everyone. It plays a key role in decision making by enabling individuals to deal with changing and complex situations. Plans are used either formally or informally, for guiding any activity that has not been entirely automated. Planning mechanisms intervene in situations where a response cannot be obtained from rules triggered by current environment information [Hoc 1988].

In the field of production management, it is important to make a distinction between the tasks of designing and operating a production system, although both of them represent different forms of planning. Production system design is to discuss and translate the intended production strategy into a set of decisions, forming the structure that will manage the different activities [Slack, Chambers, and Johnston 2004]. It means that the production system design should create appropriate conditions for control and improvement [Ballard et al. 2001]. At the strategic level, designing a production system should involve not only on-site production, but also suppliers and consumers. In operational terms, the concern is to devise the layout as well as the material and information flows in order to create favorable conditions for a high-performance production system [Slack, Chambers, and Johnston 2004].

Production system operation is concerned with production planning, monitoring, and correcting. Indeed, Hayes-Roth and Hayes-Roth (1979) defined planning as the first stage of a two-stage problem-solving process named "planning and control," in which planning is the predetermination of a course of action aimed at achieving a certain goal, and control consists of monitoring and guiding the execution of the plan to a successful conclusion. Hoc (1988) emphasizes that planning is concerned with making a decision based on extrapolation from past events.

In this chapter, planning is defined as the process of setting goals and establishing the procedures to attain them, only being effective if intertwined with the process of controlling activity execution. Therefore, a planning and control system aims to match company output and logistic performance to the customer demands: its role is to plan, initiate, and control product delivery [Wiendahl, von Cieminski, and Wiendahl 2005]. It means that such a system should monitor production and, in the case of unforeseen deviations, readjust the order progress or the production plans.

In SCM, a planning and control system should cross company boundaries, taking into account the delivery performance of the suppliers as well as the demand behavior of the customers [Wiendahl, von Cieminski, and Wiendahl 2005].

3.3.2   Main Flaws in Production Planning and Control

Several reasons for the ineffectiveness of production planning and control in construction have been pointed out in the literature:

(a) Production planning is not usually regarded as a managerial process, but simply as the application of techniques for preparing plans. In general, little effort is dedicated to the gathering of reliable data and to the dissemination of information [Laufer and Tucker 1987]. The fact that several different organizations are involved in a single project makes data collection and planning output diffusion even more difficult.

(b) Control is usually based on the exchange of verbal information between the engineer and the foreman. It is usually focused on short-term decisions and has no link to long-term plans [Formoso 1991], creating many problems regarding the management of long lead resources, especially engineered-to-order components.

(c) Many construction companies tend to emphasize the control related to global project aims, and fulfillment of contracts, rather than production control. In this context, spotting problems in the production system and defining corrective lines of action often become problematic [Ballard and Howell 1997].

(d) Uncertainty about the future is a common feature of most problems involved in construction planning. However, it is often neglected, and the necessary actions to minimize it or eliminate its effects are not usually undertaken [Cohenca, Laufer, and Ledbetter 1989]. This is often related to the fact that production planning systems do not provide appropriate mechanisms for managing reciprocal dependences. Moreover, much effort is often spent in the production of long-term plans that are too detailed. The high level of details allied to uncertainty requires frequent time-consuming updates of plans [Laufer and Tucker 1988].

(e) Information technology has had only a limited impact in production planning and control. Computer systems are often acquired and introduced in an organizational environment without the previous identification of their users' needs. This may result in the production of a large amount of useless data that indicate only the deviation from the goals to be achieved and not the causes for such deviation [Sanvido and Paulson 1992]. Moreover such systems are usually implemented in an isolated fashion in the company without considering the necessary integration to other existing systems. Lack of systematic training related to the use of such systems is another problem that has been detected [Turner 1993]. Similar problems have also been reported outside the construction industry [Wiendahl, von Cieminski, and Wiendahl 2005]

(f) Construction managers tend to be action oriented. Most of them make decisions mostly based on their intuition and common sense, rather than on data systematically collected and analyzed [Lantelme and Formoso 2000].

As construction facilities become more complex and the markets more dynamic and fragmented, the levels of outsourcing and subcontracting tend to increase, boosting the number of organizations involved in the project supply chains, including subcontractors, designers, material suppliers, and consultants, among others. This creates additional difficulties for coordinating supply chains, such as: (a) the number of planning alternatives increases dramatically [Fleischmann, Meyr, and Wagner 2000]; (b) divergent stakeholder interests need to be managed [Wiendahl, von Cieminski, and Wiendahl 2005]; and (c) lack of overall understanding of the project by different participants [Formoso, Tzortzopoulos, and Liedtke 2002].

Both uncertainty and complexity of construction projects have a strong influence on the way projects should be coordinated and, consequently, on the requirements for planning and control systems [Baccarini 1996]. It is often necessary to use outsourcing strategies that transfer to suppliers the coordination of flow dependences between sequential tasks, such as designing, fabrication, and installation. In this context, the planning process should be decentralized, and more responsibility should be given to supply chain members. Consistent objectives must be defined and the responsibilities for fulfilling them need to be communicated clearly [Wiendahl, von Cieminski, and Wiendahl 2005].

3.3.3   Recent Advances in Production Planning and Control in the Construction Industry

In recent years, some important advances in construction management, both in theory and practice, have been made by understanding and adapting some core operations management concepts and principles to the management of production systems in construction [Koskela 2000]. This was one of the main motivations for the creation of the International Group for Lean Construction [IGLC 2007].

In IGLC annual conferences, many papers have reported the use of the Last Planner™ system [Ballard 2000] in construction projects, indicating that this system has been successfully implemented in a large number of projects from different countries, such as the United States, Brazil, Chile, Ecuador, England, Finland, and Denmark, among others [IGLC 2007]. This system is able to increase the reliability of short-term planning by shielding planned work from upstream variation, and by seeking conscious and reliable commitment of labor resources by the leaders of the work teams involved [Ballard and Howell 1997]. At the medium-term level, constraints are identified and removed, ensuring that the necessary materials, information, and equipment are available [Ballard 1997].

Although there is a lack of quantitative studies that assess the impact of the Last Planner™ system in the performance of construction projects, there are some indications that it has many advantages over traditional CPM-based planning and control systems. The way Last Planner™ manages commitments and the stability that it creates in production systems are among the main reasons for its success [Vrijhoef, Koskela, and Howell 2001]. Also, it adopts a fairly simple planning and control approach and, similarly to the Toyota Production System [Wiendahl, von Cieminski, and Wiendahl 2005], emphasizes organizational aspects instead of the application of complex software systems.

Although the Last Planner™ System is well described in the literature [Ballard and Howell 1997; Ballard 1997, 2000], much needs to be discussed on the core ideas that underlie this system. Moreover, there is a continuing effort to further improve it, for instance, by integrating other managerial functions [Marosszeky et al. 2002; Saurin, Formoso, and Guimarães 2004], extending to other managerial levels [Ballard and Howell 2003], and developing software tools that support its implementation (see, for instance, SPS® Production Manager in Chapter 7).

3.4   Framework for Production Planning and Control

3.4.1   Overview

Hopp and Spearmann (1996) propose three steps for developing a planning framework, which is very similar to the decomposition process discussed in Section 3.2. First, the overall system is appropriately divided into a set of subtasks (e.g., processes, product categories, horizons, people, etc.) that make each piece manageable but still allow integration. The second step is to identify links between those subtasks, and the third one is to use feedback to enforce consistency. For instance, Fleischmann, Meyr, and Wagner (2000) proposed a matrix for classifying the supply chain planning tasks according to the planning horizon (long, medium, and short term) and supply chain process (procurement, production, distribution, and sales).

Different hierarchical levels are necessary because production management decisions differ greatly with regard to the length of time over which their consequences persist [Hopp and Spearman 1996]. Long-term planning is mostly related to strategic decisions, concerned with setting objectives [Laufer and Tucker 1987]. Middle management is mostly concerned with the means for achieving those objectives, involving tactical decisions, such as determining what to work on and who will work on it, within the constraints established by long-range decisions [Hopp and Spearman 1996]. Finally, at the operational level, short-term decisions address control, by moving materials and workers, adjusting processes and equipment, and taking whatever actions are required to ensure that the system continues to function towards its goal [Hopp and Spearman 1996].

Different planning horizons imply distinct planning frequencies, modeling assumptions, and levels of detail.[2] A major challenge in any planning and control system is to maintain consistency between different decision-making levels. In fact, the effectiveness of a planning and control system depends not only on how it solves subproblems but rather on how well it coordinates these with one another [Hopp and Spearman 1996].

Regarding the construction industry, Laufer and Tucker (1987) divide the production planning and control process into five planning phases: (a) planning the planning process; (b) information gathering; (c) preparation of plans; (d) information diffusion; and (e) planning process evaluation. The first and the last phase are typical of project-based industries, because each project may demand a different planning and control system configuration. For instance, the appropriate planning horizons and the required level of detail may vary according to the degree of variability and uncertainty involved.

The tasks "information gathering," "preparation of plans," "information diffusion," and "action" form a continuous production control cycle that must be undertaken at different planning levels. Figure 3.2 illustrates that control cycle. The preparation of plans involves setting targets, forecasting some information that is necessary for producing plans, allocating resources (materials, labor, equipment), and deciding the tasks to be carried out. Action is concerned with performing production tasks, but also procuring or scheduling the delivery of resources. Finally, information gathering involves data collection, processing and analysis, and creating learning opportunities.

Another important aspect of production planning and control is to define which processes will be pushed and which ones will be pulled.[3] According to Hopp and Spearman

---

[2] Chapter 2 presents a hierarchy of decision categories related to SCM from the perspective of organizations. In this chapter planning hierarchical levels are concerned with project supply chains.

[3] Push and pull systems are also discussed in Chapter 6 and Chapter 7.

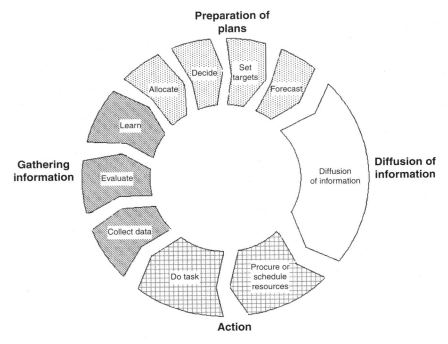

FIGURE 3.2   Production control cycle. (Adapted from Wiendahl, H.H., von Cieminski, G., and Wiendahl, H.P., *Production Planning and Control*, 16 (7), 634–51, 2005. With permission.)

(1996), a push system schedules the release of work based on demand, while a pull system authorizes the release of work based on system status. It means that a push system releases a job based on an exogenous schedule, and the release time is not modified according to what is happening in the plant. In contrast, a pull system only allows a job onto the floor when a signal generated in line status calls for it. These are, in fact, two different methods for coordinating flow dependences. Most real-world production systems are a combination of push and pull systems [Hopp and Spearman 1996].

Regarding the content of construction plans, most traditional WBS-based production plans adopt what Koskela (1992) calls the conversion model, which assumes that process improvement can be mostly achieved by improving each of its parts (subprocesses). The conversion model has, to some extent, contributed to the lack of transparency, since it abstracts away the flows between the conversion activities, and does not encourage the clear identification of internal and external clients in each process [Koskela 1992]. The focus of control in the conversion activities is a major cause of uncertainty in process management, increasing the share of non-value-adding activities [Alarcón 1997]. Therefore, production plans should make explicit not only conversion (or value adding) activities, but also production and information flows, so that these are adequately monitored and controlled. There are several techniques that can be used for doing that, such as the line of balance and process mapping.

3.4.2 Description of the Framework

In the construction industry, the production planning and control process often plays the role of designing, at least partially, the production system. This is because the production system in construction is usually implicit or taken for granted [Ballard et al. 2001]. More recently, some studies have proposed ways of comprehensively and explicitly designing production systems, before the beginning of the production stage [see, for instance, Tsao et al. 2004; Schramm, Costa, and Formoso 2004; Schramm, Rodrigues, and Formoso 2006].

Table 3.1 presents an overview of the proposed production planning and control framework, which is divided into three hierarchical levels. It shows the main categories of decisions involved and the main participants in the production of plans, as well as typical planning horizons and control cycle time. Some of the decisions included in this framework should be first established at the level of production system design, such as overall installation sequence (initial breakdown of the work into subtasks), rhythm of key processes, site layout, and resource capacity [Schramm, Costa, and Formoso 2004]. The role of long-term planning should be to refine or adjust those decisions, establishing a production environment capable of meeting the production system's overall goals [Hopp and Spearman 1996].

The number of hierarchical planning levels may vary, depending on the organizational structure available and also on the size and complexity of the project. For instance, Ballard and Howell (2003) proposed an additional planning level between master planning and look-ahead planning named phase scheduling, which is mostly concerned with the definition of the overall installation sequence. However, most of the companies involved in the case studies carried out at NORIE-UFRGS adopted the three planning levels presented in Table 3.1.

3.4.2.1 Long-term Planning

Long-term (or master) plans establish the general goals to be achieved during the execution of the project. For that reason, they should not be revised frequently. In many companies, changes in the master plan are typically event-related. It means that a new plan is not drawn up at regular intervals but only in case of an important event [Fleischmann, Meyr, and Wagner 2000].

The master plan horizon is normally the total duration of the production stage of the project. It strongly depends on project finance: the duration of the project as well as the timing of some of its milestones are generally determined by the project cash flow.

Long-term planning plays an important role in project SCM since it should synchronize key on-site production processes and also off-site production, such as the production of prefabricated components, based on a *takt* time[4] defined according to project milestones. As a result of synchronization, it is possible to reduce inventory and work

---

[4] *Takt* is a German word for rhythm or meter. In the Toyota Production System, it is the rate at which the customer is buying products. It is used to set the pace of production and alert workers whenever they are getting ahead or behind [Liker 2004]. Differently from the car industry, the demand is dependent in construction. For that reason, the *takt* time may be established according to project milestones, rather than directly from customer demand [Bulhões, Picchi, and Granja, 2005].

TABLE 3.1 Overview of the Production Planning and Control Framework

Hierarchical Level	Main Output	Main Decision Categories	Typical Planning Horizon	Typical Control Cycle Time	Main Participants
Long term	Master plan	– Production milestones based on cash flow forecast – Overall installation sequence – Overall site layout – Rhythm (*takt* time) of key processes – Capacity planning – Scheduling long lead time (Class 1) resources	Production phase	Event based	– Production director – Project manager – Design manager – Site engineer
Medium term	Look-ahead plan	– Breaking down work packages – Constraint identification and removal – Releasing work packages for short-term planning – Scheduling resources (Class 2) – Pulling design information – Design of critical processes – Site layout revision and preparation of work spaces	5–13 weeks	2–4 weeks	– Project manager – Site engineer – Key subcontractors – Foreman – Purchasing department representative – Safety specialist – Quality manager – Design manager
Short term	Commitment plan	– Refining work packages – Assigning tasks to crews – Pulling short lead time resources (Class 3)	1 week	1 week	– Site engineer – Foreman – All subcontractors – Safety specialist – Representatives of crews

in progress levels. It is necessary to establish an adequate use of capacity for the supply chain members involved in order to synchronize material flows effectively [Rohde and Wagner 2000].

Based on the master plan, long lead time resources are procured—these have been named in the proposed framework as Class 1 resources. These are usually purchased in one batch, and their delivery time might last several months (e.g., building elevators).

Several different techniques may be used for producing and controlling the master plan, such as bar charts, lines of balance, and CPM networks. Due to the large number of activities involved and the need to control project duration, a software tool is often used. Many companies find it useful to have a version of the master plan that is easy for everyone to understand, containing some key information such as the rhythm of key processes, milestones, and work flows. In repetitive projects the line of balance can play this role (see example of Figure 3.3).

In general, middle- and higher-level managers should be involved in the generation of the master plan. If necessary, some supporting staff may participate in this process, such as a planning engineer. In case there is overlapping between the design and production stages, the participation of the design manager or representatives of the design team is necessary in order to synchronize the delivery of design drawings and production activities.

Considering the overall installation sequence and the site layout, the facility to be built should be divided into work zones (e.g., floors, apartments, rooms, etc.), which are important for defining production batch sizes and hand-offs between crews. For instance, Figure 3.4 shows a plan from one of the projects investigated at NORIE-UFRGS indicating the division of a facility into work zones, which was used as a reference for planning the flows of four different crews.

3.4.2.2   Medium-term Planning

Medium-term planning establishes a link between the long-term plan and the operational plans, which are the ones that effectively guide execution on site. At this level, the long-term plan work packages are divided into smaller packages, based on the work zones previously established. In this readjustment of work packages, the milestones and the *takt* time established at the master plan must be considered.

The medium-term planning has a rolling horizon, typically from 5 to 13 weeks. The control cycle time usually varies from one to four weeks, which means that the same week needs to be replanned a few times before execution. The planning horizon and control cycle time depend on the speed and on the degree of complexity and uncertainty involved—the faster, the higher the uncertainty, and the more complex the project, the shorter the medium-term planning horizon and its control cycle tend to be. In the case studies carried out at NORIE-UFRGS, for instance, fast refurbishment industrial building projects normally had 5–9-week look-ahead plans that were updated weekly. By contrast, slower and less complex projects, such as new residential building, typically had a three-month medium-term planning horizon, and an update cycle of one month.

In the Last Planner™ System, the main role of look-ahead plans is the identification and removal of constraints in the production environment, such as procuring materials and lining up subcontractors [Ballard 2000]. Once constraints are identified, the necessary actions to make work ready must be carried out, such as producing design details, ordering materials, or hiring labor. All work packages from the look-ahead plan must be systematically screened and the ones that have no constraints may be released to the short-term plan. This was named by Tommelein and Ballard (1997) as the screening and pulling mechanism: those resources and information are pulled from the look-ahead plan, rather than pushed by long-term plans.

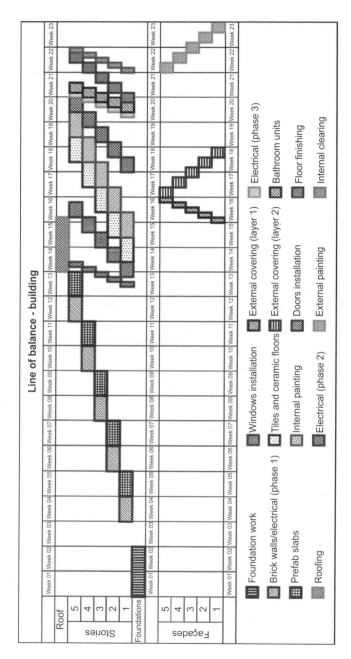

FIGURE 3.3   Example of a line of balance that makes some key decisions transparent to everyone.

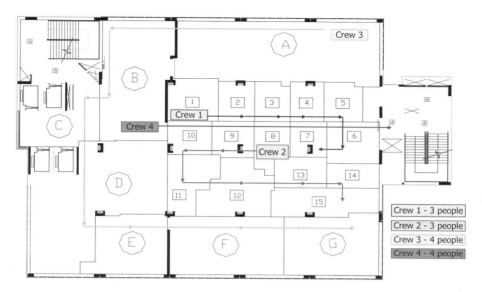

FIGURE 3.4   Example of plan showing the division of a facility into work zones.

The resources that are procured at the look-ahead planning level are classified in the proposed framework as Class 2. These generally have a lead time shorter than the medium-term planning horizon, usually less than 30 days, and are often delivered in more than one batch. Of course, the classification of resources in Class 1 or Class 2 depends on the procurement practices of each company and also on the planning horizons that are adopted in each project.

In some projects, preparing a look-ahead plan may be time consuming due to the large number of activities and constraints involved, sometimes lasting for more than three hours. Several people should participate in the look-ahead planning meeting, including production managers, a representative of the purchasing department, the design manager, the safety specialist, and representatives of key subcontractors, among others. Their participation is usually important, since they are all capable of contributing in terms of identifying and, sometimes, removing constraints. In one of the case studies carried out at NORIE-UFRGS, even a representative of the client was invited to take part in those meetings, because during a certain stage of the project the client organization had a major role in removing constraints. A common mistake made by site managers is to prepare look-ahead plans by themselves: such plans are generally ineffective because very rarely is one single person capable of identifying all look-ahead constraints in a construction project.

Another important role of the medium-term planning level is to revise the site layout and the preparation of work spaces, such as inventory spaces, installation of safety safeguards, and temporary access for equipment. This type of action should be based on the look-ahead plan because, on the one hand, it needs reliable information that is not available from the master plan and, on the other hand, it requires some time to be implemented. If they are left to the short-term planning level, this may cause some delays. Explicitly planning and controlling the physical flows in construction sites is an important step to reduce the share of non-value-adding activities [Alves and Formoso 2000]. If necessary, the sequencing and pace of work must be adjusted, in order to reduce the congestion of people and materials.

3.4.2.3 Short-term Planning

The main role of short-term (or operational) planning is to control production directly. The work packages that have had all constraints removed are filtered and may be assigned to different gangs. If the necessary resources are not available, a work package must be rescheduled for a later date [Ballard 2000]. This is another key element of the Last Planner™ system: production is shielded from workflow uncertainty, increasing the reliability of short-term planning, and enabling production units to improve their own productivity [Ballard and Howell 1997].

The operational plan is produced in a meeting in which representatives of different gangs, including the subcontracted ones, agree upon the production short-term goals. The aim is to obtain a conscious and reliable commitment of labor resources by the leaders of the work teams involved, rather than simply delivering preprepared plans to them, as is often done in traditional planning. For that reason this level of planning is called commitment planning [Ballard and Howell 1997]. This meeting is an opportunity to solve conflicts and negotiate common goals between different gangs, since a large number of dependences in construction are reciprocal. A common mistake made by site managers is to prepare the weekly plan by undertaking separate meetings with different subcontractors: this often results in poor management of the linkages among distinct subcontractors.[5]

The planning horizon at this level is typically one week. For that reason, this level of planning is also called weekly planning. Very rarely, it might be necessary to reduce the planning cycle to one day—this happens in projects that are too fast and have a very high level of uncertainty.

Ballard and Howell (1997) proposed a set of quality criteria that must be considered in the definition of each work package:

(a) *Definition*: Work packages must be specific enough in terms of type and quantity of materials to be used. It must be possible to clearly identify at the end of the week whether the assignment has been completed.

(b) *Soundness*: The necessary resources must be available whenever they are asked for.

(c) *Sequence*: Work packages must be selected considering the sequence that is demanded by the clients and the constructability of the production process.

(d) *Size*: The size of a work package must correspond to the productive capacity of each production team.

---

[5] See more about the use of the Last Planner™ system for managing subcontractors in Chapter 7.

(e) Learning: The work packages that were not completed within the week must be tracked and the real causes for their delay must be analyzed, in order to define corrective actions and to identify the packages that are likely to be affected.

During the production of the short-term plan, a workable backlog might be established. This backlog consists of a set of activities that fulfill the short-term plan quality requirements, but have not been identified as priorities in the long-term plan [Ballard 2000]. It works as a contingency plan that might be carried out to guarantee the continuity of the work of the production teams in case a problem occurs [Ballard and Howell 1997].

At the end of the short-term planning cycle, an analysis of planning effectiveness is carried out, by using an indicator named PPC (Percent Plan Complete), proposed by Ballard and Howell (1997). This indicator is the rate between the number of assignments concluded and the total number of scheduled work packages. The root reasons for noncompletion of work packages must be identified, so that corrective measures can be implemented.

At the short-term level, Class 3 resources may be pulled directly from site management, based on inventory control. These are typically made-to-stock low-cost items, and have a very short lead time.[6]

3.5 Discussion

As was mentioned before, from the point of view of the main contractor, outsourcing often results in some level of blindness regarding the tasks that are transferred to suppliers. Since it is no longer possible to directly control the execution of tasks, management progressively shifts from monitoring based on the observation of actions towards the alignment of actors' objectives [Grandori 1997].

This shift brings fairly subtle but important changes. Firstly, the analytical approach to processing control, based on observation of action and its comparison with expected standards, becomes ineffective since economic boundaries will now hide the details of process execution. Secondly, the importance of achieving cooperation of the parties involved increases with the increment of their autonomy in process execution. Finally, in order to coordinate several actors simultaneously, it is of critical importance to well define the project objectives for each project supply chain member as well as the sequence in which each of those actors will intervene in production in order to guide their individual actions towards a common direction. That seems to be one of the roles played by production planning and control in the coordination of construction project supply chains. This happens especially at the long-term level, in which some project goals are defined, such as milestones, overall installation sequence, and *takt* time.

Another important role of production planning and control is the management of commitments, pointed out by Macomber and Howell (2003) as one of the core practices of the Last Planner™ system.

---

[6] See Chapter 8 for some good practices on the management of this type of resource.

Each of the hierarchical planning and control levels plays a distinct role in terms of the management of the commitments. Look-ahead planning addresses mainly the commitments regarding the procurement or scheduling of resources and information delivery, occasionally including the negotiation among subcontractors or different crews about shareable and nonconsumable resources, such as, for example, the use of space, in order to avoid interferences among different workstations or conflicts concerned with storage space needs for distinct supplies.

Nonetheless, most of the ability of the Last Planner™ system in managing commitments effectively along the supply chain and also across the hierarchical levels of the planning and control system relies on the short-term planning level. This includes a clear statement by the project manager on the work packages that are required to be executed in the following period, a good definition of work packages in terms of their soundness, a fair negotiation with the crews and subcontractors in order to get realistic assignments, and their explicit commitment to the accomplishment of these assignments. It is also important that the crew leaders and subcontractors are aware of the need to get their teams committed to the assignments that were agreed.

At the end of each short-term planning cycle, subcontractors and crew leaders will be asked if they have concluded their assignments, and the production (or site) manager will state whether he/she is satisfied, by accepting or refusing each of those deliveries. At that point, workflow loops are closed and the cycle starts again with the request of subsequent work packages.

3.6 Conclusion

This chapter presented an overview of a production planning and control framework that was devised for small- and medium-sized companies. This framework adopted several elements of the Last Planner™ system, which has been successfully adopted in construction companies from several different countries.

Two important roles of production planning and control in the coordination of project supply chains were discussed: the definition of goals for guiding actions of individual supply chain members, and the management of commitments. The first role is mostly played by the master plan, which should make transparent to project supply chain members some major goals to be achieved in the construction stage of the project, such as milestones, the rhythm of some key processes (*takt* time), and the overall installation sequence. The second role depends strongly on the set of meetings that are carried out as part of the planning and control process, especially at the look-ahead and operational levels.

The implementation of the Last Planner™ system in practice seems to indicate that the management of commitments still relies strongly on short-term planning, because there are formal mechanisms for closing workflow loops at this level. Despite the critical role played by look-ahead planning in the management of project supply chains, this seems to be the least developed level of planning. Further research is necessary on how to make its implementation more successful and the management of commitments at this level more effective.

From a practical perspective, this chapter proposes a number of practical guidelines for devising production planning and control, which are summarized below:

(a) Planning should be divided into hierarchical levels: this is an important mechanism for dealing with uncertainty and also for adequately supporting decision making at different managerial levels.

(b) Each managerial planning level has different roles: The distinction between different planning levels is not limited to the amount of detail made available. Each level has a different role in the planning and control process and involves a set of decision categories, as suggested in the proposed framework.

(c) Different projects have distinct planning needs: The production planning and control process should be defined according to the characteristics of each project, such as its size, degree of complexity, the level of uncertainty involved, and the organizational structure available. For instance, the faster, the higher the uncertainty, and the more complex the project, the shorter the medium-term planning horizon and its control cycle tend to be.

(d) Resource procurement and scheduling must be connected to the planning process: Resources must be classified according to their lead time and the type of coordination method to be used (e.g., pushing or pulling). Then, each class of resources should be clearly linked to a level of production planning and control.

(e) Consistency between different planning levels should be maintained: the effectiveness of a planning and control system depends not only on the performance of each subprocess, but also on whether there are mechanisms to keep them consistent.

(f) Feedback and learning must be effective: Consistency between planning levels depends on the effectiveness of feedback [Hopp and Spearman 1996]. It is necessary to devise a set of metrics that are easy to collect and analyze—in fact, this is a major strength of PPC. It is also important to have opportunities for discussing and reflecting on the information available in order to stimulate learning.

References

Alarcón, L. F. 1997. Tools for the identification and reduction of waste in construction projects. In: Alarcón, L. F. (ed.). *Lean construction*. Rotterdam: A.A. Balkema, 365–77.

Alves, T. C. L. and Formoso, C. T. 2000. Guidelines for managing physical flows in construction sites. *Proceedings of the 8th Annual Conference of the International Group for Lean Construction*, Brighton, England.

Azambuja, M. M. B., Isatto, E. L., Marder, T. S., and Formoso, C. T. 2006. The importance of commitments management to the integration of make-to-order supply chains in construction industry. In: Sacks, R. and Bertelsen, S. (Org.). *Proceedings of the 14th Annual Conference of the International Group for Lean Construction*, Santiago, Chile, 609–24.

Baccarini, D. 1996. The concept of project complexity: A review. *International Journal of Project Management*, 14 (4), 201–4.

Ballard, G. 1997. *Look-ahead planning: The missing link in production control.* Berkeley, USA: Dept. of Civil and Environmental Engineering, University of California. Technical Report 97-3.

———. 2000. *The last planner system of production control.* University of Birmingham. PhD Thesis.

Ballard, G. and Howell G. 1997. Shielding production: An essential step in production control. *Journal of Construction Engineering and Management*, 124 (1), 11–17.

———. 2003. An update on the last planner. In: Martinez, J. and Formoso, C. T. (Org.). *Proceedings of the 11th Annual Conference of the International Group for Lean Construction*, Blacksburg, East Virginia, USA.

Ballard, G., Koskela, L., Howell, G., and Zabelle, T. 2001. Production system design in construction. *Proceedings of the 9th Annual Conference of the International Group for Lean Construction*, Singapore.

Ballou, R., Gilbert, S. M., and Mukherjee, A. 2000. New managerial challenges from supply chain opportunities. *Industrial Marketing Management*, 29, 7–18.

Bernardes, M. M. S. and Formoso, C. T. 2002. Contributions to the evaluation of production planning and control systems in building companies. In: Formoso, C. T. and Ballard, G. (Org.). *Proceedings of the 10th Annual Conference of the International Group for Lean Construction*, Gramado, Brazil, 489–99.

Bortolazza, R. C. and Formoso, C. T. 2006. A quantitative analysis of data collected from the last planner system in Brazil. In: Sacks, R. and Bertelsen, S. (Org.). *Proceedings of the 14th Annual Conference of the International Group for Lean Construction*, Santiago, Chile, 625–38.

Bresnen, M. 1996. An organisational perspective on changing buyer–supplier relations: A critical review of the evidence. *Organisation*, 1 (3), 121–46.

Bulhões, I. R., Picchi, F. A., and Granja, A. D. 2005. Achieving continuous flow in construction: An exploratory case research. In: Marosszeki, M. and Kenley, R. (Org.). *Proceedings of the 13th Annual Conference of the International Conference on Lean Construction*, Sydney, Australia.

Cherns, A. and Bryant, D. 1984. Studying the clients role in construction management. *Construction Management and Economics*, 2, 177–184.

Christopher, M. 2000. The agile supply chain: Competing in volatile markets. *Industrial Marketing Management*, 29, 37–44.

Cohenca, D., Laufer, A., and Ledbetter F. 1989. Factors affecting construction planning efforts. *Journal of Construction Engineering and Management, ASCE*, 115 (2), 70–89.

Cox, A. and Ireland, P. 2002. Managing construction supply chains: The common sense approach. *Engineering, Construction and Architectural Management*, 9, 409–18.

Crowston, K. 1991. *Towards a coordination cookbook: Recipes for multi-agent action.* Cambridge, EUA, MIT, Sloan School of Management. PhD Dissertation.

Denning, P. J. and Medina-Mora, R. 1995. Completing the loops. *Interfaces*, 25, 42–57.

Fleischmann, B., Meyr, H., and Wagner, M. 2000. Advanced planning. In: Stadtler, H. and Kilger, C. (Org.). *Supply chain management and advanced planning*, Berlin: Springer, 57–71.

Flores, F. 1982. *Management and communication in the office of the future*. University of California, Berkeley, USA. PhD Dissertation.

Formoso, C. 1991. *A knowledge based framework for planning house building projects*. University of Salford, UK. PhD Thesis.

Formoso, C. T., Tzortzopoulos, P., and Liedtke, R. 2002. A model for managing the product development process in house building. *Engineering Construction and Architectural Management*, 9 (5/6), 419–32.

Grandori, A. 1997. An organizational assessment of interfirm coordination modes. *Organization Studies*, 18 (6), 897–925.

Hayes-Roth, B. and Hayes-Roth, F. 1979. A cognitive model of planning. *Cognitive Science*, 3, 275–310.

Hoc, J. M. 1988. *The cognitive psychology of planning*. London: Academic Press. Translated by C. Greenbaum.

Hopp, W. J. and Spearman, M. L. 1996. *Factory physics: Foundations of manufacturing management*. Boston, MA: McGraw-Hill.

IGLC. *See* International Group for Lean Construction.

International Group for Lean Construction. 2007. www.iglc.net. Accessed on the 20th Dec. 2007.

Isatto, E. L. 2005. *Proposition of a theoretical-descriptive model for the inter-organizational coordination of construction project supply chains*. Federal University of Rio Grande do Sul, Porto Alegre, Brazil. Doctorate Thesis (in Portuguese).

Koskela, L. 1992. *Application of the new production philosophy to construction*. CIFE, Stanford University, USA. Technical Report #72.

———. 2000. *An exploration towards a production theory and its application to construction*. Technical Research Centre of Finland, Espoo. PhD Thesis.

Lambert, D. M. and Cooper, M. C. 2000. Issues in supply chain management. *Industrial Marketing Management*, 29, 65–83.

Lantelme, E. M. V. and Formoso, C. T. 2000. Improving performance though measurement: The application of lean production and organizational learning principles. *Proceedings of the 8th Annual Conference of the International Group for Lean Construction*, Brighton, England.

Laufer, A. and Tucker R. L. 1987. Is construction planning really doing its job? A critical examination of focus, role and process. *Construction Management and Economics*, 5 (3), 243–66.

———. 1988. Competence and timing dilemma in construction planning. *Construction Management and Economics*, 6 (4), 339–55.

Liker, J. K. 2004. *The Toyota way: 14 management principles form the world's greatest manufacturer*. New York: McGraw-Hill.

Macomber, H. and Howell, G. 2003. Linguistic action: Contributing to the theory of lean construction. In: Martinez, J. and Formoso, C. T. (Org.). *Proceedings of the 11th Annual Conference of the International Group for Lean Construction*, Blacksburg, East Virginia, USA.

Malone, T. W. and Crowston, K. 1994. The interdisciplinary theory of coordination. *ACM Computing Surveys*, 26, 87–119.

March, J. G. and Simon, H. A. 1958. *Organizations*. John Wiley and Sons, New York.

Marosszeky, M., Thomas, R., Karim, K., David, S., and McGeorge, D. 2002. Quality management tools for lean production: Moving from enforcement to empowerment. In: Formoso, C. T. and Ballard, G. (Org.). *Proceedings of the 10th Annual Conference of the International Group for Lean Construction*, Gramado, Brazil, 87–100.

O'Brien, W. J. 2002. A call for cost and reference models for construction supply chains. In: Martinez, J. and Formoso, C. T. (Org.). *Proceedings of the 11th Annual Conference of the International Group for Lean Construction*, Blacksburg, East Virginia, USA.

Olander, S. and Landin, A. 2005. Evaluation of stakeholder influence in the implementation of construction projects. *International Journal of Project Management*, 23, 321–28.

Rohde, J. and Wagner, M. 2000. Master planning. In: Stadtler, H. and Kilger, C. (Org.). *Supply chain management and advanced planning*, Berlin: Springer, 57–71.

Sanvido, V. and Paulson, B. 1992. Site-level construction information system. *Journal of Construction Engineering and Management*, ASCE, 118 (4), 701–15.

Saurin, T. A., Formoso, C. T. and Guimarães, L. B. M. 2004. Safety and production: An integrated planning and control model. *Construction Management and Economics*, 22, 159–69.

Schramm, F. K., Costa, D. B., and Formoso, C. T. 2004. The design of production systems for low-income housing projects. In: Formoso, C. T. and Bertelsen, S. (Org.). *Proceedings of the 12th Conference of the International Group for Lean Construction*, Helsingor, Denmark, 317–29.

Schramm, F. K., Rodrigues, A., and Formoso, C. T. 2006. The role of production system design in the management of complex projects. In: Sacks, R. and Bertelsen, S. (Org.). *Proceedings of the 14th Annual Conference of the International Group for Lean Construction*, Santiago, Chile, 227–40.

Slack, N., Chambers, S. and Johnston R. 2004. *Operations management*. Harlow: Prentice.

Soares, A. C., Bernardes, M. M. S., and Formoso, C. T. 2002. Improving the production planning and control system in a building company: Contributions after stabilization. In: Formoso, C. T. and Ballard, G. (Org.). *Proceedings of the 10th Annual Conference of the International Group for Lean Construction*, Gramado, Brazil, 477–87.

Stadtler, H. 2000. Introduction. In: Stadtler, H. and Kilger, C. (Org.). *Supply chain management and advanced planning*, Berlin: Springer, 1–4.

———. 2005. Supply chain management and advanced planning: Basics, overview and challenges. *European Journal of Operational Research*, 163, 575–88.

Thompson, J. D. 1967. *Organizations in action: Social science bases of administrative theory*. New York: McGraw-Hill.

Tommelein, I. D. and Ballard, G. 1997. Look-ahead planning: Screening and pulling. *Technical Report No. 97-9*, Construction Engineering and Management Program, Civil and Environmental Engineering Department, University of California, Berkeley, CA.

Tsao, C. C. Y., Tommelein, I. D., Swanlund, E. S., and Howell, G. A. 2004. Work structuring to achieve integrated product-process design. *Journal of Construction Engineering and Management, ASCE,* 130 (6), 780–89.

Turner, R. 1993. *The handbook of project-based management.* London: McGraw-Hill.

Venkatesan, R. 1992. Strategic source: To make or not to make. *Harvard Business Review,* Nov/Dec, 98–107.

Vrijhoef, R., Koskela, L., and Howell, G. 2001. Understanding construction supply chains – an alternative interpretation. In: *Proceedings of the 9th International Group for Lean Construction Conference,* Singapore, National University of Singapore.

Wiendahl, H. H., von Cieminski, G., and Wiendahl, H. P. 2005. Stumbling blocks of PPC: Towards the holistic configuration of PPC systems. *Production Planning and Control,* 16 (7), 634–51.

Williamson, O. E. 1985. *The economic institutions of capitalism.* New York: Free Press.

Winograd, T. and Flores, F. 1986. *Understanding computers and cognition: A new foundation for design.* Boston, USA: Addison-Wesley.

4

Supply Chain Management in Product Development

Patricia
Tzortzopoulos
University of Salford

Mike Kagioglou
University of Salford

Rachel Cooper
Lancaster University

Erica Dyson
*Trafford National
Health Service Trust*

4.1 Introduction

The product development process (PDP) in the built environment requires the involvement of a large number of players, which are part of different businesses that need to work jointly during a project. Therefore, in construction, the PDP involves intra and interorganizational interactions throughout all its stages within a project context.

Lambert and Cooper (2000) suggest that a supply chain is not a chain of business with one-to-one, business-to-business relationships, but a network of multiple businesses and relationships. Similarly, the supply chain involved in the PDP of construction projects usually consists of a network of multiple businesses and multiple relationships. The management of such relationships is important for the quality, cost, and delivery of products and services [Olander and Landin 2005]. Furthermore, customers and suppliers must be integrated in the PDP to ensure the effectiveness of the process. Supply chain management offers the opportunity to capture the synergy of intra and intercompany integration and management [Lambert and Cooper 2000; Tan and Tracey 2007].

Most research studies on construction supply chain management have focused on physical production; that is, construction. The improved understanding of the important role of supply chain management emerging from such efforts stimulates further research concerned with the PDP. More specifically in construction, partnerships between clients, construction companies, and other suppliers are increasingly being fostered to provide appropriate means for effective product development.

There has been relatively little academic guidance to date for the management of the construction supply chains involved in product development. There is a need for theory building and the development of tools and methods for successful management of those supply chains in practice [Poirier and Quinn 2004; Lambert and Knemeyer 2004]. The empirical findings described in this chapter are part of a research effort to better understand the supply chain involved in the delivery of primary healthcare facilities in the UK. This case study is representative of a category of complex projects and procurement strategies that are becoming very common in different countries. The research has focused on developing design requirements at the front end of the PDP and understanding how the circumstances in which the supply chain operates influence the process and the final product. The front end includes the cross-functional strategic, conceptual, and planning activities that typically precede detailed design and development [Khurana and Rosenthal 2002]. Three research questions were addressed:

1. Who are the key supply chain members involved in the PDP?
2. What are the difficulties faced by the supply chain members at the product development front end?
3. How do such difficulties affect requirements definition and management?

Initially, this paper discusses theoretical issues related to the concept of the PDP in construction. The characteristics of stakeholders within the supply chain involved in product development are then discussed. Difficulties in managing supply chain members during a project are examined. Thereafter, the research method is presented and case study findings are described. Finally, the implications in terms of managing the supply chain involved in the PDP are identified and discussed.

4.2   Product Development Process

4.2.1   Definition

The PDP has been a key topic of study within industrial and manufacturing engineering, and has more recently been considered in construction [e.g., Koskela 2000; Kagioglou et al. 1998]. Most studies examine the management of new product introduction, as well as the processes used for developing an idea or need into a finished product, with the associated support and services [Reinertsen 1997].

Ulrich and Eppinger (2000) define the PDP as "the set of activities beginning with the perception of a market opportunity and ending in the production, sale and delivery of a product." The authors stress that it is necessary to bring together the main functions of the enterprise; that is, marketing, design, manufacturing, and after sales, in order to achieve successful product development. Smith and Morrow (1999) and Hales (1993) define product development as the process of converting an idea, market needs, or client requirements into the information from which a product or technical system can be produced.

Similarly, the PDP in construction is conceptualized in this chapter as "the set of activities needed for the conception and design of a built product, from the identification of a market opportunity to its delivery to the client."

Clark, Chew, and Fujimoto (1992) contend that design produces a product much as any other process produces a product, and the design product differs mainly in the sense that it is information rather than a physical asset. These authors also argue that both design and production have outputs of two types, physical and information, but the physical outputs differ in their use (e.g., prototypes are used to refine design). However, design is an intellectual, nonrepetitive activity, closely related to the customers, as it focuses on translating requirements into design solutions, thus creating value and innovation.

Koskela (2000) points out that product development and design have two key characteristics: uncertainty and interdependency. Uncertainty is related to the lack of information or to instability of information. This means that not all information necessary to conduct a design task is available when needed; and requirements change throughout design development. Thus, uncertainty is a consequence of the lack of information on matters under development. Interdependency happens due to tight links between activities; that is, the information produced in one activity is interactively needed in the development of other activities [Austin, Baldwin, and Newton 1994]. Koskela (2000) further states that iterations may also occur due to constraints of downstream stages being overlooked in upstream stages, which is a problem that could be avoided by considering all life-cycle phases simultaneously. In practice, teamwork strategies and early involvement of stakeholders are strategies used to allow for such whole life-cycle considerations.

Excessive changes in requirements and the project scope cause disruptions in the PDP. Therefore, it is important that the project vision and scope are defined carefully at early product development stages, in order to eliminate avoidable changes at later stages [Sebastian 2005; Koskela 2000; Barrett and Stanley 1999; Cooper and Press 1995]. This is one of the reasons why the process front end[1] has been recognized as being critical for the PDP success.

At the front end of product development, new product ideas gain shape, plans, and support, leading to their approval and execution [Khurana and Rosenthal 2002]. In this chapter, we adopt the term "front end" with reference to the preliminary, preproject stages of the design and construction process. Requirements management is considered to be an on-going activity throughout the PDP.

In the new product development literature, front-end activities are often divided into stages, for instance: (a) prephase zero (opportunity identification and idea generation); (b) phase zero (product concept and definition, including market and technology assessment); and (c) phase one (product definition, justification, and planning) [Cooper 1998;

---

[1] The preliminary stages of a project have been labeled fuzzy front-end of the PDP by Smith and Reinertsen (1991), and this term has since been adopted by several authors. Reinertsen (1997) and Zhang and Doll (2001) propose that the front-end activities represent the final gate before the team decides to invest in designing and building products.

Zhang and Doll 2001; Khurana and Rosenthal 2002; Van Aken and Nagel 2004]. In the construction literature, the initial phases of construction projects are often referred to as front end. Kagioglou et al. (1998), for instance, defines the construction process front end as follows:

(a) Phase 0 (demonstrating the need for the project), including the outline business case
(b) Phase 1 (conception of need), including the brief
(c) Phase 2 (outline feasibility), including feasibility studies for different design options
(d) Phase 3 (substantive feasibility study), which is the building product definition, and includes the conceptual project brief

The literature stresses that front-end success depends on the existence of foundation elements; that is, a clear product strategy and a product development organization [Khurana and Rosenthal 2002]. Product strategy elements include the formulation and communication of a strategic vision for the product, as well as its high-level requirements. The product development organization should be clearly defined in terms of structure, stakeholders, communication networks, roles, and norms. In construction, foundation elements heavily depend upon clients, since the product strategy should be defined by clients, and the project organization will be greatly determined by the procurement and contractual arrangements selected by clients. Moreover, providing products and services of value to customers is facilitated through the integration of supply chain members within the product development organization [Tan and Tracey 2007].

It has been widely acknowledged in the literature that benefits resulting from improvements in the front end are likely to exceed those that result from improving subsequent design stages [Cooper and Kleinschmidt 1996; CRISP 2001; Khurana and Rosenthal 2002]. Appropriate decisions and decision-making mechanisms at the front end may therefore help reduce uncertainty throughout the PDP [Reinertsen 1997; Van Aken and Nagel 2004].

4.2.2   Generic PDP Models

Early research on new product development has produced descriptive models and frameworks that model this process as mainly linear, and divided it into sequential and discrete stages. Generic, phased, or "high-level" models aim to provide an overview of the whole process, describing the main stages and activities [Kagioglou et al. 1998; Austin et al. 2000]. These models were usually built to be used as templates containing generic processes and a set of tools, working as checklists for the key steps in the product development activity. Such maps focused upon organizing the work to be developed and the main flows of information within an organization, and between different actors and organizations in a broad perspective.

Phase reviews, or stage-gate processes, are often included in product development models [Cooper 1998]. These define steps that require input from clients, including the evaluation of ideas early in the process, the identification of important features as the product concept is defined, measures of the importance of customer needs as the product is engineered, and accurate evaluation when the product is delivered [Dahan and Hauser 2002]. Different actors become involved at different process stages in order to support client involvement and the production of appropriate design information. This creates a network of internal client–supplier relationships.

However, in construction, relationships between stakeholders are mostly implicit [Oakland and Marossezkey 2006]. There is a long chain of events throughout the project and each stakeholder may have different internal customers at different process stages. Moreover, such relationships are not made clear through contractual arrangements—for instance, a subcontractor may perceive the contractor as the only customer, neglecting the requirements of other subcontractors or the final user of the facility. This may generate further uncertainty and increase the interdependency between tasks, and may also result in the lack of consideration of interdependencies, causing waste during design.

Such implicit relationships between supply chain members have led to the recognition that product development is not a linear process. Recently, recursive and somehow complex frameworks of the PDP have been developed, acknowledging that the PDP progresses through a series of stages, but with overlaps, feedback loops, and resulting behaviors that may not fit reductionism or linear analysis [McCarthy et al. 2006].

Similarly, recent research has shown that the design and implementation of PDP models within different organizational contexts is very complex. Empirical research results have demonstrated that the expected benefits of a PDP model can only be accomplished and enhanced if there is an appropriate strategy. This strategy should provide an overall driving force for implementation, and allow the adaptation of the model to the company and to specific projects [Tzortzopoulos 2004]. Gann and Salter (2000) and Koskela and Vrijhoef (2001) support this argument, stating that tackling the relationship between business processes (at the company level) and project processes (at the project level) is a major challenge both in practice and theory. Therefore, identifying and understanding the role of supply chain members in the PDP is challenging.

4.2.3 Complexity of the Client Organizations

Different conceptual approaches have been adopted to understand the characteristics of the construction client. Early views generally assumed that the term "client" implies a person or a well-defined group of people that act as a single entity [Newcombe 2003; Bertelsen and Emmitt 2005]. However, issues such as the separation of ownership and occupation, and the rise of corporate clients have led to confusion about the clients' identity and their interaction with the industry [Newcombe 2003].

Past research identified different types of clients. Higgin and Jessop (1965), for instance, made a distinction between *sophisticated* and *naïve* clients on the basis of their previous experience with construction. Masterman and Gameson (1994) argued that clients cannot be classified solely on the grounds that they possess previous experience; they must have experience of the particular building type in question.

Darlington and Culley (2004) described two customer types that can also be used to classify construction clients. The first type, the identifiable customer, represents the individual who has a specific design problem, such as a family needing a new house. The identifiable customer usually has a fairly clear view of the problem and the context in which it occurs, which can be discussed directly with designers. By contrast, the "virtual" customer represents a group or class of individuals. The customer's requirement management process is quite different for each type of client.

In summary, there are different approaches to describe the construction client's identity; some focus on the level of experience the client has with construction, and others on the level of complexity of the client (organization). Public sector organizations like the UK National Health System (NHS) may be classified as virtual, complex clients. This type of client must appropriately represent various stakeholder groups as well as the needs of wider society. In the primary healthcare context, Primary Care Trusts (PCTs) are newly established organizations operating against a background of change in primary care, whose employees have little or no experience with design and construction. Therefore, they can be considered as naïve or novice clients. The challenge for virtual, novice, and complex clients is, therefore, to provide appropriate information for design and construction even though they possess poor understanding of the PDP.

4.3   Suppliers' Involvement in Product Development

In a project environment, it is important to identify who the supply chain members are. The parties involved in a project are usually referred to as project stakeholders. Stakeholders have been defined by Newcombe (2003) as groups or individuals who have a stake in or an expectation of a project's performance. The same author states that this includes people inside the project (e.g., designers and contractors) and people outside the project (e.g., users and the community at large). Stakeholders are usually associated with companies, which are composed of individuals and groups who usually have differing values, needs, and goals, which often leads to conflicts [Pettigrew and Whipp 1991]. Such conflicts exist in construction, particularly at the project level [Ivory 2005]. Construction is a complex multiactor activity that requires "consensus building" in the context of often diverging interests. However, the objective of the PDP should be best value creation generated by integrated teams [Accelerating Change 2002]. Achieving such integrated teams in practice is challenging [Tan and Tracey 2007].

The involvement of the supply chain in product development often requires some kind of strategic alignment between organizations, and the integration of processes from different stakeholders. However, such integration and the management of links across all supply chain members tend to be complex, since there are a very large number of stakeholders involved. Lambert and Cooper (2000) state that managing all those links is counterproductive in most cases. These authors propose that supply chain members should be identified by defining some basis for determining which members are critical for the success of the supply chain and, thus, should be allocated managerial attention and resources.

Therefore, the supply chain involved in PDP comprises different coalitions of key individuals and interest groups within and throughout organizations, involved in a project from concept through to postconstruction activities. A distinction should be made between primary and supporting members [Lambert and Cooper 2000] in order to make project supply chains more manageable. In this chapter, primary members are defined as the companies or business units who carry out value-adding activities throughout the PDP to produce a specific output for a client or customer. Similarly, secondary members are the companies that provide resources, knowledge, or assets for the primary members of the supply chain.

Achieving collaboration and smooth communications during the PDP is not an easy task due to the nature of temporary project organizations [Koskela and Vrijhoef 2001]. In manufacturing, repetitive processes allow for the development of interpersonal understanding and the build up of relationships, which benefits communication and consequently reduces process uncertainty. It is difficult to achieve such benefits in construction. Consequently, the supply chain members involved in the PDP need to interact in every single project in order to enable appropriate definition and understanding of the project scope, as well as to establish suitable communications and appropriate flows of information. Furthermore, looking at the contribution made by individual organizations during various stages of product development is important. Suppliers may have different responsibilities at the concept stage, detail design stage, and production stage. The introduction of innovation in construction projects by one party requires the collaboration of others if it is to be successful [Ivory 2005].

Properly identifying and understanding the supply chain involved in the PDP is also complex due to membership changes. Newcombe (2003) asserts that the power basis of the main stakeholders, and indeed the stakeholders themselves, "shifts" over time as the project unfolds. Consequently, one stakeholder (group) may have varying levels of involvement throughout the process. For instance, at the front end, production management tends to provide little input. As the process unfolds, more production information and activities need to be developed, requiring a more active input from this stakeholder group.

Changes in the stakeholders, stakeholders' group membership and power over time increase uncertainty, as different objectives need to be taken into account. Product development is often conducted in a system of multiple and conflicting objectives. This means that the project strategy evolves through a bargaining process between the key stakeholders, and changes in the strategy may emerge as the power of participants changes over time [Tavistock Institute 1966; Newcombe 2003]. Consequently, it is possible to argue that stakeholders affect project decisions, and project decisions affect stakeholders during the PDP [Olander and Landin 2005].

Traditionally, the architect is the first supplier to be involved in the PDP. However, other suppliers are increasingly becoming involved earlier [Kamara and Anumba 2000]. This upstream participation in the PDP has led to changes in the management of the buyer–supplier relationship, with a tendency towards the establishment of partnerships [Twigg 1998; Emmitt 2007]. Lamming (1993) views this as delegating greater design responsibility to suppliers, as design uncertainties are resolved through joint discussions and reciprocal exchange of information. The same author describes that partnership through the PDP should involve: (a) suppliers with responsibilities for their own network of suppliers; (b) collaborative design development; and (c) stronger relationships to share expertise.

Suppliers become involved in the process at different stages according to a number of issues, such as their speciality, the type of procurement adopted, the type of client, and partnering arrangements in place. Twigg (1998) contends that the relative position

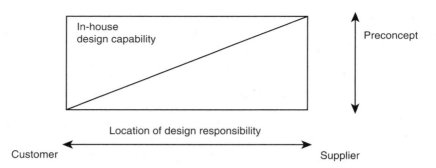

FIGURE 4.1 Location of design responsibility for contractors (Adapted from Twig (1998). With permission.)

of each supplier can be considered along a design capability continuum and their entry point in product development. The same author argues that a continuum of alternative design capabilities is available to a customer, as shown in Figure 4.1. At one extreme, a contractor may retain all necessary design capabilities in-house, whilst at the other end, a contractor may outsource all design work and act as a focal point for the coordination of the design process. There are a number of options with varying degrees of internal and external design capability. Considering such a continuum is important for providing insights into the contribution of each supplier to product development, and which suppliers may require greater coordination of activities.

In the UK construction industry, there has been a tendency for contractors to outsource most or all design work and act as a focal point for the coordination of the process. This is true for most types of existing procurement routes, for example, Design and Build (D&B) and Private Public Partnerships (PPP), such as Private Finance Initiative (PFI) or Local Finance Improvement Trust (LIFT). Even though PFI and LIFT are partnership-based models where design is undertaken as part of the partnership, the firms doing design development are normally subcontracted by the construction firm that plays the role of the main contractor. In such arrangements, the design manager needs to establish close relationships with stakeholders and facilitate coordination. Each firm may have its individual design and project management processes. However, all processes should follow the milestones determined by the customer or by the contractor as the main coordinator [Gray and Hughes 2001; Sebastian 2005].

This has imposed on contractors the need to appropriately manage the PDP, to maintain competitiveness in the marketplace, and to reduce waste both in design and in downstream construction activities [Broadbent and Laughlin 2003]. Effectively, in some cases, even the design coordination is outsourced, with the emergence of organizations specialized in this role.

In the literature, it is still unclear how contractors can effectively manage product development and design. Even though there are companies that successfully manage design, there is still a need for further research concerned with the scope of design management and the development of the necessary skills from a contractor's perspective [Tzortzopoulos and Cooper 2007]. Such a gap may be a partial consequence of the fact that design management has typically been solely approached from the perspective of the professionals involved in design [Press and Cooper 2002].

Twigg (1998) argues that distinguishing between design responsibility and design authority would provide a clearer understanding of the roles of control and expertise in managing design supply chains. The organization with design authority should be able to manage the conflicting objectives of the multiple and evolving stakeholders involved in the project.

Therefore, a broader, clearer perspective on product development and design management is needed. Such a perspective should accommodate the characteristics of the constantly evolving supply chain involved in the PDP.

4.4   Research Method

There is currently a large program of public building underway for the primary healthcare sector in the UK [DOH 1998]. The ethos of this program is to deliver health and social care jointly, so that people can have better, easier access to services through buildings that should also help regenerate deprived urban areas. This has created an opportunity to take a fresh approach to healthcare buildings, as well as an opportunity to consider the impact that the physical environment has on patients and on healthcare staff. This research focuses on a partnership developed as part of this building program.

In construction, product development is a complex process very much shaped by the context in which it takes place, as well as by the perspectives, beliefs, and motivations of the individuals involved. Consequently, this study adopted an interpretative research approach, analyzing the design front end with an emphasis on the facts, words, and meanings, in order to reach a broader understanding of supplier involvement and problems faced throughout the design front end.

The primary unit of analysis of the case study is at the level of a program of buildings for primary healthcare delivery. Information about the program and, in particular, four projects was collected through documentary evidence as well two- to three-hour semistructured interviews with key personnel from different parts of the supply chain. A total of 22 interviews were conducted. The profile of the interviewees is presented in Table 4.1.

Data analysis was based on interview summary sheets, filled in immediately after each interview [Miles and Huberman 1994]. This proforma followed the format of the interview protocol and included the following broad questions:

(a) Interviewee's background
(b) Story of the project, to identify the organizational structure in which schemes were developed, front-end activities, and the interviewee's involvement
(c) Accuracy of building requirements as perceived by the interviewee
(d) Issues and problems identified during the process and
(e) Possible improvement strategies

Data analysis was done by combining qualitative responses and documentary evidence into narratives describing the front-end process in its context. By tracing the process

TABLE 4.1   Case Study Interviewees

Interviewee Organization/Position	Supply Chain Members	Interviewee Organization/Position	Supply Chain Members
Project director of program	LIFT team	TY Architects—architect	Private sector partner
General manager of program	LIFT team	TA Architects—architect	Private sector partner
Executive Board Chair of the program	LIFT team	Joint venture's project manager	Private sector partner
Partnerships for Health— NHS executive	LIFT team	Healthcare planner	Private sector partner
Partnerships for Health	LIFT team	Salford PCT	Public sector client
Architectural advisor	Consultant	South Manchester PCT	Public sector client
Responsible for initial requirements	Consultant	Trafford South and North PCT	Public sector client
Joint venture—Chief executive	Private sector partner	Salford PCT and LIFT team	Public sector client
Contractors project manager	Private sector partner	Salford City Council	Public sector client
Contractors design manager	Private sector partner	Salford PCT	Public sector client
Contractors design manager	Private sector partner	General practitioner (doctor)	Public sector client

from the perspective of different stakeholders interviewed, a "story" was constructed [Eisenhardt 1991]. The story was used to identify stakeholder groups involved in the process, as well as issues experienced during the design front end. These are described in the next section.

4.5   Case Study: Supply Chain Involved in the PDP of Primary Healthcare Facilities

The British Government has established a new procurement method to avoid a fragmented approach to the delivery of primary healthcare facilities and to concentrate on the areas of greatest need. LIFTs are public private partnerships, set up to allow NHS PCTs and their local partner organizations to develop primary healthcare facilities. Through LIFT, a number of schemes are clustered and delivered by a single private sector partner, selected through a bidding process.

PCTs are responsible for redesigning healthcare services and are the main clients of the new facilities. As such, they are responsible for a number of activities, including identifying stakeholders, capturing requirements, and briefing designers. Such design and development activities require the involvement of a number of stakeholders from both the public and private sectors, with different interests, power, and involvement in the projects.

The LIFT program was launched in the UK in 2001. The empirical data presented here focuses on the first four schemes developed at Manchester, Salford and Trafford (MaST) LIFT.

4.5.1 Findings

The idea of developing schemes for Manchester, Salford, and Trafford through a single LIFT organization began with the recognition of the need to cluster schemes to achieve economies of scale, and simplify the delivery of long-term partnerships between PCTs and a private sector partner. Figure 4.2 represents the novel and complex MaST LIFT structure, and the overall relationships between the more than 25 public and private sector primary supply chain members involved in product development.

Partnership for Health is a wholly owned subsidiary of the Department of Health, UK, and was established to deliver a new model of investment into primary care, the NHS LIFT. At the local level (MaST), the organization started with a Strategic Part-nering Board (SPB), which was responsible for the bidding process through which the private sector partner was selected. The LIFT Company is the delivery vehicle for the projects, and it is controlled through the SPB. Under the private sector partner, there are a number of supply chain members, including:

(a) Two healthcare planners providing advice and support for the PCTs
(b) Two solicitor consultancies, representing the private and the public sector
(c) Two cost management companies (public and private sectors)
(d) One facilities management company

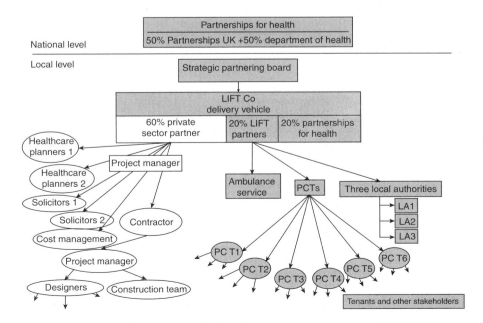

FIGURE 4.2 Stakeholder groups involved in the organizational structure of MaST LIFT.

(e) One contractor, involved with product development and design, construction and facilities management

(f) Two architectural consultancies

In addition, there are a number of other supply chain members involved within each specific scheme, as the future tenants of the facilities. These differ from scheme to scheme in number and composition, and their memberships change over time. Generally, these include the general practitioners (GPs), the local authorities (both as land owners and as tenants), community services, acute trust clinicians, dentists, pharmacists, and hospitals, among others. Such supply chain members, called user groups, were normally identified, invited for meetings, and represented by the PCTs. The user groups for the four schemes developed by MaST are described in Table 4.2.

The organizational structure put in place for MaST LIFT is complex in nature, involving many participants, both public and private sector organizations. The low level of experience of the public sector organizations further influenced project complexity. The clients (PCTs) were newly established organizations and were also inexperienced with construction. Furthermore, most PCTs appear to have allocated insufficient time and resources to LIFT, as it was perceived to be an extra activity by PCT members. The private sector partner was also inexperienced, as it was a joint venture established specifically as a response to LIFT. The design capability has been totally outsourced, being subcontracted to external consultancies.

The established partnership has dealt with a multifaceted and novel product, a new type of building that provides innovative services, and a nontraditional procurement process based on novel legal and financial systems. The large number of supply chain members involved has further increased the complexity of the initiative, and in some instances created difficulties in building up relationships and communications between partners.

TABLE 4.2 User Groups for the Four Projects

Partington, Trafford	Wythenshawe, Manchester	Benchill, Manchester	Douglas Green, Salford
• PCT	• PCT	• PCT	• PTC
• Senior GPs from 2 practices	• Wythenshawe partnership (from the Forum building that was linked with the healthcare facility)	• Willow Park Housing Trust—land owners—had plans to redevelop the area	• CHAPS— Community Health Action Program
• Local Authority (site owner)			• Roman Catholic church (neighbors)
• Healthy living center (LA)	• Metrolink—GMPTE	• Community services	• Local Authority
• General public through consultations	• 3 GP practices	• 2 dentists—1 private	• General public through consultations
	• Dental practices	• 3 GP practices	
• Dental—community services	• Local Authority	• Clinical services	
• GMAS base for a paramedic	• General public through consultations	• Podiatry	
		• Minor operations	
		• Mental health	
		• General public through consultations	

Due to such complexity and the inexperience of the teams, the organizational structure and roles and responsibilities of each party were not clear for the supply chain members. The description and diagrams presented were developed by the research team and were later validated by interviewees. No such comprehensive review had been undertaken before the research took place. As a consequence of these factors, delays have occurred throughout the process.

The tenants for each scheme had not been properly defined at the beginning of the PDP. This was due to different reasons, the primary one being that the service delivery system was being redesigned, and consequently future building tenants or users were unknown. Other reasons included poor knowledge on affordability, the search for third-party revenue, and conflicts between the needs of different stakeholder groups.

Consequently, there were difficulties in managing client requirements. The PCTs organized meetings to establish requirements and discuss design. However, different stakeholders attended each meeting, demonstrating a lack of commitment to the program. Problems were caused by stakeholders being absent from specific meetings. Each stakeholder had varying needs that had to be taken into account at different design stages. This happened without any ranking of the importance of the new requirements or of the proposing stakeholder (group), and usually without justification of costs associated with the proposed changes.

Although requirement changes are expected within any design process as the project vision evolves and the knowledge about the problem increases, it appears that in the early MaST schemes such changes were excessive and poorly controlled, therefore being disruptive and costly. In some schemes, changes in the strategic direction of the projects arose in the middle or end of design development. This was partially due to the poor commitment of the suppliers involved. The excessive amount of changes was also a consequence of the fact that neither the PCTs nor the architects appeared to have enough authority to determine which requirements should be considered and which should be disregarded. PCTs as clients did not have the necessary expertise to do so. By contrast, architects felt they were not empowered to do so due to contractual relationships. As the architects were selected by the main contractor, they perceived the contractor as their main client and therefore responsible for managing requirements.

Therefore, there were difficulties in decision making due to political pressures and sometimes divergent interests surrounding the partnership. Tensions were identified between the private and public sector partners. While the private sector was focusing on developing design and defining contractual issues as fast as possible, the public sector was poorly resourced and under pressure to define an effective and appropriate new healthcare delivery system. As such, they needed time to gather information and to make decisions. The lack of clarity on how each group influenced (or should have influenced) the process further aggravated this problem.

The public sector culture has been partially blamed for this problem. In the public sector, decision making tends to be lengthy due to the involvement of many parties and the need for public accountability. However, such tensions have also had a negative impact on the public sector. Some PCTs perceived that the private partner was "pushing" decision making at very early stages. This fact has been described by interviewees as a failure of the contractors to understand the nature of healthcare delivery.

Finally, conflicts between suppliers' interests also led to problems in the definition of contracts. In order to solve these problems, a number of solicitors were involved, generating high legal fees and, more importantly, poor trust and lack of openness.

4.6 Discussion

The case study presented is an example of the environment in which new product development often takes place in the construction industry nowadays. The supply chain involved in the PDP of primary healthcare facilities is highly complex, and requires innovative procurement schemes. Stakeholders from within and outside the project play different roles, being involved at different process stages, demanding the establishment of new types of partnerships. Such stakeholders often have conflicting interests in the project, and achieving consensus is challenging. In this context, sufficient learning needs to take place in order to develop mature partnerships among supply chain members. In fact, such early schemes should be considered as prototypes in the product development of primary healthcare facilities in the UK.

The study also indicated difficulties in identifying who the supply chain members are at different points in time throughout product development. There were changes in the supply chain organization as well as in the membership of the different organizations involved in the process. Furthermore, there was little control over inputs in the process. This increased process uncertainty, and resulted in delays and additional costs.

A generic supply chain membership for healthcare projects emerged from the case study, and is presented in Figure 4.3. It should be noted that for each supply chain member

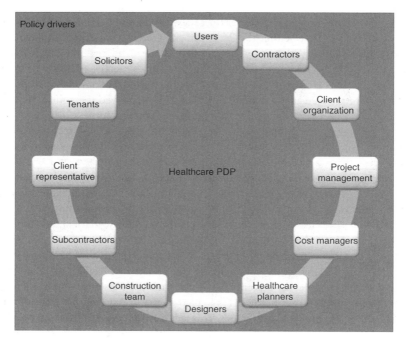

FIGURE 4.3   Product development supply chain involved in the delivery of healthcare facilities.

type described, there may be a number of different organizations involved. Also, the relationships between the supply chain members are not organized or linear, and these tend to change over time.

The supply chain representation for primary healthcare projects looks similar to traditional construction supply chain configurations [see, for instance, Anumba and Evoumwan 1997]. However, the partnership arrangements and the type of procurement process used in such projects needs to be different, adding complexity in terms of the number of organizations involved and their relationships throughout the PDP. Also, there is a direct influence from policy in terms of the structure of the supply chain and on the overall benefits to be achieved, which may not be present in other types of supply chains. Differences between public and private sector organizational cultures also add complexities in the supply chain organization and dynamics.

Furthermore, even though the project strategy should emerge and evolve through a bargaining process between supply chain members, this is not easily achieved in practice due to frequent changes of stakeholders. This results in changes in terms of the overall design strategy and the subsequent design and development process. One example was the fact that several design changes were made in one of the schemes due to one organization wanting to have a separate entrance in the building, and the PCT wanting the building to have one single entrance. Consequently, one supply chain member, a future building tenant, abandoned the project.

There are additional difficulties associated with the upstream involvement of suppliers. Such upstream involvement generates difficulties in terms of achieving consensus due to the increased number of people involved in the project. Consequently, there are greater difficulties in establishing and managing conflicting requirements, and in making sure the project's main strategic objectives are actually considered throughout all design decisions. The larger the number of stakeholders involved, the more difficult it is to achieve appropriate communications and reduce uncertainty. Therefore, the need to clearly classify primary and secondary supply chain members [as proposed by Lambert and Cooper 2000] became clear in the case study.

Finally, the design capability has been totally outsourced in the case study. Even though the design authority (in terms of expertise) was clear, there was poor clarity with regards to design responsibility (in terms of control). Appropriate design management depends on the clear definition of who takes responsibility for requirements and stakeholder inputs (e.g., time, sign offs) throughout product development. The appropriate and clear location of design responsibility is vital to project success.

4.7   Conclusions

Managing the supply chain involved in the PDP of primary healthcare facilities is demanding due to the number of stakeholders involved, the need for joint problem solving, integrated work, collaboration, coordination and appropriate communications, high task uncertainty, and the frequent information transfers during design and development.

Project goals should be defined through consensus between the different supply chain members. Such goals may range from solving individual problems of the smallest element to the most strategic elements. Conflicts over immediate goals may lead to questioning the whole purpose of the project. These may occur in order to improve long-term goals, but may also occur as a consequence of badly managed requirements being established by stakeholders.

Next, some proposed recommendations for improving supply chain management in such projects are presented.

It is important to clearly identify the supply chain involved in the PDP at an early stage. Empirical evidence suggests that problems occur during design development due to a lack of clarity about who the supply chain members are. Also, as different supply chain members have different levels of involvement during a project, it is necessary to understand and plan their involvement at the different stages of the PDP. Such involvement should be aimed at supporting value generation without increasing complexity by involving too many stakeholders at any specific project stage.

A strategy that may support the identification and communication of the different supply chain members is the design of a process model for the project, including the definition of the high-level activities, responsibilities, milestones, and major decision-making points. Such a model would be helpful in clearly defining roles and responsibilities, and in providing an overall process perspective to all involved. Improvements such as phase reviews and design freeze could also be introduced. The design of the "best" possible way to managing the supply chain should consider good practices alongside the structure of physical, political, and cultural settings of product development action at the specific project context. The appropriate skills mix of the supply chain members should be identified alongside the process definition.

Furthermore, there is a clear need to put in place collaborative arrangements across the supply chain in order to achieve an integrated PDP. Collaboration between the different stakeholders involved in the process is necessary to allow the delivery of value to the costumers, for instance, Chapter 5 presents a collaborative value-based approach for improving decision making in the construction design chain. Appropriate collaboration allows that the envisaged organizational benefits that should accrue from the project, the building vision, and requirements are suitably developed throughout the PDP.

Additionally, the adoption of classification methods may support the development and management of building requirements during the PDP. Requirements proposed by primary and secondary supply chain members could be ranked both according to their importance in the overall project and according to the importance of the supply chain member that proposed them, therefore providing a framework to organize decision making and support innovation.

Learning and commitment across the supply chain should be stimulated to allow for the establishment of mature partnerships. Participatory decision making in both process- and product-related issues should support suppliers' commitment during the PDP, and enable better communications, knowledge transfer, and learning.

Managerial focus should be on earlier stages in the PDP, especially on the design relationships that a company creates with its primary suppliers. Each relationship may be considered as part of a design chain within a network of firms [Twigg 1998]. Consequently, the partial transfer of project management responsibility to suppliers is probably a key focus for effective management of the supply chain involved in the PDP.

Finally, there is still a need for a better understanding of the types of suppliers involved, and the scope of their participation in interfirm product development and design activities. Similarly, further research is needed to establish who is better able to take design authority and responsibility within the supply chain involved in the PDP, and how to empower such stakeholders so that they are able to effectively support design management.

References

Accelerating Change. 2002. *Accelerating change: Strategic forum for construction.* London: Rethinking Construction. http://www.strategicforum.org.uk/pdf/report_sept02.pdf.

Austin, S., Baldwin, A., and Newton, A. 1994. Manipulating the flow of design information to improve the programming of building design. *Construction Management and Economics*, 5, 445–55.

Anumba, C., and Evoumwan, M. 1997. Concurrent engineering in design-built projects. *Construction Management and Economics*, 15, 271–81.

Austin, S., Baldwin, A., Li, B., and Waskett, P. 2000. Analytical design planning technique (ADePT): A dependency structure matrix tool to schedule the building design process. *Construction Management and Economics*, 18, 173–82.

Barrett, P., and Stanley, C. 1999. *Better construction briefing.* London: Blackwell Science.

Bertelsen, S., and Emmitt, S. 2005. The client as a complex system. *Proceedings of the 13th Annual Conference of the International Group for Lean Construction.* 19–21 July, Australia.

Broadbent, J., and Laughlin, R. 2003. Public private partnerships: An introduction. *Accounting, Auditing and Accountability Journal*, 16 (3), 332–41.

Clark, K. B., Chew, B., and Fujimoto, T. 1992. Manufacturing for design: Beyond the production/R&D dichotomy. In *Integrating design and manufacturing for competitive advantage*, ed. G. I. Susman, 178–204. New York: Oxford University Press.

Cooper, R. G. 1998. *Product leadership: Creating and launching superior new products.* Reading, MA: Perseus Books.

Cooper, R., and Kleinschmidt, E. 1996. Winning businesses in new product development: The critical success factors. *Research Technology Management*, 39 (4), 18–29.

Cooper, R., and Press, M. 1995. *The design agenda: A guide to successful design management.* Chichester, UK: John Wiley & Sons.

CRISP. 2001. *Issues on the early stages of construction projects: CRISP Commission 00/8.* University of Bristol and Halcrow Group.

Dahan, E., and Hauser, J. 2002. The virtual customer. *Journal of Product Innovation Management*, 19, 332–53.

Darlington, M., and Culley, S. 2004. A model of factors influencing the design requirement. *Design Studies*, 25, 329–50.

DOH. 1998. *Modernising healthcare and social services: National priorities guidance 1990/2000–2001/2002.* Department of Health publications. September.

Eisenhardt, K. 1991. Better stories and better constructs: The case for rigour and comparative logic. *Academy of Management Review*, 16 (3), 620–27.

Emmitt, S. 2007. *Architectural design management.* London: Blackwell.

Gann, D., and Salter, A. 2000. Innovation in project-based, service-enhanced firms: The construction of complex products and systems. *Research Policy,* 29, 955–72.

Gray, C., and Hughes, W. 2001. *Building design management.* London: Butterworth Heinemann.

Hales, C. 1993. *Managing engineering design.* Harlow, UK: Longman Scientific and Technical.

Higgin, G., and Jessop, N. 1965. *Communications in the building industry: The report of a pilot study.* London: Tavistock Publications.

Ivory, C. 2005. The cult of customer responsiveness: Is design innovation the price of a client-focused construction industry? *Construction Management and Economics,* 23, 861–70.

Kagioglou, M., Cooper, R., Aouad, G., Hinks, J., Sexton, M., and Sheath, D. 1998. *Final report: Generic design and construction process protocol.* University of Salford, UK.

Kamara, J. M., and Anumba, C. J. 2000. Process model for client requirement processing in construction. *Business Process Management Journal,* 6 (3), 251–79.

Khurana, A., and Rosenthal, S. R. 2002. Integrating the fuzzy front-end of new product development. In *Innovation: Driving product, process, and market change. MIT Sloan Management Review,* ed. E. B. Roberts, 47–85. Cambridge, MA: MIT Press.

Koskela, L. 2000. An exploration towards a production theory and its application to construction. PhD diss., Helsinki University of Technology. VTT Publications 408. Espoo, Sweden: VTT.

Koskela, L., and Vrijhoef, R. 2001. Is the current theory of construction a hindrance to innovation? *Building Research & Information,* 29 (3), 197–207.

Lambert, D. M., and Cooper, M. C. 2000. Issues in supply chain management. *Industrial Marketing and Management,* 29 (1), 65–83.

Lambert, D. M., and Knemeyer, A. M. 2004. We're in this together. *Harvard Business Review,* 82 (12), 114–22.

Lamming, R. C. 1993. *Beyond partnership: Strategies for innovation and lean supply.* London: Prentice-Hall.

Masterman, J. W. E., and Gameson, R. N. 1994. Client characteristics and needs in relation to their selection of building procurement systems. *Proceedings of CIB W96 Symposium,* East meets West, Hong Kong, 79–87.

McCarthy, I. P., Tsinopoulos, C., Allen, P., and Rose-Anderssen, C. 2006. New product development as a complex adaptive system of decisions. *Journal of Product Innovation Management,* 23 (5), 437–56.

Miles, M. B., and Huberman, A. M. 1994. *Qualitative data analysis. An expanded sourcebook.* San Diego, CA: Sage Publications.

Newcombe, R. 2003. From client to project stakeholders: A stakeholder mapping approach. *Construction Management and Economics,* 21 (1), 841–48.

Oakland, J., and Marossezkey, M. 2006. *Total quality in the construction supply chain.* London: Elsevier.

Olander, S., and Landin, A. 2005. Evaluation of stakeholder influence in the implementation of projects. *International Journal of Project Management,* 23, 321–28.

Pettigrew, A. M., and Whipp, R. 1991. *Managing change for competitive success.* Cambridge, MA: Blackwell.

Poirier, C. C., and Quinn, F. J. 2004. How are we doing? A survey of supply chain progress. *Supply Chain Management Review*, 8 (8), 24–31.

Press, M., and Cooper, R. 2002. The design experience: The Role of Design and Designers in the Twenty-First Century. Ashgate Publishing, UK.

Reinertsen, D. 1997. *Managing the design factory: A product developer toolkit.* New York: The Free Press.

Sebastian, R. 2005. The interface between design and management. *Design Issues*, 21 (1), 81–93.

Smith, R. P., and Morrow, J. A. 1999. Product development process modelling. *Design Studies*, 20, 237–61.

Tan, C. L., and Tracey, M. 2007. Collaborative new product development environments: Implications for supply chain management. *The Journal of Supply Chain Management*, Summer, 1–15.

Tavistock Institute. 1966. *Independence and uncertainty.* London: Tavistock.

Twigg, D. 1998. Managing product development with a design chain. *International Journal of Operations and Production Management*, 18 (5), 508–24.

Tzortzopoulos, P. 2004. The design and implementation of product development process models in construction companies. PhD thesis, University of Salford.

Tzortzopoulos, P., and Cooper, R. 2007. Design management from a contractor's perspective: The need for clarity. *Architectural Engineering and Design Management*, 3, 17–28.

Ulrich, K. T., and Eppinger, S. D. 2000. *Product design and development.* 2nd ed. Boston MA: Irwin/McGraw-Hill.

Van Aken, J. E., and Nagel, A. P. 2004. *Organising and managing the fuzzy front-end of new product development.* Eindhoven Centre for Innovation Studies, Working paper 04.12, Technische Universiteit Eindhoven, The Netherlands.

Zhang, Q., and Doll, W. J. 2001. The fuzzy front-end and success of new product development: A causal model. *European Journal of Innovation Management*, 4 (2), 95–112.

5

Collaboration and Communication in the Design Chain: A Value-Based Approach

Stephen Emmitt
Loughborough University

Anders Kirk
Christoffersen
*NIRAS Consulting
Engineers*

5.1 Introduction

Like many construction sectors around the world, the Danish construction industry was criticized during the 1990s for its poor performance and failure to deliver value to the customer, although parallel research found that some innovative practices were being implemented successfully [Kristiansen, Emmitt, and Bonke 2005]. In the past 10 years or so, a number of initiatives have been taken to: (i) try and improve the value delivered to clients, users, and society; and (ii) implement more efficient production processes. These have helped to emphasize the importance of supply chain management at all stages of projects, with concepts such as lean construction, project partnering, and more recently, value management being promoted as a way to improve performance and deliver better value. The current situation in Denmark is a dynamic market, with actors claiming to use different approaches to distinguish themselves from the competition. However, the common theme is concerned with better integrated supply chains and more effective (leaner) operating methods to deliver better value to construction clients. One manifestation of this has been the establishment of Lean Construction Denmark, comprising a community of clients, consultants, contractors, user organizations, and universities dedicated to construction supply chain improvement. This has led to some of the major contractors and consultants in Denmark implementing lean construction.

The aim of the work carried out by consulting engineers featured in this chapter was to develop a better approach to the design and construction of buildings through attention to the entire supply chain. Central to this was the inclusion of the client in the design chain and a focus on value. The case study organization regard value as the end-goal of all construction projects, and therefore the discussion and agreement of value parameters is fundamental to the achievement of improved productivity and client/user satisfaction. Value creation and value delivery are crucial components of the approach, explored and developed in what the case study organization call the "value universe," and implemented via a bespoke value-based design management model. This relies on the use of facilitated workshops and attention to the process.

The value-based model was first developed through a series of trials, starting in the mid to late 1990s with the HABITAT consortium managed by NIRAS [reported in Bertelsen 2000], and further evolved through the design and construction of an urban renewal project in central Copenhagen and the William Demant Dormitory in Lyngby [see Bertelsen, Fuhr Petersen, and Davidsen 2002; Christoffersen 2003a]. The HABITAT consortium consisted of a number of partners representing the entire supply chain, from the client to the contractors and suppliers. The aim was to improve the productivity in the Danish housing sector by focusing on both the product and the process. The approach was to clarify client needs, and to include participation by all partners throughout the entire process. The houses were designed as a modular system (based on 3–4 basic units), produced off site, delivered, and fixed using a pull system methodology. HABITAT was based on long-term relationships, working with the philosophy of continuous improvement, resulting in the construction of over 250 houses and apartments in a variety of projects.

The William Demant Dormitory comprised the design and construction of 100 student rooms with common facilities at the Technical University of Denmark in Lyngby, a suburb of Copenhagen, with a build cost of 4.5 million euro. This project used the value-based model, in which customer values were central to the development of the design and the realization of a successful building. The value-based model was developed further via this project and projects running parallel to it (urban renewal projects in central Copenhagen). The incorporation of lean thinking was important in helping to translate client value to the delivery phase. A modified version of the Last Planner System [Ballard 2000] was used in the detailed design phase, and a full version of it in the construction phase.

All projects were completed within the time frame (the William Demant building was handed over two months ahead of schedule), and to budget. These projects were also rated very highly by the client, project participants, and end users as measured in satisfaction surveys. More recently, the value-based model has been further developed in a number of consecutive projects within the greater Copenhagen area, including both new build and refurbishment residential projects, a new biochemical plant, laboratory facilities, and urban renewal of street lighting and paving for one of the municipal districts. This methodology is also being implemented on a major project, which aims to improve urban planning methods, from early political decision making through to specific projects. This represents an extension of the value universe model to include cultural values (including social, spiritual, and technical values).

This chapter provides an insight into this value-based approach to supply chain integration. The data for the case study was collected through a variety of methods, including direct experience of the authors and independent research, which included nonparticipant observation of the workshops, interviews with participants, and analysis of project documentation. Before the value-based method is described, the fundamental theoretical and philosophical issues underpinning the approach taken by the case study organization are explained. This is followed by articulation of the method, discussion of the facilitator's role, and reflection on the method. The chapter concludes with a number of recommended strategies based on the authors' experience of implementing a value-based approach.

5.2 Background and Conceptual Framework

The early work was based on approaching supply chain management from a logistics perspective [e.g., Christopher 1998], followed by supply chain management from a lean production stance, drawing on the popular management literature of Womack, Jones, and Roos (1991), and Womack and Jones (1996). The seminal work of Koskela (2000) was also important in helping to emphasize the importance of the process. This eventually resulted in the use of the Last Planner System [Ballard 2000] and the application of lean thinking to the design process. This has further evolved through a focus on value chains [Porter 1985] and value management [Kelly and Male 1993], which is an important part of supply chain management. Focusing on value has also led to increased interest in how project participants interact and communicate. These underpinning principles can be explored in a little more detail under three headings.

5.2.1 Value and Values

Establishing value frameworks is an important principle behind integrated collaborative design [Austin et al. 2001]. Similarly, the establishment of common objectives and common values are important objectives in the drive for greater cooperation and reduced conflict in construction projects [e.g., Kelly and Male 1993]. Value is what an individual or organization places on a process and the outcome of that process, in this case a building project and the resultant building [Christoffersen 2003b]. Value is often related to price, although other factors relating to utility, aesthetics, cultural significance, and market are also relevant. Values are our core beliefs, morals, and ideals, which are reflected in our attitude and behavior, and shaped through our social relations. Our values are not absolute, existing only in relation to the values held by others, and as such in constant transformation. Perception of value is individual and personal, and is therefore subjective. Indeed, agreement of an objective best value for a group will differ from the individuals' perception of value.

At the level of the individual construction project it may be very difficult to improve working methods, even when all participants and organizations agree to some common values. Maister (1993) has argued that many firms do not share values within the organization, and also fail to adequately discuss values with clients early in the appointment process. The implication is that the sharing of values is a challenge for organizations and temporary project groupings. The challenge is not exclusively with the implementation of tools to streamline the process, but it is about the interaction of organizations, or more specifically, the efficacy of relationships between the actors participating in the temporary project coalition. This social interaction needs to be managed and someone should take responsibility for leading the process.

5.2.2   Applying Lean Thinking to the Design Chain

Value to the end customer is an important aspect of the lean thinking philosophy [Womack, Jones, and Roos 1991; Womack and Jones 1996]. Lean thinking and techniques borrowed and adapted from lean manufacturing can provide a useful array of tools through which the value of the design can be enhanced, and waste reduced. Although developed specifically for manufacturing and mass-produced products, the philosophy is relatively robust and can, with some interpretation, be applied to a project environment. In the context of this study, the five principles of lean thinking have been interpreted and adapted to design management to:

- Specify value: clearly and precisely identify the client's values and requirements, and then identify the specific functions required to deliver a solution
- Identify the value stream: identify the most appropriate processes to deliver the building through the integration of the functions identified when specifying value
- Enable value to flow: remove any unnecessary or redundant cost items from the design to get to the optimal solution (as agreed by the major project stakeholders)
- Establish the "pull" of value: this means frequently listening to the client and other key stakeholders during the project and responding iteratively
- Pursue perfection: incorporate process improvement methods into the organizational culture and practices of the project participants' firms

These five principles underpin the workshop method, starting with the definition of value and continuing through the entire process, as described below. Lean thinking can be applied at different levels in the product development process, from the entire project to distinct phases and substages, which can assist the planning and scheduling of the various work packages. Approaching design from a lean thinking perspective also helps to emphasize the need for designers to understand how design value is physically realized and the associated production costs; that is, they need to understand the supply chain. Again this is addressed within the workshops. Depending on the type of project and the approach adopted by the design team, this may involve a greater understanding of craft techniques, or manufacturing production techniques, and the associated cost and time parameters. The five principles also underpin the development of the 7 Cs model, which is used by the case study organization and described below.

5.2.3   Interaction

The philosophy behind the approach used is that to explore values and implement lean thinking requires face-to-face interaction within the project team, and the primary mechanism for this is a series of facilitated workshops. The goal is to improve the integration and realization of project values, with the ultimate goal of delivering better value to the customer. This is not a new idea, for example, architects Konrad Wachsmann and Walter Gropius introduced a teamwork method for the development of complex building concepts in the 1940s [Gropius and Harkness 1966], and Caudill (1971) promoted architecture by team in a book of the same title. What has changed is that groupwork and teamwork have taken on more significance with the promotion of relational forms of contracting and integrated supply chain management.

Integration of supply chain members in the product development process brings to the project the skills, knowledge, and experience of a wide range of specialists, often working together as a virtual team from different physical locations. This requires social parity between actors, which means that professional arrogance, stereotypical views of professionals, and issues of status have to be put to one side or confronted through the early discussion of values. To do this effectively, all actors must engage in dialogue to explore and then confirm a set of values that form the basis of the project. The most effective way of doing this is through face-to-face meetings that recognize the value of group process [Luft 1984]. Interactions within groups, power relationships, leadership, and decision making are extremely complex matters, and contradictory views exist as to the ability of a group to reach its defined goals [e.g., Stroop 1932; Yoshida, Fentond, and Maxwell 1978; Emmitt and Gorse 2007]. However, the authors of this chapter strongly believe that unless interaction is addressed from the very start of projects in a professional and ordered manner, then it can be very difficult to achieve very high value in the design chain.

5.3   Value-Based Model

The focus on value and attempts to explore the values held by members of the supply chain has led to a method that relies on facilitating the interaction of the project stakeholders. This is achieved through a series of workshops in which the focus is on the client's "value universe," and the interaction practices within interdisciplinary groups [see Emmitt and Gorse 2007].

The literature on value management and value engineering overlaps, therefore it is necessary to state how the two terms are interpreted and applied in relation to this case study. Value-based management attempts to control values, primarily through value management [see Kelly and Male 1993] to "create" value in the early stages of the project. Value engineering techniques [see Miles 1972] are used to "deliver" value in the production stage. Value-based management uses face-to-face workshops as a forum to allow actors to discuss, explore, challenge, disagree, and eventually agree to commonly shared project values. These values are then defined in a written document as a set of value parameters, and prioritized in order of importance to the project team. This forms part of the project briefing (also known as architectural programming) documentation. Getting to know each other, and thus establishing common values and/or knowing why values differ between the stakeholders is crucial to the method. It is about how to work together, and how to keep agreements between the client and the delivery team.

In Denmark, it is also common to differentiate between the values of the client (external values), and values of the delivery team (internal values), and these are not to be confused [Christoffersen 2003b]. External values are further separated into: (a) process values, and (b) product values. Process values comprise both "soft" and "hard" values. The soft values include work ethics, communication, conflict solving, trust, etc., between the client and the delivery team. These are intangible and difficult to measure objectively. The hard values include the delivery team's ability to keep agreed time limits, cost estimates, quality of the product, workers' safety, etc. These are tangible and can be measured objectively to assess project performance. Primary product values comprise beauty, functionality, durability, suitability for the site and community, sustainability, and buildability. As the understanding of values improves and evolves through the design process, we are dealing with a learning process that relies on the development of trust and effective interpersonal communication.

5.3.1   Design Chain: Value Design and Value Delivery

The design process is separated in two main phases (Figure 5.1), which are completely different in their aims and management:

- The value design phase is where the client's wishes and requirements are determined and specified. These values are developed into a number of conceptual design alternatives before entering the value delivery phase of the process. Management should be focused on stimulating creativity and determining maximum value in the project; that is, establishing needs before solutions.
- The value delivery phase is where the best design alternative, which maximizes the client/customer value, is transformed through production. The aim is to deliver the specified product in the best way and with minimum waste, using value chain mapping and value engineering techniques. Value delivery comprises the final (detail) design and the construction of the project. Knowledge from contractors, subcontractors, and suppliers as well as knowledge and experience from using (consuming) the building, or similar buildings, is incorporated via facilitated workshops. Management is concerned with keeping time, budgets, and quality in a more traditional construction management context.

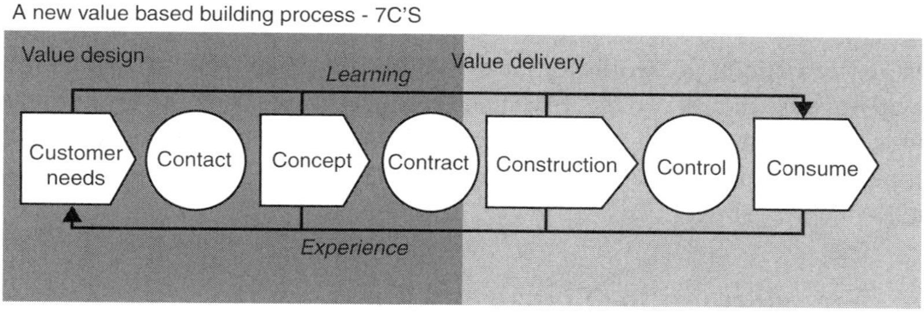

A new value based building process - 7C'S

FIGURE 5.1 The 7 C's model. (Developed from Bertelsen, S., Fuhr Petersen, K., and Davidsen, H., *Bygherren som forandringsagent – på vej mod en ny byggekultur* [The client as agent for change – towards a new culture in building]. Byggecentrum, Denmark, 2002. From Christoffersen 2003b.)

Transition between value design and value delivery is through the formal contract phase. However, it should be noted that these phases often overlap in practice. The "7 Cs" model (Figure 5.1) developed by Bertelsen, Fuhr Petersen, and Davidsen (2002), and subsequently developed further by Christoffersen (2003b), further breaks down these phases. This model forms the framework that was used in the facilitated workshops.

5.3.2 Value Design Phase

5.3.2.1 Customer Needs

Here it is important to address the client organization and make a stakeholder analysis, in order to map the interests in the project and organize it effectively. Stakeholder analysis is carried out in a facilitated process involving the key (primary) stakeholders to identify other stakeholders and categorize them as either primary, secondary, or tertiary. Then decisions can be made about who should be involved and when. In this phase, the basic values of the client organization and other identified stakeholders can be mapped, together with the contractual framework represented as time and cost budgets. The mapping process helps to identify values, which then form the basis of a value-based design brief. This design brief should not be seen as a static document, since it may evolve as issues are discussed, and values redefined through the workshop process. However, this document is important because it is used as the first specification of the client needs, and is used to communicate client values to the professional delivery team members.

5.3.2.2 Contact

Client contact with the delivery team reflects the preferred organization of the project. All stakeholders, including representatives from the owner, the user, the operation and management organizations, and society (typically represented by the authorities) should have already been identified, and should preferably be present. This helps to ensure that the appropriate specialist knowledge is represented in the delivery team (architects, engineers, contractors, and suppliers, etc.).

5.3.2.3 Concept

In this phase, a number of workshops are used to help guide the participants through the value universe in order to identify client needs and values before suggesting design solutions. The initial design brief is reevaluated as values are explored and made explicit. A number of conceptual designs are produced based on these values, and evaluated via the facilitated workshops.

5.3.3 Transformation (Transition) Phase

5.3.3.1 Contract

The construction contract is signed when the value design work is deemed to be complete; that is, when everybody agrees that no more/no better value can come out of the project within the time available. Establishing and agreeing the "point of no return," where the creative value design phase is replaced by the more pragmatic value delivery phase is crucial for preventing rework and waste. The transformation point varies between projects and it is essential that all participants are aware of this shift in emphasis, and respect it.

5.3.4 Value Delivery Phase

5.3.4.1 Construction

The client usually plays a less active role in this phase because the requirements have already been clearly identified. Much decision making still remains, although this is primarily related to production activities. These are dealt with by the main contractor, working closely with the subcontractors and suppliers. The client role (supported by professional advisors) is to deliver detailed decisions as scheduled, and to check that the specified value is delivered by the contractor. In order to achieve effective communication between the participants in the delivery team, a series of production workshops is used, focusing on waste reduction in the process as well as in the product by value engineering activities, and by introducing logistic planning tools based on the underlying ideas of the Last Planner System. This involves the establishment of a process plan for production activities, and looking two to five weeks ahead using the decision list to identify possible obstructions to the flow of activities, and mitigate them accordingly.

5.3.4.2 Control

There are many control activities within projects. However, in this context the "control activity" relates to the hand over of the property to the client and users. This represents the completion of the project. Control is executed with two goals in mind. The first is to check that the product is error free, as far as can be determined from the information available. The second one is to check whether the product fulfils the client's value specification, as agreed in the contract, and specified in the design and associated value-based documentation.

5.3.4.3 Consume

In this phase the agreed values are brought into use, and product values are experienced by building users, with facility and maintenance management coming to the fore. This is not discussed in this chapter, other than to note the importance of regular feedback from the consume phase in the drive for continual improvement. A systematic approach to the gathering of experiences is used as part of the stakeholder knowledge for incorporation in future projects.

5.4 Creative Workshops

Interaction in a facilitated workshop forum helps to establish common values, and enables actors to better understand why their values differ. The creative workshops start with the agreement of common process values, followed by discussion of client intentions and abstract ideals. Then, work proceeds to produce a complete set of production information prior to the commencement of construction activities (see below). The workshops continue into the production phase, in which the main contractor gets the main subcontractors involved. In fact, each workshop phase may comprise a series of facilitated workshops that deal with a particular issue or value stage, which continue until agreement has been reached. Workshops are "value generators" (or value drivers) and are concerned with problem framing. Part of a typical sequence is shown in Figure 5.2.

It is a demand of the value-based approach that the entire panel of participants is in place from the start of the project to its completion. For that reason the workshops tend to involve quite large numbers of people. Numbers present in the meetings vary between projects and stages, typically ranging between 15 and 30 people. The organizational format of the workshops can be changed to accommodate more people if necessary by

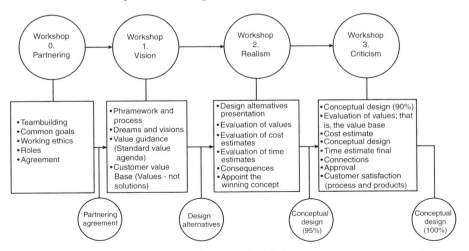

FIGURE 5.2 Sequence of workshops for the value design phase.

dividing into subgroups. The number of workshops varies depending on the size and complexity of the project. Typically, workshops last for a half or a full day, but they never last for longer than one working day. Some flexibility in programming is required to accommodate the inherent uncertainty in knowing exactly how many workshops will be required to reach agreement. The experience of the facilitator is crucial here in accurately predicting the number of workshops necessary. When problems with understanding and attitudes persist, additional workshops are convened to explore the underlying values, and tease out creative input. In extreme cases, if participants are unwilling to discuss and hence share values, they are asked to leave the process and are replaced by a new participant. Experience has shown that incompatibility usually manifests in the first few workshops. Thus, from the start of the project the whole process should be consensus-based, and participants should have a shared vision and goals. The facilitator's role is to stimulate discussion, thus helping to identify areas of agreement and conflicting interests. It is important that he or she remains objective and neutral, allowing the participants to make the decisions within a facilitated and supportive environment, as discussed below.

A standard value agenda is used as a framework for decision making in the workshops. This is referred to as the "basic value structure for buildings," and is based on the six product values (beauty, functionality, durability, suitability for the site and the community, sustainability, and buildability). This value hierarchy addresses the primary project objectives and breaks them down further into subobjectives as part of an iterative process carried out within the workshops. Common value management tools, such as the value tree [see, e.g., Dallas 2006], and quality function deployment (QFD) are used to weight options (values) in a decision matrix to help find the solution that provides the best value. The process facilitator guides participants through the discussion of values in a systematic and objective way.

5.4.1 Workshop Sequence in the Value Design Phase

5.4.1.1 Workshop 0: (Partnering) Building Effective Relationships

The function of the preliminary workshop is to bring various actors together to engage in socializing and team-building activities. The intention is to build the communication structures for the project, thus allowing actors to engage in open and effective communication during the life of the project. In addition to setting the stage for the events that follow the outcome of the first workshop is the signing of a partnering agreement, which confirms the process values for cooperation. Early workshops are also concerned with the selection of the most appropriate consultants, based on their ability to contribute to the project (their "fit") rather than the lowest fee bid. Collective dialogue helps to explore and develop relationships that can (or conversely cannot) develop into effective and efficient working alliances. Early workshops are also designed to help build a certain amount of trust and understanding before work commences.

5.4.1.2 Workshop 1: Vision

It is not possible to know values in depth at the start of a project, so workshops are primarily concerned with exploring values and establishing a common vision. Knowledge and experience from other projects are brought into the workshop, for example, from facilities management. The main focus of the effort is the establishment of client values (value-based parameters) on the basis that the better these are known and clearly identified, the better the team can deliver. An example could be functionality subdivided into optimal layouts to suit different users, for example, office workers, visitors, cleaners, and maintenance staff. Critical connections between decision making are explored so that everyone is certain of roles and responsibilities. The result of Workshop 1 is the establishment of basic values for the project; a very pragmatic document of prioritized values, which does not contain any drawings.

5.4.1.3 Workshop 2: Realism

Workshop 2 addresses how the basic project values may be fulfilled by presenting various design alternatives and looking at how they meet the basic value parameters. The contractual framework of the project is also addressed. Project economy is introduced here, along with constraints associated with authorities, codes, and regulations. Design proposals are worked through and ranked according to value. Architects are encouraged to produce at least three schemes that can be presented and discussed. Two to three workshops are normally required at this stage because there is much to discuss. Basic project values and project economy should be respected in this process, and any changes justified within the value parameters. The outcome of the realism phase is the selection of the best suited design proposal.

5.4.1.4 Workshop 3: Criticism

The presentation of the design proposals and the criticism are undertaken in two different workshops because it helps to encourage creativity and innovative solutions. In this workshop the proposed design solution is analyzed and criticized to see if it really is the best solution and whether it could be improved. Discussion is centered on the chosen design solution and its potential for improvement within the agreed value parameters. There is usually some pressure at this stage to get the scheme into production quickly, and this has to be balanced against reducing uncertainty in the design before entering the production phases. Stakeholder satisfaction with the process value and the product value is measured on the basis of the partnering agreement and the basic product value parameters. This is done using key performance indicators at various stages in the process to measure the participants perceived satisfaction. Then the project is approved for production, and the contractual delivery specifications are fixed.

5.4.2 Workshop Sequence in the Value Delivery Phase (Production Design Phase)

5.4.2.1 Workshop 4: Design Planning

As the abstract design work turns into production information there is a shift in thinking. The value management techniques are supplemented with "harder" value engineering exercises. A process management tool is introduced to support process planning and define goals. This is currently based on a modified version of the Last Planner System [Ballard 2000], which takes several issues into consideration, and aims to give participants a clear view of what needs to be done, by whom, and when. A design process plan maps activities and relationships, and is an important coordination tool. Value engineering and value mapping exercises are conducted in order to identify, and hence reduce waste. Many of the decisions are related to production activities, which are dealt with by interaction with the main contractor, working closely with the subcontractors. Supply chain issues are planned and the first steps toward a production plan in the construction phase are taken in order to identify critical supply lines, and accommodate any impact within the detailed design schedule.

5.4.2.2 Workshop 5: Buildability

Here the focus is on improving the buildability (or constructability) of the project, while trying to reduce waste in the detailed design and construction phases. The foremen and craftsmen meet with the designers to help discuss the efficient and safe realization of the design. This often leads to some reconsideration of the design, and revision of detailed designs to aid manufacturability and assembly. Changes may also be agreed to suit the available production capability and capacity. Once buildability issues have been resolved and agreed, it is possible to move onto the final workshop stage where the construction work is planned.

5.4.2.3 Workshop 6: Planning for Execution

These workshops involve interaction between the main contractor and the subcontractors to design and plan the control process. A process plan is produced that helps to map the various production activities. This also helps to identify any missing or erroneous information. The Last Planner System is usually applied by the main contractor at this stage. Due to the high level of interaction during the earlier phases, many of the problems and uncertainties have been resolved, although the inclusion of subcontractors allows for discussion about alternative ways of realizing design value. On completion of the construction schedule, the information should be complete, thus providing the contractor and subcontractors with a high degree of certainty in the production phases.

5.5   Facilitator's Role

It is common for the client to employ the process facilitator directly to represent their interests. Alternatively, the contractor might pay for the facilitation role because the early resolution of problems and rapid development of trust within the team appears to be cost-effective over the course of a project, although this is difficult to prove in absolute terms. Regardless of who pays for the service, the process facilitator plays a key role in scheduling and facilitating the meetings. He or she usually has no contractual responsibilities and is not at liberty to contribute to the discussions, merely to try and ensure that all participants have equal participation rights. Thus, the facilitator acts as an informal leader, charged with creating an effective social system that can drive the project forward based on consensus. The responsibility of the facilitator extends only to the process, not the output of the process, which remains the responsibility of the team. The facilitator has no influence on the program running alongside the workshops, other than to discuss and coordinate workshops with the project manager.

During the early meetings, the facilitator is primarily concerned with creating a harmonious atmosphere within the workshops so that actors are able to communicate and share values, with the hope of reaching agreement. Negative conflict is managed to ensure that any disagreements are dealt with in a positive manner. Positive conflict and criticism is sometimes encouraged to try and prevent the manifest of groupthink [Janis 1982], and hence try and prevent the group from making poor decisions. The facilitator's role changes as the workshops proceed, with priority given to keeping the team together during difficult discussions in the later stages when cost and time tend to dominate the discussions (and when conflict is more likely to manifest). With no formal power, the facilitator has to build trust and respect within the project team to enable the workshops to function effectively. Moral support from formal managers, for example, the project manager and the design manager, as well as the client is essential in this regard, helping the process facilitator to function as an effective informal leader. Needless to say, interpersonal communication between these parties must be effective and based on trust to allow the process to function. The process facilitator must possess excellent interpersonal skills, and have sufficient knowledge of construction to be able to guide the process, allowing sufficient time for focused discussion on the task, and time for socioemotional interchanges that promote a team ethos. The aim is to encourage the formation and retention of interpersonal relationships.

Success of the facilitated workshops will be colored to a large extent by the experience and skill of the facilitator. However, the actions of the participants are also a determining factor. Observations of meetings have revealed instances when participants have come to the meeting unprepared (for example cost information was not circulated before the meeting). This can cause a certain amount of turbulence, and sometimes this can result in the need for an additional workshop. In such situations, the facilitator speaks to the "problem" participant(s) outside of the workshop environment to try and encourage better performance in future meetings. Observation has also revealed that a great deal of informal communication takes place before and after the formal workshop sessions.

There is a widely held view in the Danish construction sector that females make better facilitators than their male counterparts. This perception appears to come from strongly held stereotypes about specific roles in a very conservative construction sector. There has not been any research undertaken that may help to confirm or dispel such views. Women are perceived to be better at communicating and to have better social skills than males. Scientific research based on some form of comparison between male and female facilitators would be helpful to see if differences between facilitators are related to gender or other factors.

5.6   Discussion and Reflection

Bringing people together in facilitated workshops is time-consuming and expensive. However, the workshops have proven to be an essential forum for discussing differences and reaching agreement. Workshops have also helped to encourage open communication and knowledge sharing, with learning as a group contributing to the clarification and confirmation of project values. There are some challenges associated with this method. Most importantly, the value-based model has to be implemented very early in the project, and all key actors must sign up to the approach. There have been some instances where the facilitated workshop method has been applied late in the process, and these "insertions" have proven to be ineffective despite the efforts of the facilitators (since the problems have already manifested). The proposed approach only works when all participants engage in open communication, and this takes a shift in thinking for many of the participants who are more familiar with adversarial practices and closed (defensive) communication.

Scheduling the process accurately to coincide with project management programs and specific milestones can be a challenge. The number of workshops required to ensure that all participants reach agreement on the project value parameters (or at least establish areas in which consensus is not reached and why) can sometimes exceed that planned. For projects with very tight schedules such uncertainty can present problems for project management teams that are not familiar with the approach. Experienced facilitators are able to bring people together quickly, and usually conduct the workshops efficiently. Sometimes it is simply a matter of discussing and agreeing on whether or not time is the crucial constraint for the project. The schedule of meetings may be extensive on a large project, and there is a concern that the cost of the meetings may outweigh the value realized through them. There is also the constant danger of holding too many workshops and the participants becoming jaded though overfamiliarization. All participants need to constantly monitor the effectiveness of the workshops and critically assess their added value through the use of various benchmarking tools. Although the workshops act as informal control gates, there are no formal gates (unlike some other process models). Some consideration of more formal procedures in line with total quality management could help the process facilitator and project managers to coordinate programs a little better.

The Danish Building Research Institute (SBi) has independently evaluated the delivery design phase of one project. Although a small and limited investigation, they found improved performance across a whole range of performance parameters [By og Byg 2004; SBi 2005]. In addition to this, the members of the projects have consistently evaluated the process highly, finding it an enjoyable and productive way of working, although as stakeholders in the project, one would expect this sort of response because it is difficult to criticize one's own contribution. Further work is required to investigate the effectiveness of, for example, the workshop method in terms of the realization of group goals. The literature on effective groups and teams suggest that the group size should be just large enough to include individuals with the relevant skills and knowledge to solve the problem; this is the principle of least group size [Thelen 1949]. The optimum size is considered to be around five or six people [Hare 1976], which is considerably smaller than the typical groups experienced in the case study projects. A related issue concerns the role of the workshop method in promoting and delivering creative solutions; and this would be a logical extension of this case study. Other areas of potential research relate to the skills and competences of the process facilitator, not just in facilitating the meetings, but also as a socializing function of project management. Some investigation of interpersonal communication skills (task-based and social–emotional) may also be useful avenues to explore in terms of educating/training process facilitators.

5.7   Recommended Strategies

It is possible to put forward a number of strategies for others to take and apply to their specific context. It is the authors' experience that failure to start the process correctly will probably result in ineffective projects. Therefore, our recommended strategies form a series of steps:

Step 1: Establish the project strategic plan; that is, establish the framework/vision, and appoint a qualified and experienced process facilitator before starting the project. The model is very sensitive to the social skills and experience of the process facilitator, so the appointment of a suitable (for a specific project context) process facilitator should be given serious consideration.

Step 2: Ensure that all key project stakeholders are present at the start of the project, and that they are committed to collaborative working and open communication.

Step 3: Make sure that everyone is educated in the process method, and that they understand their role in relation to other participants before starting the project.

Step 4: Make sure that the stakeholder organization is clearly defined, and that decision-making responsibilities are determined at the outset.

Step 5: Be specific in separating the determination of the client needs and the corresponding solutions (remembering "needs before solutions"). Client needs must be established through an effective briefing process before design solutions are presented. Do not be tempted to rush into (premature) solutions, since this has been found to generate waste.

Step 6: Ensure client needs are structured using common value management tools, such as the value tree and QFD.

Step 7: Make sure that whenever a design alternative is presented to the client, it is supported with detailed and realistic cost estimates and a value evaluation. This will help the client and other key stakeholders to make informed decisions, and hence move forward with confidence.

Step 8: Try to ensure that differences between individual company interests and personal interests do not compromise the outcome of the project. Further research on the application of human resource techniques, and the importance of social skills in design and construction projects will be useful here.

Step 9: Make sure that the transition point between value design and value delivery, the point of no return, is determined and agreed by all project participants.

Step 10: Be aware of the different values, professional cultures, and attitudes of the various actors in the delivery team and their reasons for participating in the project. Hidden agendas often prove to be a determining factor in undermining the success of projects.

5.8   Concluding Comments

Our experience is that the way in which people interact within the project environment and with their colleagues in their respective organizations will have a major influence on the success of individual projects, and the profitability of the participating organizations. Both the metaphorical and physical space between the organizations participating in projects will influence interaction practices, and hence the effectiveness of the project outcomes. This means that (considerable) effort is required in trying to manage interpersonal relationships to the benefit of clients' projects and also to the profitability of the organizations contributing to them. We strongly believe that emphasis should be on maximizing value through improved interaction, communication, and learning within the entire supply chain.

The approach described in this chapter does work extremely well for some clients and their delivery teams in a Danish (democratic) context. It is important to recognize that it might not suit all clients, or all societies. Clients are extremely complex entities, usually representing a multitude of stakeholders with differing sets of values, and different levels of interest in the project. Thus, exploring and mapping values is complicated and trying to satisfy all stakeholders is, on reflection, rather ambitious. We also feel that it is important to emphasize that this approach has evolved over a number of years, and that it is one of many approaches to try to improve supply chain management in the Danish construction sector. As intimated above, it is a challenge to implement the value-based model in practice, requiring considerable effort and determination on behalf of the process facilitator, and commitment from the project participants. The value-based method is continually evolving—it is getting simpler—and the recent award of a research grant has provided the resources to investigate some of the issues reported here in greater detail.

References

Austin, S., Baldwin, A., Hammond, J., Murray, M., Root, D., Thomson, D., and Thorpe, A. 2001. *Design chains: A handbook for integrated collaborative design.* Tonbridge: Thomas Telford.

Ballard, G. 2000. The Last Planner System. PhD thesis, University of Birmingham.

Bertelsen, S. 2000. *The habitat handbook.* Copenhagen: Danish Ministry of Industry and Commerce.

Bertelsen, S., Fuhr Petersen, K., and Davidsen, H. 2002. *Bygherren som forandringsagent – på vej mod en ny byggekultur* [The client as agent for change – towards a new culture in building]. Denmark: Byggecentrum.

By og Byg. 2004. *Evaluering af forsøg med trimmet projektering og trimmet byggeri.* Report number 421-047, January. Horsholm: Statens Byggeforskningsinstitut.

Caudill, W. W. 1971. *Architecture by team: A new concept for the practice of architecture.* New York: Van Nostrand Reinhold.

Christoffersen, A. K. 2003a. *Report on the William Demant Dormitory process – methodology and results.* Copenhagen: The Danish Ministry of Commerce and Economy.

———. 2003b. *State of the art report: Working group value management.* Byggeriets Evaluerings Center, August.

Christopher, M. 1998. *Logistics and supply chain management: Strategies for reducing costs and improving services,* 2nd ed. London: Pitman.

Dallas, M. F. 2006. *Value & risk management: A guide to best practice.* Oxford: Blackwell Publishing.

Emmitt, S., and Gorse, C. 2007. *Communication in construction teams.* Abingdon: Spon Research, Taylor & Francis.

Gropius, W., and Harkness, S. P. eds. 1966. *The architect's collaborative, 1945–1965.* London: Tiranti.

Hare, A. P. 1976. *Handbook of small group research,* 2nd ed. New York: The Free Press.

Janis, I. 1982. *Victims of groupthink: A psychological study of foreign policy decisions and fiascos,* 2nd ed. Boston, MA: Houghton Mifflin.

Kelly, J., and Male, S. 1993. *Value management in design and construction: The economic management of projects.* London: E & FN Spon.

Koskela, L. 2000. *An exploration towards a production theory and its application to construction.* Espoo, Finland: VTT.

Kristiansen, K., Emmitt, S., and Bonke, S. 2005. Changes in the Danish construction sector: The need for a new focus. *Engineering Construction and Architectural Management,* **12** (5), 502–11.

Luft, J. 1984. *Group process: An introduction to group dynamics.* Palo Alto, CA: Mayfield.

Maister, D. 1993. *Managing the professional service firm.* New York: The Free Press.

Miles, L. D. 1972. *Techniques of value analysis and engineering.* New York: McGraw Hill.

Porter, M. E. 1985. *Competitive advantage: Creating and sustaining superior performance.* New York: The Free Press.

SBi. 2005. Journal no. 421-042, May. Statens Byggeforskningsinstitut (Danish Building Research Institute, SBi). Hørsholm, Denmark.

Stroop, J. R. 1932. Is the judgement of the group better than that of the average member of the group? *Journal of Experimental Psychology,* **15**, 550–62.

Thelen, H. A. 1949. Group dynamics in instruction: Principle of least group size. *School Review,* **57**, 139–48.

Womack, J., and Jones, D. 1996. *Lean thinking: Banish waste and create wealth in your corporation.* New York: Simon & Schuster.

Womack, J. P., Jones, D. T., and Roos, D. 1991. *The machine that changed the world: The story of lean production.* New York: Harper Business.

Yoshida, R. K., Fentond, K., and Maxwell, J. 1978. Group decision making in the planning team process: Myth or reality? *Journal of School Psychology,* **16**, 237–44.

6

Supply Chain Management for Lean Project Delivery

Iris D. Tommelein
University of California

Glenn Ballard
University of California

Philip Kaminsky
University of California

6.1 Introduction

6.1.1 What is a Supply Chain? What is Supply Chain Management?

The term "supply chain" refers to a series of interdependent steps of activities or processes (sometimes sequential and sometimes overlapping) as well as flows between them, supported by infrastructure (people, equipment, buildings, software, etc.) [e.g., Simchi-Levi et al. 2007; Fine and Whitney 1996]. These flows express real or forecast customer demand

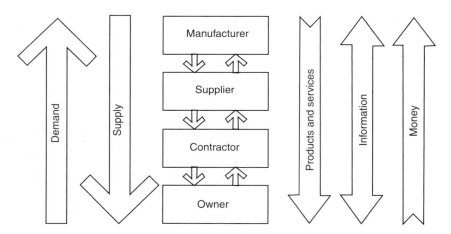

FIGURE 6.1 Example of supply and demand with flow of products/services, information, and money in a supply chain.

going in one direction, and supply going in the other direction in order to fulfill that demand. Figure 6.1 outlines the general directions of these flows. Demand and supply flow in opposite directions but may follow different routes (they are not necessarily one-on-one opposites of each other). Information flows both ways. Products and services also may flow both ways (e.g., a fabricator may ship products to a galvanizer and then incorporate returned products into larger assemblies). Accordingly, the term supply "network" might be a better characterization of this system than supply "chain" is, but the latter term is used more commonly and will thus be used throughout this chapter.

Supply chain management (SCM) refers to managing the flows of physical products and services, information, and money between the activities or process steps that companies perform, while aiming for customer service as the goal (i.e., get the right product to the right place at the right time for the right cost). Defined in this way, SCM applies to the delivery of capital projects (so-called "project supply chains") as it does to the delivery of products or services in other industries[1] (supply chains that deliver products are sometimes referred to as "product supply chains").

6.1.2 SCM in Project Settings

A project constitutes steps to design, make, and then deliver a product or service to a customer (or customers). In order to do so, the project may acquire goods and services from a combination of preexistent and custom-made supply chains (SCs). One difference between SCM at large vs. SCM in project delivery settings is that some project supply chains are relatively short-lived: they must be established, configured rapidly,

---

[1] We disagree in this regard with Fernie and Thorpe (2007), who equate SCM with partnering and claim that SCM does not apply in the construction industry.

and remain flexible to match demands that vary over the course of project execution.[2] Another difference is that in project production systems, owners tend to be involved throughout project delivery and influence their project supply chains directly (they are "prosumers"[3]). In contrast, in many manufacturing production systems (especially for commodities that are made-to-stock to production targets that are set to meet forecast demand), retail customers remain anonymous until receipt of the final product (they are consumers) though they may influence supply chains indirectly. Notwithstanding such differences, SCM and new product development practices have been pushing for more individual customization (which also has been the pursuit in "mass customization" efforts, for example, Davis 1989; Pine 1993, 1999; Gilmore and Pine 1997) so that several concepts and techniques used to manage various manufacturing SCs are now akin to those used to manage project SCs.

Project supply chains may be parts of existing, longer-lived supply chains that operate regardless of whether or not any one specific project exists. Alternatively, project supply chains may be established specifically to meet one project's or several projects' needs. For example, a formwork contractor may tap into an existing supply chain to purchase lumber on an as-needed basis from a local reseller and get partial truckloads shipped from the reseller's storage location to the contractor's yard. Alternatively, while ordering from the same reseller, full truckloads may get shipped directly from the mill to the contractor's yard, bypassing the reseller's storage location. As illustrated, the supply chain from the mill to the reseller exists to meet the demands from a pool of customers, whether or not this one contractor places any order with that reseller; based on order size (to take advantage of economies of scale in transportation and handling) that supply chain flexes to suit the magnitude of the specific demand. Such flexibility is not uncommon in construction supply chains because materials and shipments are often bulky, heavy, or of exceptional size (e.g., 30 m long precast piles), and these characteristics may weigh in considerably in SC performance metrics (e.g., economics expressed as total cost installed or total cost of ownership).

6.1.3 Narrow and Broad Views on SCM

The goals of SCM (meeting customer service/cost objectives) can be viewed and optimized from a project, enterprise, or industry perspective. Tommelein, Walsh, and Hershauer (2003) noted that "while SCM may be practiced on a single project, its greatest benefits come when it (a) is practiced across all projects in a company, (b) involves

---

[2] Such flexibility has been characterized as "agility" by some, but it is what we expect of "lean-ness" on projects and their supply chains (e.g., Preiss (2006) compares and contrasts agility with lean-ness). Others have used the term "leagility" to refer to the combination of lean-ness with agility [e.g., van Hoek 2001]. A key distinction to make when comparing those terms used in the literature is whether or not they apply only to making (manufacturing) or also include designing. In our view, lean includes both.

[3] "Prosuming" means involving the customer in production [e.g., van Hoek 2001, 163]. In contrast, other traditional supply chains serve consumers.

multiple companies, and (c) is applied consistently over time. In today's marketplace, companies no longer compete one-on-one; their supply chains do."

Historically speaking, SCM has evolved from materials management by broadening its scope. SCM thus includes procurement (sourcing and purchasing) and logistics (warehousing and transportation). As the scope of SCM continues to broaden, operations and production are also included, so SCM includes the design and execution of activities or process steps themselves, as well as the design and management of the system they make up, in order to deliver value to the owner. By considering more functions in an integrated fashion, SCM is increasingly better positioned to shape product and service flows so as to more optimally meet management objectives.

This broad interpretation resonates with the Construction Industry Institute (CII)'s definition [Tommelein, Walsh, and Hershauer 2003]: "SCM is the practice of a group of companies and individuals working collaboratively in a network of interrelated processes structured to best satisfy end-customer needs while rewarding all members of the chain." Other views on SCM amplify considerations different from those presented here. For example, Cox (2001) has highlighted power relationships as a means to gain leverage in SCs. While competing definitions of SCM exist, as is clear from other chapters in this book and the voluminous body of literature on this subject [e.g., Hershauer et al. 2005], many authors state that delivering optimal customer service is the goal of SCM. The remainder of this chapter focuses specifically on the application of SCM in lean project delivery settings.

6.2 SCM in "Lean" Project Delivery

6.2.1 Toyota Production System

"Lean production" is a term coined by John Krafcik to characterize the Toyota Production System (TPS) [Womack, Jones, and Roos 1990]. Toyota, like other automobile manufacturers, produces cars on a large scale, but it uses a different[4] way of designing and making them [e.g., Liker 2003; Liker and Meier 2005]. The philosophy Toyota has developed and the culture it instills through company-wide use of "lean" practices has enabled it to become a world leader in automobile manufacturing. In a nutshell, this philosophy promotes "doing what the customer wants, in no time, with nothing in stores" [Womack and Jones 1996]. It focuses on value streams (recognizing how, where, and when value gets created in the process of transforming raw materials into finished goods) and shaping them to reduce waste [Rother and Shook 1999].

The lean philosophy supports Toyota's entire business enterprise, including not only manufacturing and production [e.g., Liker 2003; Liker and Meier 2005] but also new product development (a type of project production system with characteristics different from those of a manufacturing production system) [e.g., Morgan and Liker 2006], accounting, supplier relationships, strategic planning, etc. Toyota's production-systems thinking can also be applied to project settings, for example, those encountered in

---

[4] Many companies have been trying to copy Toyota's system, but it remains different from other production systems in many regards.

architecture-engineering-construction (AEC) project delivery. Accordingly, a theory of how to deliver projects in a lean fashion is emerging, though it is yet to be fully articulated. The term "lean construction" refers to this theoretical development.

6.2.2 Lean Construction and Lean Project Delivery

6.2.2.1 Transformation-Flow-Value Theory as a Foundation for Lean Construction

The theory of lean construction recognizes that three schools of thought have emerged in production management and that these views are orthogonal yet complementary: one adopts the transformation view "T", the second the flow view "F", and the third the value view "V" [Koskela 1992, 2000].

Koskela et al. (2002, 213–15) characterize these views as follows:

> In the transformation view, production is conceptualized as a transformation of inputs to outputs. [The] ...principles by which production is managed... suggest, for example, decomposing the total transformation hierarchically into smaller transformations, called tasks, and minimizing the cost of each task independently of the others. The conventional template of production has been based on this transformation view, as well as the doctrine of operations management. ... However, this foundation of production is an idealization, and in complex production settings the associated idealization error becomes unacceptably large. The transformation view of production has two main deficiencies: first, it fails to recognize that there are phenomena in production other than transformations, and second, it fails to recognize that it is not the transformation itself that makes the output valuable, but, instead, that there is value in having the output conform to the customer's requirements. The transformation view is instrumental in discovering which tasks are needed in a production undertaking and in getting them realized, however, it is not especially helpful in figuring out how to avoid wasting resources or how to ensure that customer requirements are met in the best possible manner. Production managed in the conventional manner therefore tends to become inefficient and ineffective.
>
> The early framework of industrial engineering introduced another view on production, namely that of production as flow. ... The flow view is embodied in "lean production," a term coined ... to characterize Toyota's manufacturing practices. In the flow view, the basic thrust is to eliminate waste from flow processes. Thus, such principles as lead time reduction, variability reduction, and simplification are promoted. In a breakthrough book, Hopp and Spearman (2000) show that by means of queuing theory, various insights that have been used as heuristics in the framework of JIT can be mathematically proven.
>
> A third view on production was articulated in the 1930s, namely that of production as value generation. In the value generation view, the basic goal is to reach the best possible value from the point of the customer. The value generation view was initiated by Shewhart (1931). It was further refined in the framework of the quality movement but also in other circles. Principles related to rigorous requirements analysis and systematized flowdown of requirements, for example, are forwarded.

Cook (1997) recently presented a synthesis of a production theory based on this view.

Thus, there are three major views on production. ... These three views do not present alternative, competing theories of production, but rather theories that are partial and complementary.

Lean construction is a TFV theory that acknowledges that all three views weigh in on production system management. In contrast, traditional construction and project management practices have amplified the transformation view, while demoting "F" and sacrificing some "V". SCM practices have amplified the flow view (as highlighted by Figure 6.1), while demoting "T" and sacrificing some "V." Lean construction applies TFV specifically to project settings such as—but not limited to—those encountered in the AEC industry. Lean SCM, an integral part of lean construction, thus differs from SCM at large in that it aims to balance all three.

Lean construction adopts a holistic and systemic view of project delivery, recognizing that the project delivery system may be viewed at different levels with TFV pervasive in all, namely (a) the physics of the task (how work at the lowest level actually gets done), (b) production (how work relates to other work), (c) organizations (how people and the relationships between them affect how work gets done), and (d) formal and informal contracts (how incentives motivate people to behave or production to be organized in one way or another).

At the production level this system includes five phases (Figure 6.2): (a) project definition, (b) lean design, (c) lean supply, (d) lean assembly, and (e) use. Spanning these phases are production control and work structuring. Production system design, operation, and improvement are driven by work structuring and production control.

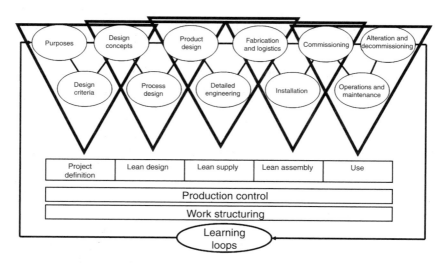

FIGURE 6.2 Lean Project Delivery System™ (LPDS). (From Ballard, G. Lean Project Delivery System™. White Paper-8 (Rev. 1). Lean Construction Institute, Ketchum, ID, 23 September, 2000a. With permission.)

6.2.2.2 Work Structuring

Lean work structuring is project production system design (process design integrated with product design to deliver a project) and extends in scope from an entire production system down to the operations performed on materials and information within that system [after Ballard et al. 2002]. Work structuring means "developing a project's process design while trying to align engineering design, supply chain, resource allocation, and assembly efforts" [Howell and Ballard 1999].

"The goal of work structuring is to make work-flow more reliable and quick, while delivering value to the customer" [Howell and Ballard 1999]. In particular, work structuring views a project as consisting of production units and work chunks [Ballard 1999]. A production unit is an individual or group performing production tasks. Production units are recipients of work assignments. A work chunk is an output of a production task that is handed off from one production unit to the next. In the process of performing a production task, each production unit may or may not make changes to the boundaries of the work chunk before handing it off to the next production unit. While performing tasks, production units typically will add value to work chunks. In turn, these chunks morph while moving through the production system until they become a completed product.

Work structuring involves determining: (a) in what chunks will work be assigned to production units, (b) how chunks will be sequenced, (c) how chunks will be released from one production unit to the next, (d) where decoupling buffers will be needed and how they should be sized [Howell, Laufer, and Ballard 1993], (e) when different chunks will be done, and (f) whether consecutive production units will execute work in a continuous flow process or whether work will be decoupled [Tsao, Tommelein, and Howell 2000; Tsao 2005]. These determinations are fundamental to production system design, be it project production systems or SCs. Nevertheless, these determinations are not explicitly and routinely made in practice today. In contrast, current work structuring practices—if it is appropriate to call them that—mostly focus on local performance and are driven by contracts, the history of trades, and the traditions of craft. Hampered by these drivers, decision makers rarely take the liberty to consider how to optimize the entire production process. The resulting work breakdown structures more-often-than-not prevent the smooth flow of work and hamper performance effectiveness. In contrast, work structuring, as defined here, adds a production system's view to the other views, while aiming to reveal such and other optimization opportunities through adoption of a holistic view on project delivery, ranging from project definition through use of a capital facility (Figure 6.2).

6.2.2.3 Production Control

Production control (as also discussed in Chapter 2 of this book) means shaping work and planning it at successive levels of detail, covering with greater accuracy increasingly shorter time periods into the future as time for action approaches, while making adjustments as needed to steer the project towards best meeting system objectives during project execution. Its objective is to maximize the likelihood of getting the work done according to project objectives. Plan reliability—a key objective in lean project delivery—can be managed by means of the Last Planner™ system [Ballard 2000b]. This system embraces a methodology to shield planned work from upstream variability, thereby allowing performance improvement to take place behind the shield [Ballard and Howell 1998] and work flow to be stabilized upstream from the shield[5] [Ballard and Howell 1994].

6.3   Project Supply Chains

Projects are undertaken to create a unique product, service, or result [PMI 2004, 5]. They can be conceptualized as temporary production systems. This means that, in contrast to ongoing business operations, which take place in a manufacturing or service facility, a project has a start time and an end time. A project would not be designated as such if it went on forever. A project production system of course does not exist in a vacuum: it is established in context and draws upon existing supply chains (that support permanent production systems) to "feed" its needs.

Figure 6.3 shows a work-structure model, highlighting primary flows that must be managed in a construction project. The "Con" (construction) processes flow to an intersection, an assembly point <Con C, 1> (where processes merge and products get matched[6]), of which there are obviously very many in a project. For each construction installation such as <Con A, 1>, there is a prior act of engineering/design such as <Eng A, 1> and procurement (purchasing, fabrication) such as <Proc A, 1>. The arrows connecting the <Eng X, i> boxes indicate that they must be coherent with one another,

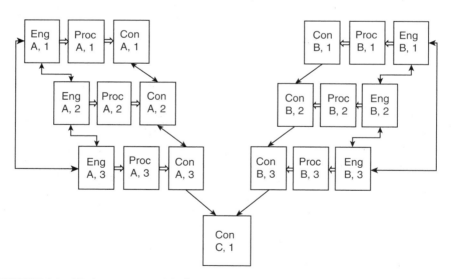

FIGURE 6.3   Work structure model of a project production system.

---

[5] More details about the Last Planner[TM] system are presented in Chapter 3.
[6] Tommelein (1998) presents a computer model to illustrate how uncertainty involving matching problems may be accounted for in project planning and execution.

amounting ultimately to a description of a system—structural, mechanical, electrical, etc. The model could be expanded to show product supply chains and the corresponding flows that support fabrication and site deliveries as well.

Figure 6.4 expands the view on the project work-structure model in that it also includes product supply chains. The wide arrow shows project delivery as a progression through time from start to completion, through the phases of lean project delivery (as detailed in Figure 6.1). This arrow is akin to a so-called "development chain" in SCM [Simchi-Levi, Kaminsky, and Simchi-Levi 2007], involving project participants such as the owner, designers, contractors, and other providers of specialist services who contribute to a new product's development.

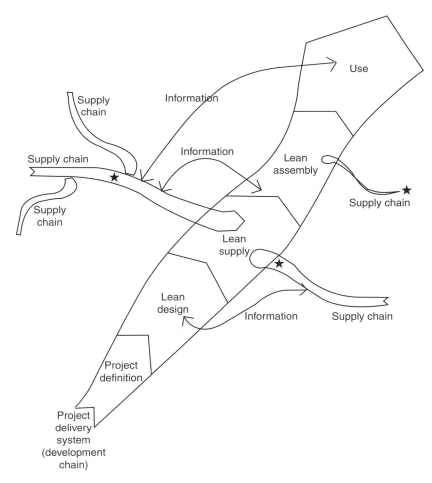

FIGURE 6.4 Project supply chain hinging on preexistent and custom-made supply chains. *Note:* Stars denote customer order decoupling points (CODPs).

In a narrow and traditional sense, supply refers to a supplier handing-off a "black box" product to a project participant. "Black box" means that the supplier designs and makes the product while treating the project participant as a consumer. The delineation appears to be defined by the customer or product on one end, and by the choice where to "black box" supply on the other; that is, to aspire to only transactional relationships. Accordingly, suppliers are not traditionally thought to be project participants.[7]

In contrast, lean project delivery includes suppliers in the team, recognizing that they may offer not only "black box"[8] products but could also deploy their production system to suit the project. Indeed, a project is a set of resources structured to achieve the project's objectives, and hinges on preestablished production systems from which goods and services are acquired. The project hinges on supply chains in the Lean Supply triad (Figure 6.2). A variable is the extent to which those preestablished production systems are objects of coordination and shared fortune for the contractor or owner, as opposed to existing entirely independently of the contractor or owner. Consequently, project SCM could be understood as the design, execution, and improvement of SCs independent of any one project, and the design, execution, and improvement of the project supply chain.

Figure 6.4 shows several narrow arrows, each exemplifying a supply chain that "feeds" the project. There typically will be many more of these project-feeding supply chains, but only a few are shown. In the course of work structuring, these supply chains were selected from among many that exist independently of whether or not a particular project materializes (i.e., the project exists in a "universe" of supply chains). Some supply chains may be used "as-is," to hand-off a product to the project. Others may have their production system tailored to meet a specific project's requirements. Custom-tailoring is shown in Figure 6.4 by the stars that indicate where in the supply chain the product or service becomes customer-specific (this star marks the "customer order decoupling point," a concept that will be further detailed later). The double-headed arrows illustrate how information may flow in the system, from suppliers to project participants in different phases of project delivery, and vice versa. It is through integration of such various supply chains that project supply chain objectives are pursued.

6.4   Selected Lean Production System Design Concepts and Principles

We next define concepts and principles to be used when designing lean (project) production systems. These consider work structuring to go hand-in-hand with production control and establishment of feedback loops to promote learning. Production system design is said to be "lean" when it is done in pursuit of TFV goals [Ballard et al. 2002].

---

[7] To illustrate this point, note that in the first decade or so of functioning of the CII at University of Texas, Austin in Texas, CII membership included only owners and contractors. Only in more recent years have suppliers been invited to the table.

[8] Toyota has developed different SCM practices based on the degree to which its systems can be decomposed, and accordingly refers to "white box," "gray box," and "black box" items being procured from suppliers [e.g., Ward et al. 1998].

6.4.1 Customer Order Decoupling Points and Push–pull Boundaries in Supply Chains

In the process of work structuring, the process steps and wait times a product is subjected to while being transformed from raw materials into a final product can be mapped out. This map may take the form of a "value stream" or a "cross-functional diagram," which shows tasks and hand-offs across organizational boundaries. In any case, along the supply timeline it depicts, the customer order decoupling point (CODP) marks the point where a product is customized to meet a specific customer's needs (shown by means of a star in Figure 6.5).

Based on this CODP concept, Wortmann, Muntslag, and Timmermans (1997) developed a typology of manufacturing and supply approaches differentiating products that can be made-to-stock from those that have to be made-to-order (Figure 6.5). To the left of this point, upstream to "Raw materials," production is driven based on forecast demand. To the right of this point, downstream all the way to "Product delivered to customer," production is driven based on actual demand. The CODP defines the "push–pull boundary" because forecasting means pushing products through the supply process without knowing exactly who the customer will be, whereas sales reflect the pull of a specific customer. Because forecasts are speculative—they are always wrong (due to their stochastic nature) and even more so when projecting further out into the future—lean production systems are designed to produce products and deliver services based on customer pull where possible.

Products that remain undifferentiated until they are sold "off the shelf" are said to be "made to stock" (MTS). MTS products can be produced based on either forecast need or a specific customer order, but it is often the former, as they are mass produced to reap the benefits of economies of scale. For example, SCs for lumber, drywall, bathroom plumbing, and many light fixtures are like this (e.g., a fixture sold in a retail store is custom-wired during installation).

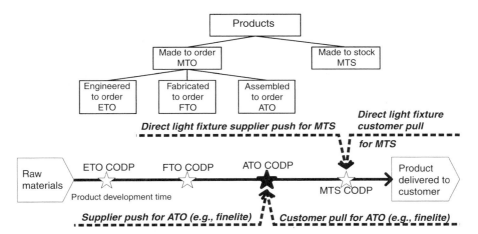

FIGURE 6.5 Typology of product supply approaches recognizing customer order decoupling points. (After Wortmann, J. C., Muntslag, D. R., and Timmermans, P. J. M. (eds). *Customer-driven Manufacturing*. London: Chapman & Hall, 1997. With permission.)

Alternatively, products may be "assembled to order" (ATO) by putting together off-the-shelf parts to suit a customer's desired configuration prior to delivery to the customer. Configuration and assembly is done at a location either on- or off-site, but not at the location of final installation. As a result, ATO production systems incur a lead-time penalty (i.e., assembly is done some time before final installation). Lean producers strive to continuously reduce such lead times so that other performance metrics may off-set the costs of ATO production, for example, the lead-time penalty may be outweighed by a reduction in installation time or an increase in safety, handling efficiency, or qual-ity. For example, SCs for light fixtures may be structured for ATO production [e.g., Tsao and Tommelein 2001].

Furthermore, products may be "fabricated to order" (FTO) (e.g., by taking off-the-shelf parts and cutting, drilling, or welding them) or "engineered to order" (ETO) (e.g., using engineering design and analysis to determine which components are needed and how to configure them). For example, SCs for pipe supports may be structured for FTO or for ETO production [Tommelein, Walsh, and Hershauer 2003; Arbulu et al. 2002, 2003]. Figure 6.5 illustrates the corresponding CODPs and shows increasing lead-time penalties for ATO, FTO, and ETO products.

While Figure 6.5 marks the CODP as a single point (star) on the timeline, "the concept does not only center around deciding at what level in the chain postponement is to be applied, it is also a matter to what degree is it applied" [van Hoek 2000]. So, work struc-turing is employed to decide which suppliers to work with, which products or services to get from whom and when, and where to position the CODP. From a TFV perspective, the location of the CODP affects where customer requirements (information flows) need to be injected in the supply time line and what lead time will be required downstream from that CODP to the point of customer hand-off. It thus affects the extent to which a supplier can deliver value both in terms of product specificity as well as responsiveness in meeting varying project demands.

6.4.2   Lean Principle of Continuous Flow

Note that CODPs exist whether or not a supply system is lean. Lean production systems are designed to use customer pull where possible to set the rhythm for production. Pro-ducing products at the so-determined customer-demand rate, lean systems also strive to achieve "continuous flow"; that is, get raw materials to proceed through all production steps without undue inventory or other waste. The lean production literature describes these concepts in detail [e.g., Rother and Shook 1999; Rother and Harris 2001; Harris, Harris, and Wilson 2003; Smalley 2004; Womack and Jones 2002]. In a nutshell, con-tinuous flow may be achieved through batch sizing (ideally, batches of one to achieve one-piece flow) and synchronization of production steps, combined with pull. Example applications of pull used in construction SCM are *kanban* systems for delivery of ready-mix concrete [e.g., Tommelein and Li 1999] and precast panels [e.g., Arbulu, Ballard, and Harper 2003], and constrained work in progress for rebar cages [e.g., Arbulu 2006].

Where pull is not feasible, production system steps may be decoupled using buf-fers. The use of buffers is necessary in systems subject to uncertainty or variability, as these are particularly detrimental to performance [Hopp and Spearman 2000].

Correspondingly, the lean approach to production system design is to first root out all unwanted variability,[9] and then accommodate the remaining variability in its design.

6.4.3 Positioning and Sizing of Buffers

Buffers are holding places for products, resources (people or equipment), or information, or time delays in between steps in a production system. If continuous flow were always possible, buffers would not be needed. Until that lean ideal has been achieved, buffers are needed and their locations and sizes must be judiciously determined in the course of production system design. Buffers can be used to serve a variety of functions.

6.4.3.1 Use of Inventory Buffers in Project Production Systems

Inventory buffers comprise raw materials, work in progress, or finished goods. Based on Schmenner's (1993) work, the major functions of raw materials inventory can be characterized as:

- To protect (buffer) against the unreliable delivery of needed raw materials.
- To hold costs down if possible by buying in large quantities or by buying at propitious times.

The same author characterized the major functions of **work-in-progress inventory** as:

- To permit one segment of the process to operate under a different production plan and at a different rhythm from another segment of the process (e.g. Howell, Laufer, and Ballard (1993) called this "decoupling inventory").
- To permit individual work stations or machine centers to produce parts, assemblies, or materials in sizable batches, rather than individually (the lean, ideal "one-piece-flow") or in smaller batches. Such "cycle inventory" acts to tide the process over until the next setup.
- To protect (as a buffer) against the unreliable (a) delivery of materials from elsewhere in the production process, (b) completion of prerequisite work, or (c) release of information.

He also characterized the major functions of **finished goods inventory** as:

- To supply the product quickly to the consumer. Made-to-stock products have zero lead time.
- To protect (as a buffer) against the uncertainties of customer demand. Buffers are thus a substitute for information, that is, if one had perfect information about future downstream customer demand (and likewise, of upstream supply), one could reduce buffer sizes.

---

[9] Hopp and Spearman (2000) distinguish two types of variability: (a) bad variability is the result of unplanned outages, quality problems, accidental shortages, or human skill; and (b) good variability may stem from the purposeful introduction of differences in product or process characteristics as a means to increase the system's ability to match or create market demand (customer value). Bad variability is always unwanted.

- To smooth (through the accumulation of finished goods inventory) demand on the process even while demand is erratic or temporarily depressed (balance capacity against demand).
- To lower costs for shipping and handling (optimize the batch process).

In addition to inventory buffers, the systems' view on project production also includes so-called "capacity buffers," "time buffers," and "plan buffers" among the variables in the design, execution, and improvement of the project-cum-supply chain.

6.4.3.2 Use of Capacity Buffers in Project Production Systems

A capacity buffer is a resource that intentionally is not fully utilized. In a system subject to uncertainty, lean practice is to schedule resources (e.g., people and equipment) at less than 100% utilization because the manifestation of that uncertainty might jeopardize system performance (e.g., if work is planned to make full use of a resource, that work will not get completed in the anticipated time period if a glitch occurs). There is value in having resources on standby—the result of such underloading—to deal with the unexpected, as it yields greater plan reliability (nevertheless, providing excess capacity is contrary to the project management wisdom that says resources must be kept busy all the time). Lean production recognizes that reliable flow is more important for increasing throughput (customer value) than high resource utilization is, and therefore relies on capacity buffers. Most production systems are not balanced anyway, so high utilization cannot be obtained for all resources in each and every step.

6.4.3.3 Use of Time Buffers and Merge Bias in Project Production Systems

A time buffer is a delay or a lag added at the end of a sequence of steps that has an uncertain finish time, in order to guarantee the start time of the immediately succeeding step.

Time buffers may be particularly useful in production system design when SCs merge, thereby creating a network. Merge bias occurs when several inputs must all be present in order to start a process step or task, that is, a shortage of any one of them will prevent the start of their successor. When the arrival of one (or several) of these inputs has some degree of variability, the likelihood of successor delay increases. Furthermore, when the number of inputs increases, each one arriving independently of the other, the likelihood of successor delay multiplies; this system characteristic is known as the merge bias. Merge bias affects the time it will take for a product to flow through the system.

Koskela (2004) noted that in construction, many tasks get started even when some inputs are in short supply, on the presumption that the supply of those inputs will catch up with the production need for it. Of course, there is no guarantee that this will happen, unless the supply system is designed and controlled to make it happen. He called this the "make-do" mind-set and characterized it as a type of waste, because all too often supply is unable to deliver and as a result steps fall short of being completed as planned. When pressured to make progress, workers use what is readily available and thereby forego the best-laid plans. They end up working in potentially unsafe conditions, produce faulty work, and make early progress, but at the expense of their own follow-on work[10] or that of others, and fail to meet the customer's needs. Thus, if the supply system is not managed well, completion of such steps becomes uncertain, thereby injecting unreliability into the work flow, which in turn hampers performance.

6.4.3.4   Use of Plan Buffers in Project Production Systems

A plan buffer is a step or task that is part of a plan but that is (1) not yet scheduled or (2) scheduled but can be rescheduled. Flexibility exists in reordering those steps or tasks (i.e., work flow). Plan buffers are hidden when hierarchical plans are created because the master plan (master schedule) typically defines tasks at an abstract level. Those tasks are broken into smaller ones when look-ahead plans and weekly work plans are created. In the process of breaking down tasks, decisions must be made regarding their definition and sequencing while adhering to the original plan. Thus, smaller tasks make up a buffer from which selection needs to be made in order to yield a good, overall system performance.

6.4.4   Lean Principles of Muri, Mura, Muda

We next summarize three objectives Toyota pursues at the same time, because they apply to project production system design as well as to SCM. They are (a) to appropriately use resources, (b) to balance loads, and (c) to eliminate waste (captured, respectively, by the Japanese words "muri," "mura," and "muda") [Kitano 1997; Liker 2003].

Muda refers to the elimination of waste, which Ohno (1988) classified using seven types: (a) defects in products; (b) overproduction of goods not needed; (c) inventories of goods awaiting further processing or consumption; (d) unnecessary processing; (e) unnecessary movement of people; (f) unnecessary transport of goods; and (g) waiting by employees for process equipment to finish its work or for an upstream activity to be completed. Womack and Jones (1996) added "design of goods and services that fail to meet user's needs" and Koskela (2004) furthermore added "make-do" as another type of waste.

Mura refers to load balancing. Mura is reflected in just-in-time system concepts (supplying the production process with the right part, at the right time, in the right amount, using small buffers, using pull [Kitano 1997]). In the context of SCM this includes striving for level production (supply) (e.g., through use of one-piece flow, lead-time reduction in order to be able to react to changing project needs, and modularization) as well as demand (e.g., using the Last Planner™ system).

Muri refers to using resources appropriately. It may be achieved, for example, using standardized work, setting a reasonable customer-demand rate to drive production, and defining a logical flow of work [Kitano 1997].

6.5   Lean Supply

With these concepts and objectives for project production system design and SCM in mind, we now return to the Lean Project Delivery System™ (LPDS) (Figure 6.2) and

---

[10] It is well known that the last x% of work on an activity tends to take disproportionately more time than remains on the schedule to complete that activity.

focus on lean supply. Lean supply in the LPDS includes "product design," "detailed engineering," and "fabrication & logistics." Through product design, lean supply connects to lean design. Through fabrication and logistics, it connects to lean assembly. As mentioned, the Lean Supply triad is the hinge between the project production system and SCs. Viewed from this perspective we describe various SCM tools and techniques, and state how they contribute to improvements with regards to TFV.

6.5.1   Product Design for Lean Supply

Cross-functional teams: "Lean" thinking strongly advocates the inclusion of suppliers in design, so suppliers will be part of the lean project delivery team. Lean design pursues "design for X" (DFX), where X stands for criteria to assess TFV (such as source-ability, constructability, maintainability, sustainability) including SC performance. For example, suppliers may advise designers on standard and easy-to-install/maintain products, or on their process capability, or transportation and storage means, in order to curtail product and process variability. Suppliers, fabricators, procurement specialists, logistics services providers, and production units can inform a design team about the realities of execution possibilities, requirements, and constraints, thereby helping the team to generate value and eliminate waste in the process of making more informed decisions, especially when considering SC strategies. A challenge is to bring suppliers in early enough so they can fully contribute to the team, while rewarding them for their engagement even if no product sale is guaranteed.

Supplier alliances:[11] Supplier involvement in a project does not need to be initiated when that project starts and end when that project is completed. A supplier alliance is a long-term relationship between a buyer (e.g., owner) and a seller (e.g., supplier) that spans multiple projects and thus results in a more permanent supply chain to meet a customer's needs. It tends to focus on specific product families (e.g., precast concrete elements or engineered-steel buildings) or services (e.g., software support). Due to its multiproject nature, alliances can address opportunities and needs at the enterprise level. On occasion, suppliers may get even more deeply entrenched with owners and support the development of their business case.

An alliance is a substitute for the many one-on-one transactions that otherwise are developed when specific project needs arise. Advantages of alliances are efficiencies stemming from longer-term relationships including collaboration on joint product and process development (value creation through such means as standardization, target costing, risk pooling, demand leveling, and increasing demand predictability further in the future) and the incentives it brings for alliance participants to invest in developing such efficiencies. Disadvantages are the trust and investment needed to develop them, and potential loss of market competition.

---

[11] In some contexts, "alliance" refers to a particular contractual arrangement. In contrast, here we simply use this term to denote a longer-term agreement between a buyer and a seller.

6.5.2 Detailed Engineering for Lean Supply

A question of work structuring in supply chains is "Which party will detail the design?" While detailing may be thought of as the last step of completion of design, instead we view it as the start of construction [Tommelein and Ballard 1997]. Fabricators and installers engaged in detailed engineering for lean supply can add significant value to the production system (e.g., Zabelle and Fisher 1999), because they are familiar with intricacies and alternative means for doing the work and the execution environment (e.g., transportation, handling, trade interference, skill availability).

Lead time reduction: When products have long lead times (e.g., ETO and FTO products in Figure 6.5), it is hard to achieve one-piece flow and use pull mechanisms. This is particularly true when submittals and approvals are required in addition to custom-making, and uncertainty can manifest itself at many occasions. Accordingly, fabricators and installers are well positioned to reduce product cycle times and improve through-put. They can strive to reduce batch sizes and eliminate multitasking practices. Through their early involvement in design they can also alleviate—if not eliminate—the design–bid–*redesign* cycle. These practices help to achieve flow (F) in the production system.

Shorter lead times also enable designers to keep their options open longer. Use of a technique called "postponement" affords designers more time to explore alternative solutions with other project participants, suppliers, and stakeholders. This practice reduces waste in the project because it avoids rework (e.g., a solution that gets selected early on based on one view, may prove to be infeasible based on other views or when additional information becomes available over time) and generates value to the project because the assessment of alternatives can be gauged by combining values ascribed from various views.

Standardization: Suppliers may also recommend that the team use standard products and processes (or develop them if no existing ones are satisfactory) (muri), rather than custom-design everything from scratch. Standardization reduces the workload pertaining to submittals and approvals. Furthermore, use of multiples of the same product helps to alleviate matching problems [Tommelein 1998, 2006], simplifies all handling, eases installation, allows for risk pooling, and promotes learning. Lean practitioners develop standard products and processes in order to be able to gauge deviation from those standards. This helps not only to control production, but also to experiment with new ways of doing things (e.g., kaizen), which in turn leads them to develop better standards.

Information transparency: A lean practice is to make system status information available to those who need it, so that there is no need for guesswork or speculation (waste) to know actual demand or system status. System-wide transparency in SCs helps to avoid the Bullwhip Effect that results from people otherwise speculating what customer demand might be, one or several steps removed from them in the SC, and thereby injecting variability into the system [Forrester 1961; Lee, Padmanabhan, and Whang 1997].

6.5.3 Fabrication and Logistics for Lean Supply

Provide materials, tools, and information to workers at point of use: Several products may be combined into a single handling (packaging) unit (the process is sometimes referred to as "kitting" or "bagging and tagging"). Parts can be similar or dissimilar but all parts in the unit will be used together, for example, installed in each other's vicinity or handled by a single person or crew. The advantage of providing materials, tools, and information to workers at point of use is that workers will not need to spend extra time to locate and count the needed parts (mura). Kitting can be done away from congested work areas, at a location where it can be managed better and less expensively. This practice consists of creating a CODP and moving it upstream in the SC. Pushing the "matching problem" upstream in this way will be successful only when the final demand for the parts in the kit and the timing of need for the contents of the kit are well known. In conditions of uncertainty, the value of kitting is diminished and it can even be counterproductive.

Control of transportation means: Many construction products are shipped by the supplier (seller) who often takes responsibility for the load while it is in transit, but as a result may use large batch sizes (full truckloads) and impose long lead times. The buyer may offer incentives to the supplier to make supply reliable and responsive to the project needs, for example, to get just-in-time deliveries. Alternatively, the contractor (buyer) can take on this responsibility, thereby gaining control over the supply.

Load consolidation: Goods shipped by suppliers to one project may be combined with goods for other projects in order to save on shipping costs—an opportunity for suppliers to apply inter/intraproject SCM. This can create a win–win situation; for example, some commodity suppliers offer supplier-managed-inventory services with daily replenishment in regions where they have multiple customers.

An unfortunate consequence of loading trucks to capacity is that extra time may be needed to achieve this load, thereby delaying the timing of the shipment. This time–cost tradeoff must be considered when one is aiming for reducing cycle time in the supply chain.

Conversely, a project could send out a vehicle to pick up loads at different supplier locations, or large companies with ongoing work in a region could pick up loads for multiple projects from one or multiple suppliers.

Third-party logistics providers: Third-party logistics providers are more common in the manufacturing industry than they are in construction. They arrange not only for transportation of goods, but also package goods as needed for easy distribution, thereby saving time later on for locating and retrieving goods. For example, some drywall suppliers also stage pallets with drywall during off-hours in each room as specified by installation crews.

Logistics centers: Logistics centers are places for handling one or more operations pertaining to the delivery of products or services to projects. These centers generally are not places where goods are produced [Baudin 2004], but some fabrication may take place there. Logistics centers can be configured to provide a wide range of functions such as: receiving, storage, break-bulk, sorting, assembly, cargo consolidation, transport, distribution (direct shipment, shipment with milk runs, etc.), distribution network management including vehicle routing, package tracking, and delivery, e-commerce services, etc. These functions can be catered to suit the requirements of one or several SCs [after Hamzeh et al. 2007], summing the demand for certain products and thereby offering the benefit of risk pooling.

6.6   Conclusions

This chapter presented a production-systems view on SCM and the corresponding principles for lean production system design that consider transformation, flow, and value when delivering customer service. According to this view, a project becomes a part of a supply chain by design, acquiring goods and services from a combination of preexistent and custom-made supply chains, each providing goods and services to the project customer, who in turn may use the facilities provided to produce goods and services for others, ad infinitum. SCM therefore is an integral and important part of lean product delivery. It concerns not only the "lean supply" triad in LPDS™ but more broadly supports "work structuring" (design) of the overall production system. Lean project SCM includes selecting and shaping existing SCs, or constituting new SCs as needed to meet production system requirements of one or several projects. It encourages suppliers to not only transact products but also to consider tailoring their production system to suit a project's needs and set up relational agreements in order to enhance performance and maximize value in the delivery of the specific project(s) at hand.

Contractual relationships can offer incentives or disincentives for SC participants to view the project holistically and strive to meet TFV objectives; they can drive or stifle work structuring efforts. A failure to understand the entire SC results in local decision making that is often contrary to the optimal function of the entire chain.

Acknowledgments

The authors' recent research on project production systems and SCM has been funded in part through gifts and in-kind contributions made to the Project Production Systems Laboratory (P²SL) at the University of California, Berkeley (http://p2sl.berkeley.edu/). All support is gratefully acknowledged. Any opinions, findings, conclusions, or recommendations expressed in this paper are those of the writers and do not necessarily reflect the views of contributors to P²SL.

References

Arbulu, R. J. 2006. Application of pull and CONWIP in construction production systems. *Proc. 14th Ann. Conf. of the Intl. Group for Lean Constr. (IGLC-14)*, 25–27 July, Santiago, Chile.

Arbulu, R. J., Ballard, G., and Harper, N. 2003. Kanban in construction. *Proc. 11th Ann. Conf. of the Intl. Group for Lean Constr. (IGLC-11)*, Elsinore, Denmark.

Arbulu, R. J., Tommelein, I. D., Walsh, K. D., and Hershauer, J. C. 2002. Contributors to lead time in construction supply chains: Case of pipe supports used in power plants. *Proc. Winter Simulation Conference 2002 (WSC02). Exploring New Frontiers*, 8–11 December, San Diego, California, 1745–51.

———. 2003. Value stream analysis of a re-engineered construction supply chain. *Building Research and Information—Special Issue on Re-engineering Construction*, Spon Press, 31 (2), 161–71.

Ballard, G. 1999. *Work structuring.* Lean Construction Institute White Paper No. 5. www.leanconstruction.org.

———. 2000a. Lean Project Delivery System™. White Paper-8 (Rev. 1). Lean Construction Institute, Ketchum, ID, 23 September. http://www.leanconstruction.org/pdf/WP8-LPDS.pdf, 7 pp.

———. 2000b. The Last Planner™ system of production control. PhD Diss., School of Civil Engineering, The University of Birmingham, UK, 192 pp.

Ballard, G. and Howell, G. 1994. Implementing lean construction: Stabilizing work flow. *Proc. 2nd Ann. Conf. of the Intl. Group for Lean Constr.*, reprinted in Alarcon, L. 1997. *Lean Construction.* A. A. Balkema, Rotterdam, The Netherlands.

———. 1998. Shielding production: Essential step in production control. *Journal of Construction Engineering and Management, ASCE,* 124 (1), 11–17.

Ballard, G., Tommelein, I., Koskela, L., and Howell, G. 2002. Lean construction tools and techniques. In Best, R. and de Valence, G. (eds) *Design and Construction: Building in Value.* Oxford, Boston: Butterworth-Heinemann, Elsevier Science, 227–55.

Baudin, M. 2004. *Lean Logistics: The Nuts and Bolts of Delivering Materials and Goods.* Productivity Press, New York, 387 pp.

Cox, A. 2001. Managing with power: Strategies for improving value appropriation from supply relationships. *The Journal of Supply Chain Management,* Spring, 42–47.

Davis, S. M. 1989. From "future perfect": Mass customizing. *Planning Review,* March/April, 16–21.

Fernie, S. and Thorpe, A. 2007. Exploring change in construction: supply chain management. *Engineering, Construction and Architectural Management,* 14 (4), 319–333.

Fine, C. H. and Whitney, D. E. 1996. Is the make-buy decision process a core competence? Working paper. MIT, Sloan School of Management, MIT Center for Technology, Policy, and Industrial Development.

Forrester, J. W. 1961. *Industrial Dynamics.* Cambridge, MA: M.I.T. Press.

Gilmore, J. H. and Pine, B. J., II. 1997. The four faces of mass customization. *Harvard Business Review,* January–February, 91–101.

Hamzeh, F., Tommelein, I. D., Ballard, G., and Kaminsky, P. (2007). Logistics centers to support project-based production in the construction industry. In Pasquire, C.L. and Tzortzopoulos, P. (eds) *Proceedings of the 15th Annual Conference of the International Group for Lean Construction <http://www.iglc.net/>* (IGLC 15), 18–20 July 2007, East Lansing, MI.

Harris, R., Harris, C., and Wilson, E. 2003. *Making materials flow.* Lean Enterprise Institute, Cambridge, MA, 93 pp.

Hershauer, J. C., Walsh, K. D., and Tommelein, I. D. 2005. Exploring the multiple perspectives that exist regarding supply chains. In Geunes, J., Akcali, E., Pardalos, P. M., Romeijn, H. E., and Shen, Z. J. (eds) *Applications of Supply Chain Management and E-Commerce.* New York: Springer. HD38.5A67.

Hopp, W. J. and Spearman, M. L. 2000. *Factory physics.* 2nd ed. (1st ed. 1996). McGraw-Hill, Boston, 698 pp.

Howell, G. and Ballard, G. 1999. *Design of construction operations.* Lean Construction Institute White Paper No. 4. www.leanconstruction.org.

Howell, G., Laufer, A., and Ballard, G. 1993. Interaction between subcycles: One key to improved methods. *Journal of Construction Engineering and Management, ASCE*, 119 (4), 714–28.

Kitano, M. 1997. Toyota production system: "One-by-one confirmation". *Lean Manufacturing Conf.*, University of Kentucky, May 15. http://www.mfgeng.com/images/toyota.pdf.

Koskela, L. 1992. Application of the new production philosophy to the construction industry. Tech. Report No. 72, CIFE, Stanford University, CA, September, 75 pp.

———. 2000. An exploration towards a production theory and its application to construction. PhD Diss., VTT Pub. 408, VTT Building Technology, Espoo, Finland, 296 pp.

———. 2004. Making do: The eighth category of waste. *Proc. 12th Ann. Conf. of the Int'l. Group for Lean Constr.*, Elsinore, Denmark.

Koskela, L., Howell, G., Ballard, G., and Tommelein, I. 2002. The foundations of lean construction. In Best, R. and de Valence, G. (eds) *Design and Construction: Building in Value*. Oxford, UK: Butterworth-Heinemann, Elsevier Science, pp. 211–26.

Lee, H. L., Padmanabhan, V., and Whang, S. 1997. Information distortion in a supply chain: The bullwhip effect. *Management Science, Frontier Research in Manufacturing and Logistics*, 43 (4), 546–58.

Liker, J. K. 2003. *The Toyota Way: 14 Management Principles from the World's Greatest Manufacturer*. New York: McGraw-Hill, 350 pp.

Liker, J. K. and Meier, D. 2005. *The Toyota Way Fieldbook*. New York: McGraw-Hill, 476 pp.

Morgan, J. and Liker, J. 2006. *The Toyota Product Development System: Integrating People, Process and Technology*. Portland, OR: Productivity Press, 400 pp.

Ohno, T. 1988. *Toyota Production System: Beyond Large-scale Production*. Productivity Press, Portland, OR.

Pine, B. J., II. 1993. Mass customizing products and services. *Planning Review*, July/August, 6–13 and 55.

———. 1999. *Mass Customization: The New Frontier in Business Competition*. Cambridge, MA: Harvard Business School Press, 368 pp.

PMI. 2004. *A Guide to the Project Management Body of Knowledge*. 3rd ed. Project Management Institute, Newtown Square, PA, 388 pp.

Preiss, K. 2006. Agility and leanness. *Keynote presentation, International Conference on Agile Manufacturing* (ICAM), Old Dominion University, Norfolk, VA.

Rother, M. and Harris, R. 2001. *Creating Continuous Flow: An Action Guide for Managers*. Engineers & Production Associates, Lean Enterprise Institute, Cambridge, MA, 104 pp.

Rother, M. and Shook, J. 1999. *Learning to see: Value stream mapping to create value and eliminate Muda*. v.1.3. The Lean Enterprise Institute, Brookline, MA, 112 pp.

Schmenner, R. 1993. *Production/Operations Management*. 5th ed. Prentice Hall, Englewood, NJ.

Simchi-Levi, D., Kaminsky, P., and Simchi-Levi, E. 2007. *Designing and Managing the Supply Chain*. 3rd ed. New York: McGraw-Hill/Irwin, 498 pp.

Smalley, A. 2004. *Creating Level Pull*. Lean Enterprise Institute, Cambridge, MA, 114 pp.

Tommelein, I. D. 1998. Pull-driven scheduling for pipe-spool installation: Simulation of lean construction technique. *Journal of Construction Engineering and Management, ASCE,* 124 (4), 279–88.

———. 2006. Process benefits from use of standard products: Simulation experiments using the pipe spool model. *Proc. 14th Ann. Conf. of the Intl. Group for Lean Constr (IGLC-14),* 25–27 July, Santiago, Chile. http://www.ce.berkeley.edu/%7Etommelein/papers/IGLC06-038-Final-Tommelein.pdf.

Tommelein, I. D. and Ballard, G. 1997. Coordinating specialists. Technical Report No. 97-8, Construction Engineering and Management Program, Civil and Environmental Engineering Department, University of California, Berkeley, CA.

Tommelein, I. D. and Li, A. E. Y. 1999. Just-in-time concrete delivery: Mapping alternatives for vertical supply chain integration. *Proc. 7th Annual Conference of the International Group for Lean Construction (IGLC-7),* 26–28 July, Berkeley, CA, 97–108.

Tommelein, I. D., Walsh, K. D., and Hershauer, J. C. 2003. Improving capital projects supply chain performance. Research Report PT172-11, Construction Industry Institute, Austin, TX, 241 pp.

Tsao, C. C. Y. 2005. Use of work structuring to increase performance of project-based production systems. Ph.D. Diss., Civil & Envir. Engrg. Dept., Univ. of Calif., Berkeley, CA.

Tsao, C. C. Y. and Tommelein, I. D. 2001. Integrated product/process design by a light fixture manufacturer. *Proc. 9th Annual Conference of the International Group for Lean Construction (IGLC-9),* 6–8 August, Singapore.

Tsao, C. C. Y., Tommelein, I. D., and Howell, G. 2000. Case study for work structuring: Installation of metal door frames. *Proc. 8th Annual Conference of the International Group for Lean Construction,* University of Sussex, Brighton, UK.

van Hoek, R. I. 2000. The thesis of leagility revisited. *International Journal of Agile Management Systems,* 2/3, 196–201.

———. 2001. The rediscovery of postponement: A literature review and directions for research. *Journal of Operations Management,* 19, 161–84.

Ward, A., Christiano, J., Sobek, D., and Liker, J. 1998. Toyota, concurrent engineering, and set-based design. In: Liker, J. K., Ettlie, J. E., and Campbell, J. C. (eds) *Engineered in Japan: Organization and Technology.* New York: Oxford University Press.

Womack, J. P. and Jones, D. T. 1996. *Lean Thinking: Banish Waste and Create Wealth in Your Corporation.* New York: Simon & Schuster, 352 pp.

———. 2002. *Seeing the whole: Mapping the extended value stream.* Lean Enterprise Institute, Cambridge, MA, 101 pp.

Womack, J. P., Jones, D. T., and Roos, D. 1990. *The Machine That Changed the World.* HarperCollins Publishers, New York, NY, 323 pp.

Wortmann, J. C., Muntslag, D. R., and Timmermans, P. J. M. (eds). 1997. *Customer-driven Manufacturing.* London, UK: Chapman & Hall, 464 pp.

Zabelle, T. R. and Fischer, M. A. 1999. Delivering value through the use of three dimensional computer modeling. *Proc. 2nd Conference on Concurrent Engineering in Construction,* Espoo, Finland.

7

Application of Integrated Materials Management Strategies

Roberto J. Arbulú
Strategic Project Solutions

7.1 Materials Management: Traditional Practice in Construction

Construction organizations are often structured in functional silos such as planning and procurement. Traditionally, the planning department generates not only master schedules that provide an overall view of how the project should be executed, but also bills of materials that define type and quantity of products. Buyers then use schedules and bill of materials to initiate the process of acquiring materials for the project in coordination with site teams. With a commercial agreement in place, suppliers proceed to fabricate or collect from existing stocks for delivery to site. When site teams receive materials, a typical action is to find a place to store them until needed for installation. After suppliers have delivered (typically in large batches), materials usually sit on-site for long periods of time until they are completely consumed (assuming that there is no surplus—a rare occurrence in construction!). This simplified and generic description of how construction manages materials is nothing but a purchasing process rather than a materials management process. Although the deployment of control processes for issuing and using materials has become widespread in some sectors of the construction industry, such as industrial construction, pulling materials to site installation is hardly adopted. In traditional construction, the management of materials is typically limited to telephone conversations with suppliers to confirm final delivery dates. Figure 7.1 illustrates the generic process described above in a cross-functional format.

Lack of materials on-site when required, lack of the 'right' materials on-site, and accumulation of material inventories are just some of the types of waste generated by traditional practices, hampering performance through delays, low-quality workmanship, cost overruns, and poor safety levels on-site (mainly due to materials laying around working areas). The most common approach to materials management in construction is to manage variability by accumulating large inventories on-site. This approach has implications for cost and time primarily due to: (a) crews looking for materials within large stockpiles instead of working; (b) hidden cost associated with managing large materials stockpiles; (c) stockpiles potentially blocking workflow and risking material quality and safety on-site; and (d) large quantities of materials delivered early to site, causing a negative impact on project cash flow (money is allocated and released too early). The desire to benefit from economies of scale, such as cheaper by the dozen, creates an incentive to purchase and also request deliveries in large quantities. From that perspective, the larger the size of the delivery batch, the less trips are required to fulfill a complete order. Therefore, the lower the transportation cost, the cheaper the unit price.

One of the main root causes of this problem is that materials are ordered (and not managed properly) based on information coming from project schedules. The general assumption is that project schedules reflect the reality of site production at any point in time, but this is almost never the case. A common reality in construction is that schedules are disconnected from production-related activities, which means that whatever the materials management process is (if this exists) will be decoupled from

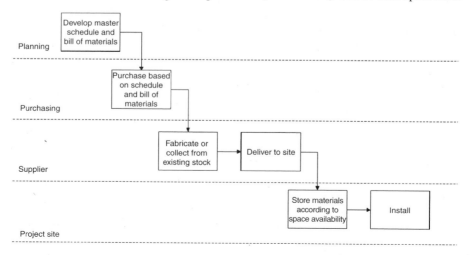

FIGURE 7.1 Simplified view of traditional materials management in construction.

(a)

(b)

FIGURE 7.2   Consequences of traditional materials management in construction.

production workflow. Figure 7.2 illustrates the consequences of traditional practices when managing materials. The value of a material stockpile and the time it has remained on-site is made explicit in Figure 7.2b.

7.2   Integrated Materials Management: A New Approach

Industries such as manufacturing and retailing have demonstrated high levels of integration in their supply chains, sometimes up to a third tier. In construction, due to its fragmentation, it is often difficult to integrate with even first-tier suppliers.[1] In this context, the term "integrated" then refers to the integration of first-tier suppliers in materials management practices to obtain better synchronization of supply and demand at the project level as a means for minimizing waste. The synchronization of supply and demand depends on several variables such as project objectives, the nature of the materials, the understanding of local conditions such as geographical conditions according to project location, and the capabilities of the first tier in the supply chain, among others. For the development of an integrated materials management strategy to be successful, it is necessary to follow a four-step process: definition, design, implementation, and operation (Figure 7.3).

The first and most important step in this development process, definition, has three main parts: (a) define project objectives including customer needs and stakeholder requirements as well as constraints on the satisfaction of needs and requirements such as any applicable requirement and local conditions at the project level; (b) identify the nature of the materials to be managed through the strategy; and (c) align the objectives of the materials management strategy with project objectives. The next steps, design,

---

[1] The definition of first-tier suppliers depends on who is looking at the supply chain. In this chapter, the perspective of an organization that delivers construction services to an owner operator has been adopted. It means that a material supplier is considered to be a first-tier supplier.

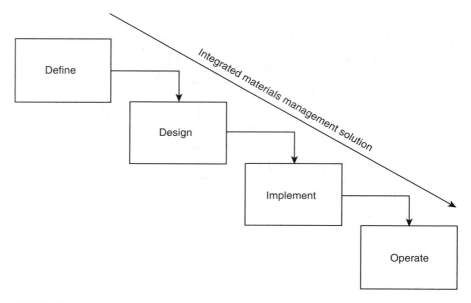

FIGURE 7.3　Development process for an integrated materials management strategy.

TABLE 7.1　Product Types for the Purpose of Materials Management

Non-Task-Specific	Task-Specific
PPE (e.g., hard hats, safety vests)	Made-to-order products (e.g., concrete)
Tools (e.g., hammers)	Engineered-to-order products (e.g., steel)
Consumables (e.g., nails, drill bits)	
Commodities (e.g., pipes, cables)	

implementation, and operation, will strongly depend on the output of definition. A key decision point for the design and implementation of a materials management strategy is to identify the nature of the products. For the purpose of materials management, there are two types of materials in construction: (a) task specific and (b) non-task-specific (Table 7.1). Task-specific materials are defined as products that are explicitly required for a unique construction task. Examples of such products include concrete, precast concrete elements, preassembled rebar, and steel. These are mostly made-to-order or engineered-to-order products. Non-task-specific materials are defined as products for which precise consumption cannot be reasonably planned ahead. These are mostly made-to-stock products. Examples of such products include safety vests, grinding discs, tying wire, nuts, and bolts. Some production teams often refer to non-task-specific products as the ones they cannot be bothered to order or plan to use in detail until those products are needed immediately. Non-task-specific materials are typical examples of relatively cheap, small items that can impact production if they are not available on-site when needed.

In this chapter, two case studies are presented in order to illustrate how to approach integrated materials management. The first case study focuses on non-task-specific materials and includes the application of *kanban* techniques to materials supply. The second case study focuses on task-specific materials, incorporating the use of real-time integrated production management and materials management systems.

7.3 Case Study 1: Management of Non-task-specific Materials

Properly managing non-task-specific materials may be challenging, especially because it is difficult to estimate consumption rates per product. Non-task-specific materials are mainly characterized by their variety, while the total amount of different non-task-specific materials depends on the project type. For instance, the variety of non-task-specific materials needed to deliver an industrial project may be greater than a housing project. Sometimes this variety implies the need to integrate several first-tier suppliers in order to supply the demand. As integrating large numbers of first-tier suppliers is not an easy task, an integrated materials management approach must focus on working with only preferred suppliers.

7.3.1 Background

The following sections provide a description of this case study including results, components of the integrated materials management strategy, and detailed approach. Readers are encouraged to adapt the ideas presented herein to their own situation. The case study was carried out during the civil phase of a £4.2 billion civil infrastructure project in the United Kingdom. Being one of the largest projects in Europe, it included about 80 different subprojects with massive activities such as earthworks, piling, tunneling, concrete placing and precast fabrication and installation, rebar fabrication, assembly and installation, among others.

The project presented specific challenges, such as traffic conditions, security, and lack of laydown area, that forced the owner to define five key principles for managing materials supply: (a) materials must be pulled from the supply network as needed at the workface; (b) materials must arrive at the right place, at the right time, in the right quantity; (c) the supply network is achieved at the best value for the customer; (d) all necessary actions are taken to minimize vehicle movements on-site; and (e) all necessary actions are taken to increase workflow reliability on-site. The development of the integrated materials management strategy followed the four-step process shown in Figure 7.3. Details of this process are presented in the following sections.

7.3.2 The Approach

The integrated materials management strategy was defined based on the above five principles established by the owner. These were translated into objectives:

- Give project teams what they want when they want it by pulling non-task-specific materials from first-tier suppliers.

- Support the reduction of material inventories (the strategy did not propose the elimination of inventories).
- Reduce the paperwork necessary to order new products or to increase stock levels.
- Enable a product rationalization process focused on reducing product variety while maintaining site needs and desired service levels.
- Reduce purchasing and supply lead times.
- Eliminate expediting, a traditional practice in construction of chasing production or purchase orders which are needed in less than the normal lead time.
- Contribute to continuous improvement.
- Act as a catalyst for change in procurement and in purchasing methods.
- Simplify site materials management processes for acquiring, storing, distributing, and disposing of non-task-specific materials by eliminating waste and reducing information processing.

With clear objectives in place, the integrated materials management solution was then designed based on five key components as shown in Figure 7.4:

(a) Marketplace or site store: A marketplace is a small warehouse or store. Typically, a marketplace varies in size according to the magnitude of the project. It requires, among other things, a detailed layout design to determine the best shelving solutions, and how materials will flow coming in and out of the marketplace. This strategy considered the marketplace basically as an inventory buffer strategically sized to shield site production from supply variability and fluctuations in demand.

(b) Collection vehicles or "milk runs": Collection vehicles were managed directly by the project, rather than by the suppliers. These vehicles visited suppliers' warehouses on a regular basis to collect materials in order to replenish marketplace stocks.

(c) Supplier *kanban*: *Kanban* is a Japanese word that means "card" or "sign." It is an approach developed by Toyota Motors Company that aims to pull materials and parts through production systems on a just-in-time basis supporting the minimization of work-in-process between workstations. In this case study, *kanbans* were used to transmit a replenishment signal to outside suppliers (first-tier suppliers), as supplier kanbans do in manufacturing [The Productivity Press Development Team 2002]. Plastic bins and cards were used as signals to trigger the replenishment of

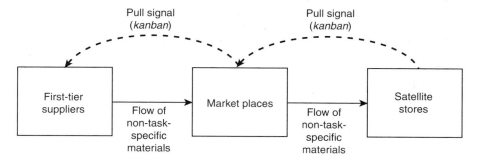

FIGURE 7.4 Non-task specific materials management strategy.

non-task-specific materials on a daily basis. More details about the use of *kanban* will be provided later in this section.

(d) Satellite stores: The project used cargo containers as satellite stores strategically located on-site. The number of satellite stores and location varied according to project progress. The type of products managed through the satellite stores was limited and defined by the project team in control of each satellite store. Satellite stores maintained two days' worth of inventory constantly replenished through the marketplace.

(e) Inventory management system and electronic ordering: Although *kanbans* triggered physical flows of materials, the overall materials management process required the use of other tools to enable the management of very large amount of information and money flows[2] between suppliers and customers on-site. The inventory management system included the use of inventory management software linked with an e-procurement system, barcode technology, and radio frequency (RF) handheld scanning devices. For instance, the inventory management system enabled the marketplace team to track all replenishments, generating transactions and billing reports that were submitted to leaders of production teams detailing each withdrawal by team member, product, and total spend by team.

The definition and design of the strategy took place in the early stages of site construction, taking approximately four months due to project complexity (mostly due to the large number of stakeholders to be engaged). Implementation took approximately two months with multiple areas under construction. The implementation process followed three main steps, which are explained below: (a) define product list; (b) select preferred suppliers; and (c) set up *kanban* replenishment system.

(a) *Define Product List*

A key decision in the materials management strategy is to clearly define which items are non-task-specific materials. Due to a fast-track project delivery approach (design being completed while construction is underway), it was difficult for production teams to provide product types and demand estimates for non-task-specific materials. Based on this situation, the marketplace implementation team took advantage of usage records and demand patterns from previous projects that had similar construction phases. Through the analysis of the data, the team identified opportunities to minimize product variety through a product rationalization process based on product function. Figure 7.5 shows an example of the product rationalization process based on data from only one supplier with a total of 2555 different non-task-specific products. From the total number of products, approximately 10% represented a value of more than £250 per year per product. The remaining 90% of products had an average demand of less than 20 items per month. It was clear that the tendency was to order a large variety of products but in small quantities. This was mainly caused by a product selection process from suppliers' catalogs with no rules or any control—that is, people ordered what they wanted based on their own preferences rather than

---

[2] Any supply chain has three flows: (a) physical; (b) information; and (c) capital or money flows.

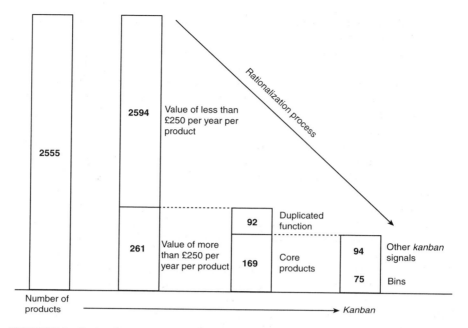

FIGURE 7.5   Rationalization process for non-task-specific materials.

from a limited and rationalized product list. Allowing this situation to occur at the project was certainly not an option; otherwise the materials management process would not work properly with site teams placing orders based on their own desires instead of real needs. The rationalization process then focused on reducing the variety of products without compromising the performance of site activities. For instance, if data showed that 20 different types of gloves with similar functions had been purchased in the past, actions were taken to limit the number of glove types available through the materials management strategy to the minimum possible.

Production teams provided input during the rationalization process in order to ensure the final product list matched their needs and the needs of the project. The rationalization process concluded with a product list of approximately 200 different products. During operations, the list was constantly increased or decreased according to site needs (the peak was around 800 different products).

(b) *Select Preferred Suppliers*

Based on a rationalized product list, preferred suppliers were clearly identified. By limiting the number of suppliers, the demand for non-task-specific products was divided amongst the selected suppliers, providing an incentive for them to actively participate in the integration effort. This, off course, does not totally eliminate resistance to change from suppliers that are not used to participating actively in integration initiatives. At the end, only five preferred suppliers shared the responsibility of supplying demand. Once in operation, preferred suppliers collaborated with the marketplace team and project teams to forecast demand figures for each product. Demand figures, typically in units per month, were converted into daily demand figures to support the implementation of *kanbans* for daily replenishments. Suppliers delivered a variety of products including personal protective equipment (PPE), hand tools, power tool consumables, general consumables, fixtures and fittings, internal and external signage, welding equipment, and related consumables.

(c) *Set Up Kanban Replenishment System*
The *kanban* system was set up including minimum and maximum inventory levels for each product in the marketplace according to demand estimates. The flexibility and speed of the *kanban* system allowed the project to adjust minimum and maximum inventory levels very quickly in support of site production. Two *kanban* signals were used for replenishment: plastic bins and cards. Three different types of plastic bins were selected to accommodate a variety of products with differences in size and shape: (a) small bin (280 mm depth × 210 mm width × 130 mm height); (b) medium bin (375 × 210 × 180 mm); and (c) large bin (375 × 420 × 180 mm). Each bin was identified by a label placed on the front of the bin that contained information about the product such as *kanban* quantity, product description, product code, bin type, and a bar code containing the same information on the label. Plastic cards (the size of credit cards) were used as *kanban* signals when a product did not physically fit in the bins. One of the goals of the *kanban* replenishment system was to keep the same amount of units per plastic bin or card at any point in time, so if there was an increase in demand, a new full bin (or several full bins) or a new card with the same number of units was added to the existing inventory. Figure 7.6 shows the arrangement of bins and cards inside the marketplace (note that cards are hung on the shelving system next to the products without bins).

   Visits to suppliers' warehouses revealed that they generally hold up to two months worth of inventory[3] based on their own analysis of market demand from multiple clients and projects. Since the intention was to integrate suppliers rather than to change their internal procedures, the materials management strategy took advantage of existing inventories at suppliers' warehouses as buffers[4] to shield the overall supply system from variability and uncertainty. Additionally, and in order to increase the speed of the replenishment operation, the materials management strategy required preferred suppliers to dedicate an area in their own warehouses to maintain exclusive inventory for the project using the same plastic bins and cards as in the marketplace. Figure 7.7 illustrates the arrangement of bins at one of the suppliers' warehouse.

---

[3] This, for example, contrasts with the food retail business, which operates supply chains where on average 95% of all supermarket products have stock levels of a maximum of 10 days, and obviously less for perishable products.

[4] An inventory buffer is one of three types of buffers that can be used to mitigate variability in a production system. The other two types of buffers are capacity and time [Hopp and Spearman 2000].

(b)

(a)

FIGURE 7.6   Use of bins and cards as *kanban* signals.

FIGURE 7.7   Dedicated inventory using bins at suppliers' warehouse.

The replenishment operation used milk run vehicles to transport empty bins from the marketplace to the supplier's warehouse. Once the milk run arrived at the warehouse, empty bins were exchanged for full bins of the same type and *kanban* quantity. Upon arrival at the marketplace, each plastic bin or card was scanned as part of the inventory control process. This was a repetitive cycle performed on a daily basis.

7.3.3 Conclusions

The preliminary results on this case study were first published by Arbulu, Ballard, and Harper (2003). A more comprehensive description of results obtained, after the implementation of this integrated materials management strategy, is presented below:

- It was estimated that at least one hour per person per day spent on purchasing transactions was saved on just chasing approvals and expediting supply. This represents cost savings of £100,000.
- It was estimated that at least five minutes per worker per day of standing time due to lack of materials or lack of the right materials was saved through the materials management strategy. This represents cost savings of £0.6 million on worker productivity alone.
- A rebate on total sales to the project from suppliers no greater than 6% was negotiated thanks to the efficiency of the overall materials management process, which enabled suppliers to better manage their own supply chain. This amount has not been disclosed.
- Daily supply reliability was higher than 97% in response to a demand originated by approximately 2000 workers on-site. Supply reliability is defined as the percentage of materials orders fulfilled in a day based on customer needs. On average, 97 of every 100 orders were delivered on time with the right quantity and quality every day.
- Traditional site stores in construction have one inventory turn per month on average. The market place achieved an average of four inventory turns per month, a reduction of inventory levels of 75% compared with traditional practices.
- Supply lead times were reduced from 1–3 days down to 1–3 hours. The agility of the supply system allowed site teams to react faster to changing circumstances on-site and to remain working on the most critical work at hand rather than allocating resources to expedite the supply of materials. In some cases, there were more than six pickups from suppliers' depots every day. As Ballard et al. (2001) highlighted, pulling requires satisfaction of the condition that material lead times fall within the window of site reliability. Reducing lead times to 1–3 hrs supported pulling.
- Integration with first-tier suppliers was achieved successfully. Five different suppliers were integrated within the materials management strategy for the delivery of more than 500 different non-task-specific products. On average, 85% of the suppliers' sales to the project were made through the integrated materials management system. The remaining 15% were orders for task-specific materials as well as for a variety of non-task-specific materials that were not included in the marketplace product list due to very low and erratic consumption volumes.
- A standardized product list was obtained through a rationalization process, especially for highly commoditized products that can be typically sourced from many different suppliers. This process contributed to a reduction in the variety of non-task-specific products simplifying implementation and operations.
- High levels of transparency between first-tier suppliers and site teams were achieved. For example, real consumption figures were shared with project

teams on a weekly basis, which allowed the teams to provide feedback about expected increments in demand due to project progress and changes in scope.

- Storage requirements on-site were low due to a combination of high supply reliability (97%) and the ability to perform frequent deliveries and collections. The materials management strategy supported production for about 30 site teams (from numerous companies) handling about 400 transactions every day. The physical area designated for marketplace operations was 180 m², of which approximately 60% was used for the stocks. On average, the marketplace turnover per month was approximately £270,000.

- Finally, the efficiency of the integrated materials management strategy allowed project leadership to implement additional initiatives involving the use of marketplace operations. This was the case of the man-packing initiative, which had the objective of providing a basic PPE package for individuals joining the project, from management to site crews. Personal information was taken at the induction center and communicated to the marketplace, and a package was assembled and given to the individual immediately upon arrival on-site. This guaranteed that every individual had sufficient PPE to start work on-site, contributing to a safer work environment.

This case study demonstrates that an integrated materials management strategy produces significant performance and financial results if it is properly aligned with project objectives. It also illustrates the application of lean production techniques, used successfully in other industries [Womack and Jones 1996], for the management of materials in project environments with the objective of generating value for the ultimate customer: production teams. Figure 7.8 illustrates the improvements that were achieved after the implementation of the materials management strategy.

7.4   Case Study 2: Management of Task-specific Materials

Managing task-specific materials has a key rule: materials management cannot be decoupled from production workflow (hence the term "task-specific"). Traditional management approaches in construction, such as the one explained at the beginning of this

(a)

(b)

FIGURE 7.8   Changes in the inventory of non-task-specific materials after implementation.

chapter, often violate this rule, generating considerable amounts of waste in the process. An effective materials management strategy for this type of material must incorporate the use of production control tools as the mechanism to trigger the need for materials deliveries to site based on real-time production status. In this scenario, production control and materials management processes must be integrated.

7.4.1 Background

This case study describes the implementation of an integrated materials management strategy for made-to-order concrete, a task-specific material, during the delivery of a £500 million civil infrastructure project in the United Kingdom, part of a £5 billion megaproject. The project consisted of several interdependent portions of work in the refurbishment and extension of a 140-year-old train station. The work content included structural modifications and the renovation in the existing station, fit-out activities, plus fairly complex civil works including an underground metro line and station. These highly interrelated tasks occurred on a highly congested site in central London. This extension is the first overland station of its kind in London in the past 100 years. The materials management strategy included the application of process reengineering as well as the use of SPS®|Production Manager (SPS|PM)[5] for production control, synchronization, and monitoring of concrete supply and demand.

7.4.2 The Approach

Similar to case study 1, the deployment of this materials management strategy followed the four-step process of definition, design, implementation, and operation shown in Figure 7.3.

7.4.2.1 Definition

The definition phase started by gaining a better understanding of specific and existing challenges related to concrete supply. Through a series of workshops with project stakeholders, the main challenges were defined and agreed as follows:

(a) Due to the complexity and dynamic nature of the production environment on-site, the concrete ordering approach was considered unsustainable and ineffective. Concrete orders were requested by production teams and submitted to

---

[5] SPS|PM [SPSINC 2007] aims to improve project delivery by enabling better and sustainable production planning and control. It systematically implements new approaches for managing projects such as Last Planner™ system, CONWIP, Last Responsible Moment (LRM), Real-Time Forecasting, and Work Structuring for Production. SPS|PM facilitates the identification of variation through a detailed approach to production control and measurement, and stabilization of production level workflow, and also enables automated updating of forecasts based on real-time completion of activities. With a stabilized project and reliable schedule forecasts, the lean concepts of resource leveling, pulling of materials, and use of standard work processes can be implemented. SPS|PM includes a standard process library which provides the foundation for implementation of best practices within a project or across multiple projects.

the on-site concrete coordinator and then fed to the concrete supplier via phone and fax.

(b) Weekly forecasts for concrete were unreliable. Concrete orders were planned one week in advance and revised one day ahead, but deliveries were made only in response to call-offs on the day required.

(c) The concrete supplier was forced to react quickly in order to supply real demand on a daily basis. The on-site concrete coordinator revised weekly look-ahead[6] plans on a daily basis and provided updates to the concrete supplier, including revised pour times and quantities. Such updates typically happened at the last moment.

(d) The lack of transparency between the production teams on-site and the concrete supplier was causing unbalanced concrete supply across the day. For instance, if no concrete was being placed in the morning, a backlog of concrete supply was created in the afternoon, with the risk of demand exceeding supplier capacity. This often led to overlapping pours late in the day, causing delays and thus overtime for the concrete crew and the supplier's batch plant.

(e) Demand reliability for concrete varied from 50 to 80%, with the supplier losing the ability to sell concrete to other projects.

(f) The existing process was highly dependent upon the concrete coordinator, who acted as a physical information link between the various parties involved, and kept much information in his head. When the concrete coordinator was absent, the process tended to falter.

(g) The existing way of working did not systematically promote learning and improvement.

The details of these challenges are further explored below through a description of the current process for concrete ordering.

Four stakeholders were involved in the execution of the concrete-ordering process in its current state: (a) construction managers/foremen; (b) engineers; (c) concrete coordinator; and (d) concrete supplier. The interaction between these four stakeholders included three repetitive loops for weekly coordination, daily coordination, and final concrete call-off (see Figure 7.9). One of the rules of the existing approach was that each Friday before 11 a.m., engineers sent the concrete coordinator the lists of concrete pours that they expected to make in the following week. In many cases, and especially when there was uncertainty on the exact day a pour was going to happen, some engineers ordered the same concrete for multiple days, forcing the concrete supplier to book their capacity to supply the expected demand. Some weekly forecasts were so unreliable that some teams were exempted from this rule, meaning their look-ahead window was reduced to less than one week.

Each morning, the provisional list of concrete orders for the following day was adjusted based on current information. A few engineers called the concrete coordinator regularly, but in most cases, the coordinator had to call the engineers to get updates. Input information was required by 11 a.m., with the exception of one team, from whom

---

[6] See more about the Last Planner™ system and look-ahead planning in Chapter 3.

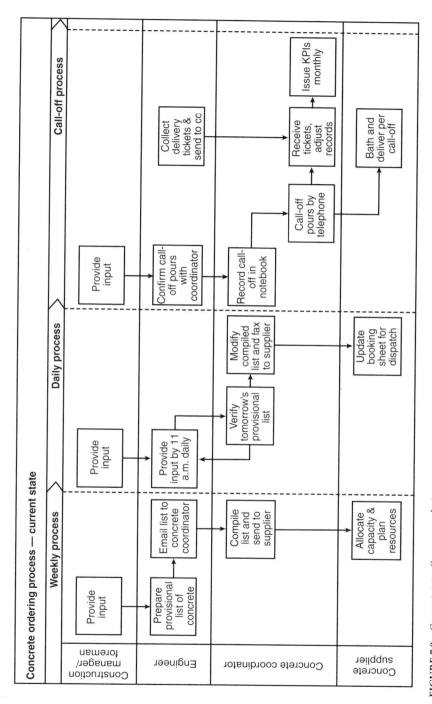

FIGURE 7.9 Current state: Concrete ordering process.

input information was required by 2 p.m. due to the level of uncertainty of their site operations requiring more time to provide more reliable information. Once the provisional list was completed, the concrete coordinator then revised, printed, and faxed it to the concrete supplier. The concrete supplier presumably used this one-day look-ahead plan to revise resource requirements and capacity allocation.

Every day, engineers called the concrete coordinator to request concrete call-offs. The coordinator then recorded each call-off in a handwritten booklet and phoned the concrete supplier to communicate the call-offs including concrete mix type, quantity, placement location, and entry gate to site. Once the concrete order was delivered, a delivery ticket found its way to the coordinator, who recorded actual deliveries from it. If the coordinator did not receive a delivery ticket in a timely manner, he requested help from the responsible engineer. If conflicts were identified from either the one-day look-ahead plan or the daily call-offs, the concrete coordinator advised managers who assigned priorities for concrete dispatches.

Monthly, the concrete coordinator produced and distributed a key performance indicator (KPI) graph showing (a) the cubic meters of concrete planned on the one-day look-ahead, (b) the cubic meters placed, (c) the cubic meters cancelled, and (d) the cubic meters of unplanned placements. These graphs indicated that the average demand reliability was 65%, the highest 83%, and the lowest 52%.

The current state of the concrete ordering process as described above is vulnerable to disruption because it is highly dependent upon the concrete coordinator, who acts as a physical conduit of information between the various parties involved and keeps a lot of information in his head. Additionally, the current process does not provide enough transparency of the overall concrete supply system for the people involved. Progressively updating the plan and having it posted on the project network and faxed to the concrete supplier is part of the current approach and allows everyone to see a more realistic picture of what is going on, but real-time information is lacking. Further, as discussed before, plans for concrete orders are realized only about 65% of the time, which basically means that the picture for the next day is only 65% accurate (not for next week, for the next day!). Finally, learning and continuous improvement are not promoted in the current approach. The concrete coordinator practiced 'name and shame' in an attempt to get engineers to improve their plan reliability. No doubt this practice may be effective when the engineer has the ability to improve and control the information necessary for doing better planning of concrete pours. However, further improvement is needed, requiring more systematic analysis of plan failures and a more thorough understanding of individual skills and the planning processes within specific teams.

7.4.2.2 Design

The design of the new integrated materials management strategy was kicked off through a series of workshops focused on incorporating direct input and feedback from key stakeholders. The aim was to avoid designing the material management strategy in isolation, knowing that people support what they help to create. Therefore, the participation of stakeholders was essential to ensure successful implementation afterwards. The definition step and the design of the strategy took approximately three months including engagement of stakeholders for commitment.

Based on the understanding of the current state, a set of rules was created to be considered for the development of the future-state concrete ordering process, and to be followed by all site teams. The rules were (a) keep at least two weeks of concrete orders in SPS|PM at all times; (b) actively manage and monitor forecast dates in SPS|PM; (c) SPS|PM will automatically close and collate all concrete orders in a report by 2 p.m. every day; (d) concrete orders will not be accepted by the concrete supplier if they do not come via SPS|PM; (e) once the concrete supplier confirms an order, concrete will be delivered to the site; (f) concrete delivery tickets must be sent within 24 hours to the concrete coordinator; (g) actual concrete delivery times and quantities must be entered into SPS|PM on a daily basis; and (h) site teams that are ready to accept delivery of concrete within one hour of scheduled time will not be preempted by deliveries to other site teams that have either placed same-day orders or have slipped more than one hour past their scheduled delivery time. If a time slot was missed, such orders were treated as second priority—this required management support and discipline, but was necessary to improve plan reliability in concrete.

With basic rules in place, the detailed development of the future-state concrete ordering process was initiated (see Figure 7.10). The future-state process started with entry by site teams of milestones, production workflows, and material orders into SPS|PM two weeks prior to planned start of work, as per the rules previously described. This allowed the concrete supplier to do rough planning of capacity and resources. Both supplier and coordinator scanned for potential conflicts as information was being generated in SPS|PM. As the start time for a workflow approached, site teams did what was needed to make the initial tasks ready to be performed (i.e., identify and release constraints), and adjusted production workflows in SPS|PM as needed. In general, the accuracy of plans tends to increase with shorter look-ahead periods, so the supplier can progressively refine capacity allocation and resource requirements based on SPS|PM data. Conflicts and overloads became more serious as the start date of the workflow approached, so both the concrete supplier and coordinator looked for potential conflicts and acted to resolve them.

The cut-off point for concrete orders was 2 p.m. the day prior. Whatever concrete deliveries were in SPS|PM at that time were considered commitments and became the basis of measuring reliability. The supplier had no choice but to treat these as real orders and so make final disposition of capacity and final specification of resources (labor, equipment, materials, overtime) accordingly. On the start date of the workflow, the supplier batched and delivered concrete only in response to call-offs from the coordinator, who acted at the request of site teams. The data loop was closed by updating SPS|PM and the financial loop was closed by collecting and sending delivery tickets to the project accounting team to be matched up with supplier invoices. Figure 7.10 shows the future state of the concrete ordering process.

In order to monitor performance during operations, project stakeholders agreed to work with standard and automatic reports in SPS|PM for concrete demand, concrete deliveries, volume of concrete ordered/delivered by day, volume of concrete

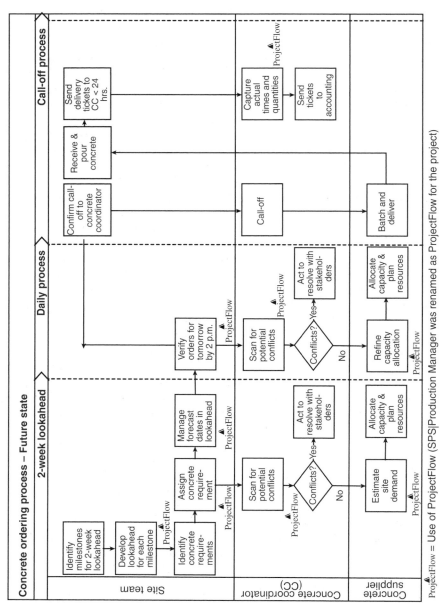

FIGURE 7.10 Future state: Concrete ordering process.

ordered/delivered by day weighted by pour quantities, and concrete demand/supply profile by hour. Samples of these reports are presented later in this chapter.

7.4.2.3 Implementation and Operation

Implementations are a sociotechnical challenge [Arbulu and Zabelle 2004]. Successful implementations, from simple new processes or procedures to complex strategies, depend on achieving the appropriate levels of capability and commitment. In this particular case, the implementation of the materials management strategy required a combination of leadership, project management, and technical support. Furthermore, realizing improvement required discipline in applying agreed rules all the way to senior management level. Without these elements, this implementation would have failed.

Implementation was kicked off through a workshop with project leadership and stakeholders involved in the concrete ordering process with the objective of providing a final review of the overall materials management approach. Separate implementation workshops were scheduled with project managers and engineers from the different site teams to review rules, roles, and responsibilities. At this time, a tentative roll-out date for the integrated materials management strategy was agreed with all site teams. Before the roll-out date, all necessary team members received training to operate under the new approach. The implementation process took approximately one month with continuous monitoring for adjustment during an additional month. Once in full operation, project leadership continuously monitored the performance of the concrete supply system in SPS|PM to ensure results were being obtained. Figure 7.11 shows a diagram of the overall concrete supply process, developed as a tool to support implementation, and to communicate overall process steps.

7.4.3 Conclusions

Within the first three months of adopting the new strategy, concrete deliveries became more reliable as the constraints between site and the supplier were better understood. This was the result of the site teams' improved planning and production control process coupled with the integration of concrete ordering and control through SPS|PM. Compared to initial project estimates, concrete productivity rates improved 19%—in the first 12 weeks of implementation there was a reduction in man-hours from 2.0 to 1.62 hrs/m^3. Based on a conservative estimate of labor rate including all costs of employment, overtime, management supervision, facilities, and equipment (£20.00/hr), this corresponds to a productivity improvement of 5000 man-hours or £100,000.

Figure 7.12 shows the improvement of the project's daily concrete reliability based on data from SPS|PM. For each day, it compares the demand as ordered through SPS|PM (left column),[7] the concrete supplied based on orders placed in SPS|PM (bottom section of right column); and the concrete supplied ad hoc (top section of the right column); that is, orders that were not placed through SPS|PM. Figure 7.12 indicates that the ratio

---

[7] This was labeled as Delivered PF in Figure 7.12, because SPS|PM was renamed as ProjectFlow (PF) in this project.

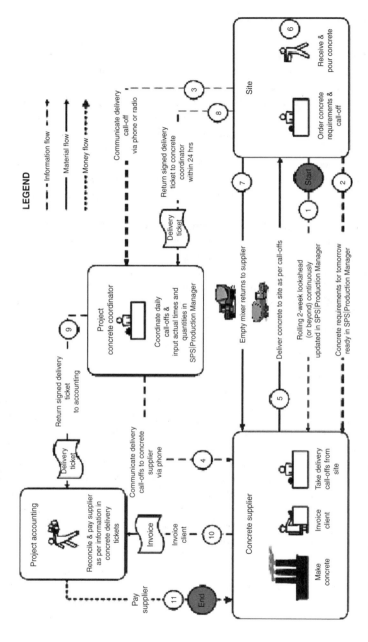

FIGURE 7.11   Graphical representation of concrete management strategy.

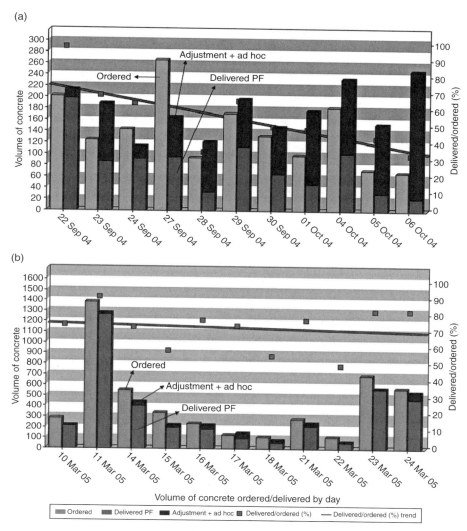

FIGURE 7.12 Daily concrete reliability: (a) before implementation; and (b) after implementation.

between concrete delivered and concrete ordered improved from 30 to 70%, more than 100% improvement, and the number of ad hoc concrete orders was reduced considerably, which illustrates the adoption and effectiveness of the new materials management approach.

Finally, Figure 7.13 compares concrete demand and supply across the hours of a day through before and after scenarios utilizing automatic reports from SPS|PM. Before implementation, the high peak indicates a time when several project teams were requesting concrete at the same time. This led to the other line extending farther to the right, representing overtime work for the concrete batch plant and concrete crew. After

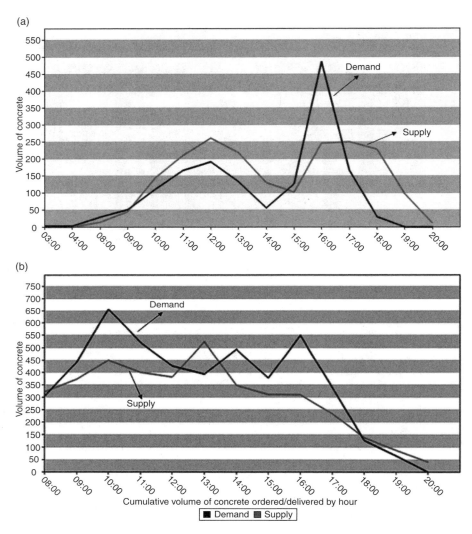

FIGURE 7.13 Concrete daily demand and supply variation: (a) before implementation; and (b) after implementation.

implementation, there was virtually no overtime work, even during the busiest phase of the project. A better balance, but certainly not perfect due to remaining variability levels, between concrete supply and demand was also obtained.

This case study demonstrates that the management of task-specific materials such as concrete can be controlled and incorporated into workflow management practices on-site, creating an integrated approach to materials management that contributes to better production management, increasing labor productivity and overall system reliability.

7.5   Summary of Key Ideas

Materials management strategies must consider the nature of the products managed through the strategy (non-task-specific, task-specific). Considering that most projects in construction are unique, we should expect an important amount of made-to-order and engineered-to-order products (mostly task-specific products). This implies that the development of materials management strategies for task-specific products, or at least for key task-specific products, should be considered a major project priority. This, of course, does not eliminate the need to develop strategies for non-task-specific products, which typically include small items, the lack of which can stop production workflows. Independently of the nature of the product, an integrated materials management strategy focuses on the synchronization of supply and demand as a means for minimizing waste. Certainly, this goes beyond traditional purchasing practices.

References

Arbulu, R. J., Ballard, G. H., and Harper, N. 2003. Kanban in construction. *Eleventh Ann. Conf. Intl. Group for Lean Construction* (IGLC-11), Blacksburg, VA, July.

Arbulu, R. J., and Zabelle, T. R. 2004. Implementing lean in construction: How to succeed. *Fourteenth Ann. Conf. Intl. Group for Lean Construction* (IGLC-14), Santiago de Chile, July.

Ballard, G. H., Koskela, L., Howell, G., and Zabelle, T. 2001. Production system design in construction. *Ninth Ann. Conf. Intl. Group for Lean Construction* (IGLC-09), Singapore, August.

Hopp, W. J., and Spearman, M. L. 2000. *Factory physics: Foundations of manufacturing management.* 2nd ed. Irwin/McGraw-Hill, Boston, 698 pp.

SPSINC. See Strategic Project Solutions.

Strategic Project Solutions. 2007. www.spsinc.net. Accessed 27 December 2007.

The Productivity Press Development Team. 2002. *Kanban for the shopfloor.* Productivity Press, Portland, OR, 102 pp.

Womack, J. P., and Jones, D. T. 1996. *Lean thinking: Banish waste and create wealth in your corporation.* Simon and Schuster, New York.

8

Production System Instability and Subcontracted Labor

Rafael Sacks

Technion—Israel
Institute of Technology

8.1 Introduction

There is an inherent mutually dependent relationship between the degree of plan stability for any construction project, on the one hand, and subcontractors' reliability in providing resources, on the other hand. The dependence of the project plan on subcontractors providing the necessary resources at the right time, in the right quantity, and with the right skills, and equipment is obvious to all who have managed projects. Construction project managers often lament that their projects were late and/or overbudget because subcontractors failed to perform as expected. Unfortunately, the culture of managing projects by subcontracting their parts often leads to construction engineers and managers ignoring the fundamentals of production planning, analysis, and control. The result is that they are not aware of the strength of the reverse relationship, which is that subcontractors are highly dependent on the predictability of construction plans in assigning resources. The less reliable the plan, the less reliable resource assignments are likely to be, and vice versa, creating a seeming vicious circle of unreliability. This chapter sheds light on this relationship, with the goal of helping project managers[1] understand the motivations of subcontractors, and thus helping encourage supply chain management approaches that strive for stability, especially when subcontracting is unavoidable.

Apart from the instability caused by internal factors, construction projects are also subject to disturbance from external sources. Design changes initiated by clients, and unforeseen statutory requirements are just two examples. The discussion that follows focuses on understanding the mechanisms of internal instability, and the construction management procedures that are recommended do not address externally driven instability. A management approach to establishing residential construction production systems that are resilient to the high uncertainty generated by late client design changes can be found in Sacks and Goldin (2007).

8.2   Subcontracting

Subcontracting is an age-old practice. The modern form of subcontracting in construction can be traced to the specialized trades employed in building the cathedrals of the Middle Ages. An early reason for subcontracting was the sophistication, and specialization of trades, which with the advent of prefabrication of some construction components, also demanded long-term investment of capital in production facilities, and equipment. The extent of subcontracting in modern construction has been documented in numerous studies [Edwards 2003; Hinze and Tracey 1994; Hsieh 1998; Maturana et al. 2007]. A 1998/1999 study of general contractors in commercial construction in the United States found that 91% of the trades were subcontracted more than 75% of the time [Costantino and Pietroforte 2002].

However, subcontracting has also developed in trades that demand little or no capital investment, and are not so specialized that they could not be performed by workers employed directly by general contractors. There are two main reasons for this. At the macroeconomic level, fluctuating regional and local demand for the services of general construction contractors demands agility in adjusting capacity. A large workforce on payroll is a burden for any company that cannot rely on long-term market demand. At the level of production management, instability and unreliability of work on site leads to decreasing labor productivity. Koskela (1992) summarized seven preconditions, each of which have inherent uncertainty for productive execution of work packages. The conditions are completion of any precedent tasks, accessibility of the space for the work, availability of labor, materials, and equipment, provision of updated and complete information, and fulfillment of external conditions (such as statutory permissions and financing). If any one of these preconditions is not fulfilled, then the work cannot proceed as planned.

As buildings become more complex products, the ability to coordinate satisfaction of all of the preconditions simultaneously using traditional management approaches is diminished, and the relevant result is the waste of workers waiting [Womack and Jones 2003]. Subcontracting per se does not eliminate this potential waste; it simply transfers the risk from the general contractor to the subcontractor.

---

[1] In the remainder of this chapter, the term "project manager" is used to denote the role of a general contractor's construction manager.

A common result of extensive subcontracting is that construction project managers and engineers abdicate their role, and responsibility for designing, and managing production systems. Research has revealed that the number of workers on site fluctuates [Sacks and Goldin 2007], handover times between trades become long as buffers of work are allowed to accumulate between them [Sakamoto, Horman, and Thomas 2002; Tommelein, Riley, and Howell 1999], and budget and schedule overruns become common. Figure 8.1 shows the number of workers provided by each of four trade subcontractors in construction of a medium-sized apartment building (28 apartments) as monitored over a period of 10 weeks during execution of the finishing works in the building. Only the plumbing subcontractor maintained a fairly stable number of workers on site. Fluctuations of this kind are common in many construction projects with subcontracted trades.

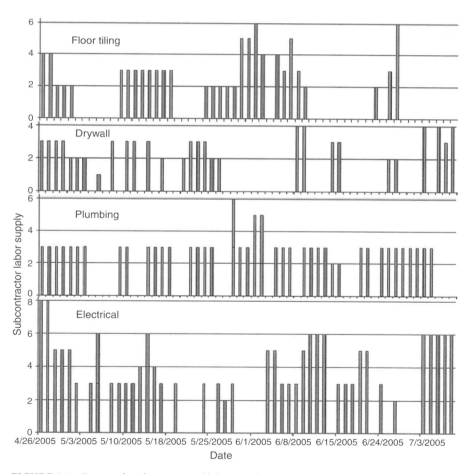

FIGURE 8.1 Four trade subcontractors' labor supply records for an apartment building.

8.3   Management Approaches

The management approaches to dealing with this situation can be classified as comprising one or both of two basic actions to improve resource allocation from the project perspective: (a) contractual changes, such as partnering, and terms for incentives or penalties; and (b) operational changes, such as application of the Last Planner™ system [Ballard 2000]. Where possible, long-term relationships with subcontractors are fostered, but this is beyond the scope of any single project.

Partnering is a contractual approach that aims to avoid short-term opportunistic behavior, and encourage cooperation between project participants to achieve common goals [Rahman and Kumaraswamy 2004]. According to Bennett and Jayes (1995), partnering is "a management approach used by two or more organisations to achieve specific business objectives by maximising the effectiveness of each participant's resources. The approach is based on mutual objectives, an agreed method of problem resolution and an active search for continuous measurable improvements." Partnering may be undertaken for a single project, or pursued as a long-term strategy. Unfortunately, partnering is often adopted by subcontractors as a short-term response to pressure from powerful clients, rather than being a fundamental cultural change [Dainty, Briscoe, and Millet 2001; Fisher and Green 2001]. If not implemented correctly, it can have detrimental effects on subcontractors [Packham, Thomas, and Miller 2001]. As Bresnen and Marshall (2000) point out, real cultural change requires an understanding of factors that dictate the basic interests of the parties involved.

One of the key features of lean construction[2] thinking is the idea that variation can be reduced, either by improving the design of construction production systems, or by managing flow using pull techniques such as the Last Planner™ system [Ballard 2000]. This system works to enhance reliability in three main ways: (a) the "look ahead planning" and "make ready" process, in which construction managers make work ready by ensuring that materials, information, and equipment are available; (b) filtering planned activities through the weekly work planning procedure to ensure that all preconditions for successful execution of activities to be assigned have been or can be satisfied; and (c) by seeking conscious and reliable commitment of labor resources by the leaders of the work teams involved.

The last facet is perhaps the most significant feature of the technique because it could be argued that the first two steps are reasonably obvious to experienced project managers and are commonly performed. It is this last step, which requires the conceptual leap from "should" to "can" (or "push" to "pull"), that requires schooling in lean thinking and lean construction. It focuses a spotlight on the inherent link between plan reliability and reliable allocation of resources, and recognizes that those actually performing the work are the most reliable predictors of labor capacity because they themselves allocate the resources.

In general, people behave according to the ways in which they are measured, and for construction contracting the ground rules are set in the contract, which is almost always

---

[2] The phrase "lean thinking" was coined by Womack and Jones (2003) to denote an approach to production whose origins lie in the Toyota Production System (TPS). The terms lean "construction" and "lean production" are defined and explained in some detail in Chapter 5, in the sections "Lean in Project Settings" and "Lean Project Delivery."

based on a set of fixed unit prices or a lump sum for the work. The Last Planner™ system deals with production at the operational level, but it is not directly concerned with the basic parameters of the commercial relationship that underlie subcontracting arrangements. Intuitively one can deduce that the more reliable a project work plan is over time, the more willing subcontractors will be to allocate resources more readily, and suffice with smaller buffers, allowing levels of work in progress (WIP) to be reduced, and flow of work to be improved. However, a rigorous understanding of this relationship is needed.

8.4 Empirical Observations

An extensive set of in-depth interviews with 57 construction project managers and subcontractors from large firms, carried out in Israel [Harel and Sacks 2006], explored the motivations of project managers and subcontractors when demanding and providing resources in the multiple project environment. One of the significant findings confirmed that subcontractors do not perceive project managers to be reliable. In response to the question "What proportion of the work promised by the project manager do you believe will actually be made ready?" more than 85% of the subcontractors believed that less than 80% of the work would be made ready. The average amount of work expected to be ready was only 60%, with a standard deviation of 19.7% (see Table 8.1). A corollary question was posed to project managers, in which they were asked: "When you demand resources from a subcontractor, to what degree do you exaggerate your report of the amount of work that will actually be available?" The responses (Table 8.2) indicate that many project managers attempt to counter subcontractors' lack of trust in them,

TABLE 8.1 Subcontractor Managers' Faith in Project Managers' Assessments of Work Ready

What Proportion of the Work Promised by the Project Manager do you Believe will Actually be Made Ready?	Number of Responses	Percentage
0–20%	1	3.7
21–40%	3	11.1
41–60%	8	29.6
61–80%	11	40.7
81–100%	4	14.8
Total	27	100

TABLE 8.2 Project Managers' Exaggeration in Reporting Work Ready

To What Degree do you Exaggerate your Report of the Amount of Work that will Actually be Available?	Number of Responses	Percentage
0–20%	15	51.7
21–40%	9	31.0
41–60%	3	10.3
61–80%	0	0
81–100%	2	6.9
Total	29	100

and consequent undersupply of resources, by exaggerating their demands: 48.3% of the respondents confirmed that they exaggerate their demands by at least 20%. These results reveal an inherent lack of trust in the relationship.

Furthermore, subcontractors claimed that in only 55.5% (on average) of situations did project managers enable them to work with the most productive crew possible for any given task. With regard to the need to expedite work on site, just over 50% of the project managers concurred that they frequently force subcontractors to begin work in projects even when the work of the preceding trade teams is not complete to a degree that allows the newcomers smooth and productive work.

The survey also explored the practice of "overbooking:" 82% of the project managers agreed or strongly agreed that subcontractors generate buffers of future available work, in other projects, to an extent where they commit themselves to multiple general contractors beyond the limits of the capacity of their own resources. Among the subcontractors, 64% concurred to a strong degree that this practice is common.

8.5   Subcontractors' Economic Motivations in Allocating Resources

There are two primary starting points for understanding the work environment that results when subcontractors are employed in construction projects. The first is that each subcontractor must perform work on multiple projects simultaneously. Unlike the project manager's perspective, from the subcontractor's point of view, the focus is not on any one specific project, but on multiple projects. In Figure 8.2, a typical situation is shown schematically: subcontractor 4, for example, who specializes in Trade III, allocates resources to projects A and C concurrently.

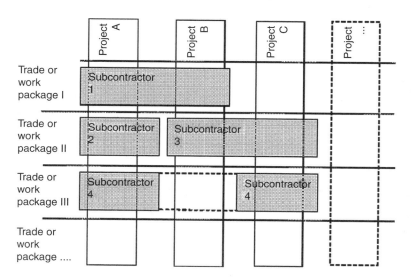

FIGURE 8.2   Metaproject subcontracted work environment.

The second starting point is the economic motivation for a subcontractor under unit price, or lump sum contracts. The income to be earned, or the loss incurred, by a subcontractor in any particular project over a single planning period can be expressed as a function of the prices and costs for each work item performed, the quantity of work planned for, the production resources actually applied, and the quantity of work actually made available. To understand the subcontractor's motivation, it is sufficient to consider a single representative work item. The appropriate expression for the net income is derived from an economic model originally presented by Sacks (2004). Its simplified form is as follows:

$$I_n \approx \min[k,q] W_D (U - C_M) - k W_D C_S, \tag{8.1}$$

where the variables are as defined in Table 8.3. The net income depends on the amount of work actually performed, and the cost of the resources incurred. The first term on the right hand side, $\min[k,q] W_D (U - C_M)$, represents the total income from the work actually performed. The work actually performed is the smaller of either the work actually made available, or the capacity of resources actually provided, which are represented, respectively, by the terms k and q. The second term, $k W_D C_S$, represents the total cost of the resources actually provided by the contractor.

At the start of any given planning period, the subcontractor manager must decide on the quantity of resources to assign to each project at hand; that is, he/she must set the value of k that will optimize its income. The subcontractor knows the unit price, the unit costs of materials and resources, and the amount of work planned (or demanded) by the general contractor's project manager.

However, the actual value of q that will occur is not known with certainty. This is because work planned will only actually become available for execution through the

TABLE 8.3   Annotation

Parameter	Definition
I_n	Net income from project I during any period T.
W_D	Quantity of work planned by the general contractor in period T.
W_A	Quantity of work that is actually made available in period T.
q	The ratio of the quantity of work actually made available to that planned during any period T. $q = W_A/W_D$. This is similar but not identical to the PPC measure of the Last Planner™ system. Unlike PPC, q may be greater than 1 because it measures the total amount of work made available, including any that may not have been planned. PPC is only concerned with the work made available that was in fact planned initially.
U	The unit price for the work set in the contract.
C_M	The cost of materials for each unit of the work.
W_D	Quantity of work planned by the general contractor in period T.
k	The ratio of the actual resources provided by the subcontractor to the quantity of resources needed to complete the full quantity of work planned in period T.
C_S	The cost of the resources per unit of work planned in period T. The total cost of resources to perform W_D is given by $W_D C_S$.

planning period when all of the seven preconditions described above are fulfilled. In this case, given that labor is assumed to be available, problems commonly encountered are that the work area is not free (precedent work teams/subcontractors have not completed their work and cleared the area), the requisite materials were not delivered on time, or information that controls the work is out of date. The quantity of work actually made available also has a second order impact on income, because productivity itself is a function of work quantity and space [O'Brien 2000]. However, here, the focus is on the subcontractors' strategy in allocating resources; for the sake of simplicity, the second order effect, material waste, and overheads are ignored.

The two starting points—multiple concurrent projects and the maximization of net income—can be jointly expressed by summing Expression 8.1 over multiple projects, thus:

$$I_n \approx \sum_{i=1}^{n} \left\{ \min[k_i, q_i] W_{D_i} (U_i - C_{M_i}) - k_i W_{D_i} C_{S_i} \right\} - C_R. \tag{8.2}$$

Expression 8.2 adds the cost of any resources not allocated to any project. As unallocated resources are only a source of cost and cannot generate income, a subcontractor will attempt to avoid this situation. Given that there is uncertainty about the values of q_i at any time, there are two possible strategies:

(a) Contract for sufficient projects to ensure that the total amount of work likely to be made available will be greater than the total capacity of resources, in which case all resources can be gainfully employed. This strategy is termed overbooking, and is common among subcontractors (see the subsection Empirical Observations).

(b) Where the total amount of work planned is less than the total capacity available, identify which projects are most reliable, and/or where the amount of work planned may be underestimated, and set k>1. Thus, situations may arise in any particular project where more resources are allocated than are needed according to the general contractor's work plan.

The parameters affecting the subcontractors' behavior can be compared using Expression 8.1.

Figure 8.3 shows the relationship between the quantity of resources allocated to any project (represented by k), and the income, I_n, and its dependence on the reliability of the project schedule (represented by q). As q declines, not only is the total income reduced, but the point at which losses are incurred occurs for increasingly smaller resource levels. Given this relationship, in order to maximize its income in any single project, a rational subcontractor must try to estimate the most likely value for q and allocate resources appropriately, (i.e., try to set $k=q$).

Two additional important perspectives can be gained regarding the influence of the labor content, and the profit margin on the importance of plan reliability, q, in setting resources. The slope of the lines on the right hand side of Figure 8.3 is $-C_S$, which is the unit cost of resources supplied.

Where the subcontractor supplies labor alone, the slope is steepest, making the subcontractor more vulnerable to unreliability in the plan; where a subcontractor supplies a

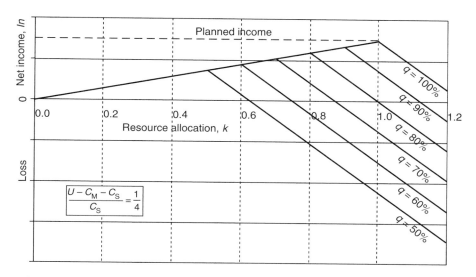

FIGURE 8.3 Relationship between net income, resource allocation, and plan reliability.

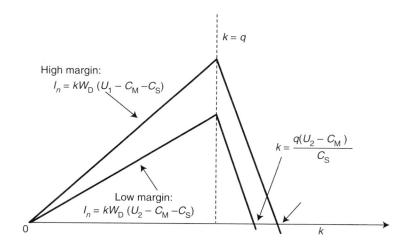

FIGURE 8.4 Influence of profit margin on resource allocation, and sensitivity to plan reliability.

significant proportion of the materials needed for the work, its vulnerability is decreased.

The impact of profit margin is similar. In Figure 8.4, two projects are compared. The difference between them is that one has a higher profit margin:

$$\frac{U_1 - C_{M_1}}{C_{S_1}} > \frac{U_2 - C_{M_2}}{C_{S_2}}. \tag{8.3}$$

The figure shows that where the profit margin is higher, not only is the income greater, but the subcontractor is less sensitive to fluctuations in the plan reliability. Positive income can be sustained for lower (more unreliable) values of q. The corollary of this is that in an industry where projects are unstable, average subcontract unit prices must be high enough to absorb a high degree of waste.

In any real project, the value of q will vary over multiple planning periods, and a frequency distribution can be compiled for its values. This distribution is a long-term measure of the reliability of the planning in the project, and using it a subcontractor could estimate expected values of q, $P[q]$. In reality, even if subcontractors do not maintain explicit records of the value of q for each project, they do build a mental picture—an impression—of the plan reliability in each project. By assuming a typical frequency distribution for any project, one can compute an expected value for the income that will be generated by a project in a planning period for any given value of k (the quantity of resources to be allocated). In this way, it can be shown that not only is the mean expected value of q important, but also its degree of dispersion. When allocating resources to two projects with equal mean values of q, the subcontractor will tend to prefer that with a smaller standard deviation.

The numerical example provided in Figure 8.5 underlines this point. Both projects A and B have the same price and costs, but project A is less reliable than project B (the probability curves for the expected values of q are shown on the left side of Figure 8.5). As a result, the optimal value for k for project B is greater than that for project A (the graph on the right shows the expected values of I_n for a range of different choices for the value of k). A rational subcontractor who has some knowledge of the reliability of each project would consistently provide fewer resources (relative to the resources called for in each project's plan) to project A than to project B.

An additional observation is that the number of circumstances in which k will exactly equal q is likely to be small, which implies that in the majority of situations, waste is present. Either capacity is underutilized, or there is excess WIP waiting for resources to be allocated to it. By corollary, the less the plan variability (the more reliable), the closer one can operate to full capacity utilization without generating excess WIP. This relationship has been described in queue theory, and discussed in the lean construction literature [Howell, Ballard, and Hall 2001].

The conclusions that can be drawn from this analysis are that a subcontractor will assign resources to any project according to its perception of the general contractor's

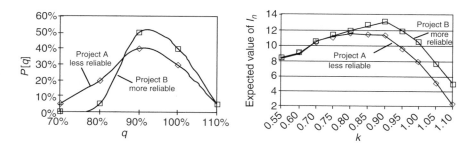

FIGURE 8.5   Relationship between plan reliability and optimal resource assignment.

reliability in supplying work at the rate that has been demanded. In the majority of cases, the capacity assigned is likely to be lower than that demanded, because the penalty of oversupply is greater than that of undersupply: the slope defined by $kW_D(U - C_M - C_S)$ in Figure 8.3, where $k < q$ is generally much less steep than that defined by $-kW_DC_S$ when $k > q$. It is better to err on the side of caution.

Indeed, if the same resources can be employed in alternative projects, then lower income in any one project due to underallocation of resources is offset by applying those resources elsewhere. Where overbooking is present, the subcontractor can achieve maximum overall income even while not achieving maximum possible income from each individual project. This is good news for subcontractors, but bad news for project managers, because lower than expected resource allocations have the undesirable consequence of further reducing plan reliability.

An alternative strategy that a subcontractor may adopt is to delay the start of work at a project beyond the start time demanded by the project manager, with the goal of ensuring that a sufficient inventory of work accumulates to buffer it from the upstream trade. In effect, and in the absence of any explicit pull mechanism, the subcontractor increases its confidence in the supply of work by simply waiting for inventory of WIP to accumulate. Common wisdom in construction project management dictates that subcontractors be required to appear on site on the date stipulated by the project manager in terms of the contract, with the result that subcontractors attempt to create a buffer through "unconventional" means.

Both of the above cases support the hypothesis presented by Howell, Ballard and Hall (2001) which states, inter alia, that independent attempts to achieve full resource utilization will result in longer wait times for assignments, ultimately reducing work flow reliability.

8.6 Subcontractor vs. Project Manager: Game Theory Model

The previous section outlined why subcontractors tend to overbook their resources, and often do not provide all the resources demanded of them by project managers as projects progress. However, the analysis disregarded the behavior of construction project managers responsible for coordinating the work of subcontractors. Managers who recognize the tendency of subcontractors to provide fewer resources than a plan calls for often counteract it by demanding more resources than needed. This has the predictable result of lowering the subcontractor's level of confidence over time, which exacerbates the problem.

These behaviors have been exemplified using a game theory model [Sacks and Harel 2006], in which a project manager demands resources, and a subcontractor allocates resources, respectively. The game theory formulation modelled the allocation of resources at the start of each planning period in a project (typically each week). The players are the project manager (PM) and the subcontractor (SUB). Each makes "moves," one after the other, through repeated cycles of the game. The project manager plans the work, as is common in traditional construction systems. In each round of the game, the PM sets the amount of work to be performed by each subcontractor (SUB) in each task i in each period on the basis of the construction master plan. In response, each SUB evaluates the demand and the amount of work they perceive will actually become available, and then supplies the resources it deems appropriate.

The extensive form of game theory analysis [Osborne and Rubinstein 1994] was used because the moves are sequential, and because both PM and SUB have imperfect knowledge about the outcome in terms of the work that will actually be accomplished (i.e., they cannot predict with certainty how much work will be made available by the upstream contractors, or whether design changes, material delays, weather conditions, or other factors will interrupt or slow work).

Figure 8.6 shows the layout of a typical round of the game. The values for q at the left of the tree represent the range of possible results for the work that will actually become available in terms of the amount of work planned. The probabilities for the values of q are described by a probability distribution, $P[q]$. For the model analysis, a range was assumed in which the cumulative probability that at least 100% will be performed is 70%, and the weighted average of the distribution is 98%.

The PM's possible moves are detailed at the second level from the left hand side of Figure 8.6 using the ratio d, which is the ratio of the work demanded, W_D, to the work the PM estimates will actually become available. The value d is also modelled by discrete values: demand resources for less work than estimated ($d=0.9$), exactly the amount estimated ($d=1$), and more than estimated ($d=1.1$). In response to the PM's request, the SUB can elect to allocate fewer resources than required for the work demanded (setting $k=0.9$), where k is the ratio of resources supplied to those demanded, exactly the amount required ($k=1$), or more than demanded ($k=1.1$). The latter value reflects a situation in which the SUB has resources available, and is willing to commit them in the hope that more work than expected will in fact become available, and that they would be utilized profitably.

The utilities for each player are calculated at the end node of each branch of the tree. The utility for the PM is assumed to be the total amount of work actually completed in

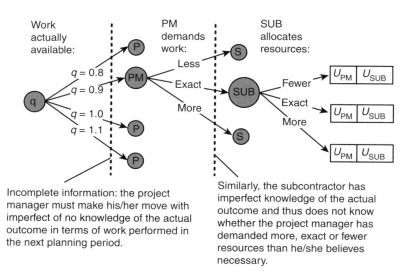

FIGURE 8.6 Extensive form game of subcontractor resource allocation.

the planning period. When insufficient resources are allocated (i.e., when $q \geq dk$), the work done is constrained by the quantity of resources available, and is proportional to dk; on the other hand, when $0 \leq q < dk$, the work done is constrained by the availability of work and is therefore directly proportional to q. Thus the utility for the PM, U_{PM}, is given by:

$$U_{\text{PM}} = \min(q, dk). \tag{8.4}$$

The SUB's utility, U_{SUB}, is defined as the total income derived from the work done in the planning period, calculated on the basis of Expression 8.1. In extensive form games with probabilistic outcomes, the utilities are replaced by expected utilities to reflect the variability possible in the outcomes. In this case, since the utilities are all functions of q, the expected utility for each combination of PM and SUB strategies is the weighted average of the utilities for each possible result for q, weighted by its probability, which is given by the distribution $P[q]$. Thus the PM's expected utility is:

$$U_{\text{PM}} \text{ expected} = \int_0^{dk} qP[q]\,dq + dk\int_{dk}^{\infty} P[q]\,dq = \int_0^{dk} qP[q]\,dq + pdk, \tag{8.5}$$

(where $p = P[q > dk]$). Similarly, since the labor cost is independent of the amount of work performed, the SUB's expected utilities are computed as:

$$U_{\text{SUB}} \text{ expected} = \left(pdk + \int_0^{dk} qP[q]\,dq \right)\left(U - C_{\text{M}}\right) - kC_{\text{S}}. \tag{8.6}$$

In practice, construction professionals would not estimate a continuous probability distribution, but rather use discrete values at significant intervals.

Project managers' and subcontractors' behavior is influenced by the degree of knowledge they have about the amount of work that will actually become available, which cannot be known with certainty at the start of each planning period, because plans are not necessarily reliable. To explore the impact of plan reliability on the expected behaviors, information sets were used. Information sets allow modelling of situations in which each player either knows or does not know the value of the "nature" variable, in this case the value of q, which measures the amount of work that will in fact become available. In all cases, the players remain ignorant of the strategy adopted by their opponent until the play is made. In reality, the quantity of work that will actually become available at a future date cannot be entirely "known" to either PM or SUB, and equally, it is extremely unlikely that either would have absolutely no knowledge of the likely value. In that sense, application or removal of information sets for each player result in extreme situations that cannot actually occur. However, they both function somewhere within the continuum between the idealized cases, and so the analysis sheds light on how they behave in the vicinity of the pole, and how their behavior changes as they move between them.

Three of four possible situations were modelled[3]:

- In **Case A**, plan reliability was assumed to be very low: information sets are applied, modelling the situation where neither player can predict the value of q. This case reduces to the normal form of the game.
- In **Case B**, plan reliability was assumed to be very high: the PM's information set was removed. The SUB's knowledge cannot be simply represented by the plan reliability because in traditional systems the SUB is called upon to provide the resources as demanded by the PM, and the SUB has imperfect knowledge about the PM's strategy (i.e., has the PM demanded more, exact or less resources than they have estimated will actually be needed?). The SUB must consider to what degree the PM can be trusted, and may attempt to gather independent information about the likely work availability in each coming period. Thus the SUB functions according to an information set applied between the PM and the SUB players. Thus, in Case B the SUB has no knowledge.
- **Case C** models the situation where plan reliability is high, and both players have full knowledge of the future value of q. All information sets are removed.

Readers concerned with the details of the solution procedures are referred to Sacks and Harel (2006). The focus of this section is on the interpretation of the results in terms of plan reliability.

Solution of the three game theory scenarios showed that the cooperative equilibrium[4] result, where exact resources are demanded, and exact resources are provided, only occurs when both PM and SUB have perfect information, as in case C. When neither participant has reliable information, both resort to defensive, competitive strategies to reach equilibrium at their "least bad" result. Because defensive behavior on the part of the SUB includes provision of fewer resources than demanded, the ability of a PM to achieve plan reliability is harmed, creating a stable, and self-perpetuating lose–lose situation. The results are summarized in Table 8.4.

8.7 Discussion

The basic economic model and the economic game theory formulation are both intended to shed light on the motivations of subcontractors in their periodic allocation of resources to construction projects, with a view to informing changes to subcontracting arrangements, and their management in ways that make project workflows smoother. Many researchers have placed emphasis on plan reliability as an indicator of work flow; the Last Planner System™ [Ballard 2000] works directly to make short-term construction

---

[3] The fourth case, where the subcontract has full knowledge and the project manager has none, is not needed.

[4] A game theory "equilibrium," as used here, represents a set of strategies employed by the players that result in a situation where neither player has anything to gain by changing only his or her own strategy, so long as the other player does not change their strategy.

TABLE 8.4 Game Theory Solution Results

Case	Description	Equilibrium Solutions
A	Neither the PM nor the SUB has any knowledge of the probability distribution of *q*; that is, neither can predict how much work will be possible.	This form has a perfect equilibrium, which is the strategy pair: **PM demand more; SUB provide fewer.** The equilibrium is insensitive to variation of the profit margin, and relatively insensitive to fines that may be applied.
B	The PM has perfect knowledge of *q*, but the SUB has none; the information set in front of the PM is removed, but that in front of the SUB remains.	This case has two significant equilibria. The first occurs for the **PM demanding more work than estimated in every situation, and the SUB providing fewer resources.** A second equilibrium, which is part of a mixed strategy, occurs when the **PM demands exact work when *q*=0.9 or *q*=1.0, and more when *q*=1.1, and the SUB provides exact resources.**
C	Both have full knowledge of the work to become available, which also implies that the SUB is fully aware of the PM's possible strategies. In this form both information sets are removed.	Here too there are two main equilibria. The first, similar to those in the previous cases, occurs when the PM demands more work in every case, and the SUB provides fewer resources in every case. The second occurs when **exact resources are both demanded and allocated in every case.** The numerical differences between them are very small.

planning reliable. Unfortunately, subcontracting has the effect of reducing the reliability of labor and other resource allocations, which is one of the most crucial of the seven inputs that are needed for reliable execution of a construction task [Koskela 2000].

The Last Planner has three impacts that are relevant in terms of the game theory model:

- The degree of plan reliability is made transparent to all participants in the process, by means of the percent plan complete (PPC) measure.
- The method implies an explicit effort to improve plan reliability.
- Subcontractors are given direct control over the work assignment process itself through the weekly work planning meeting in as far as their own supervisors are considered "last planners," which means that their knowledge of the work allocation strategy is likely to be at least as good as that of the project management function itself.

In the game theory model, improved plan reliability, and no less improved knowledge of it, remove the information set from the project manager. All three impacts, together with improved knowledge of the work allocation strategy, move a subcontractor in the direction from case B towards case C. Thus the game theory model suggests that use of the Last Planner system should result in movement to an equilibrium of resource allocation behavior that is more cooperative than the equilibrium reached without it.

Cooperative behavior in this context means increased tendency to demand exact resources required, and to provide exactly the resources demanded. In turn, cooperative behavior of this nature improves plan reliability for subsequent trades, raising the overall degree of plan reliability for the project. The mechanism of the system may thus be considered to be to facilitate movement toward stable modes of cooperative behavior.

The Last Planner functions mainly at the level of production control. Other approaches to engendering cooperative behavior, such as partnering [Fisher and Green 2001; Rahman and Kumaraswamy 2004], function mainly at the level of organizational and personal relationships. The ability of either approach to achieve sustainable improvement is dependent on maintaining cooperative behavior over time. In terms of the game theory model, as long as the basic utility functions remain unchanged, behavior within systems where partnering arrangements are initiated is likely to regress towards the original "lose–lose" equilibrium solutions inherent in the system. Examples of this effect have been reported [Lazar 2000; Packham, Thomas, and Miller 2001], in which the basic interests of subcontractors remain unchanged, and the effects of partnering are felt in the short term only. A possible implication is that if Last Planner is not implemented consistently and completely (i.e., both pull flow control using Last Planner meetings, and consistent and transparent reporting of PPC), then its impact is also likely to be transient.

In subcontracted projects, the ills common in traditional construction push work flows are exacerbated by the instability in process flow that begins when subcontractors are delayed at any point in the job, and elect to withdraw their crews to other projects where work is available. A common result is that the crews cannot then be withdrawn from the alternative projects just in time to resume work on the original project, which then increases instability even further. This can create a "snowball" effect, with impact across projects. The result is large buffers of work and long cycle times at multiple projects; even well prepared "lean thinking" general contractors are vulnerable to instability introduced from projects outside of their control.

Standard contractual arrangements between subcontractors [Barnes 2006] and general contractors make it very difficult to implement practical steps intended to improve flow according to lean construction principles. Most contracts have extensive provisions for dealing with nonconformance or nonperformance on the part of the subcontractor, but very few provisions, if any, for creating a stable work flow. Specifically, it is difficult to:

(a) Reorganize work cells across organizational boundaries (neither workers nor work packages can be moved from one subcontractor to another).
(b) Create a pull system within which subcontractors remain available for continuous periods when work is required.
(c) Improve flexibility by establishing conditions for shifting workload and/or labor and equipment between subcontractors as conditions demand at any given time.
(d) Improve stability and reduce variability in terms of the number of workers, the arrival times of crews on site, the availability of core equipment, etc. where there is no direct ownership of the equipment, and/or employment of the staff of the subcontractor on the part of the general contractor. In fact, for reasons related to liabilities for compensation, most contracts go to great lengths to avoid establishing any kind of employer–employee relationship between a general contractor, on the one hand, and the subcontractor and its employees, on the other hand.

Even while critiquing the problems subcontracting poses for lean construction management, the benefits it provides—in terms of flexibility, competitiveness, and trade specialization—should not be ignored. Subcontracting can be adapted, through new formulations of the economic incentives, to support considerations of work flow, and make application of lean construction techniques feasible. At the least, it can be used intelligently to avoid its deleterious effects. The following are specific recommendations that can be adopted by construction project managers when considering their strategies for subcontracting.

- *Shift part of the risk for lost capacity back to the general contractor*

Risk should be apportioned in accordance with the ability to control the risk. Some contracts create a safety net for subcontractors by allowing them to make claims for situations in which work does not become available as planned. This should not be left to subcontractors' claims for compensation, but built in a priori as an inherent part of each price, by splitting unit prices into two components—one to be paid for product, and the other for resource capacity. The payment for work would no longer be simply $I = WU$ (work completed multiplied by unit price), but:

$$I = W(1-\alpha)U + \alpha U_L D_A n, \quad \text{where } n = \min(n_D, n_A). \tag{8.7}$$

The term α is a measure of how much risk is shifted, and must be agreed to by both parties in advance. The second part of the right hand side of Equation 8.7 is composed of a unit labor cost per unit time, U_L, the actual duration for the work, D_A, and the lesser of the actual number of workers provided, n_A, and the number of workers demanded, n_D. The number of workers demanded should be capped for each task by the construction manager, at a value that reflects his/her confidence about the rate at which work will be made available. Where they are less certain, managers will now demand less labor, rather than more, to avoid paying for excess. On the other hand, the subcontractor has good reason to meet the demanded resource level, in order to maximize its income.

The net income for the subcontractor, based on Expression 8.5 and Expression 8.7, is now:

$$I_n \approx \min[k,q] W_D((1-\alpha)U - C_M) + \alpha U_L D_A \min[k,1] n_D - k W_D C_S. \tag{8.8}$$

In Figure 8.7, which is based on Expression 8.8, the effect of applying this type of arrangement can be seen in the shift of the projected income curves. For example, where α has been set to 50%, then for $q = 80\%$ the subcontractor loses little even when $k > q$. The general contractor's project manager begins to accept responsibility for production management rather than simply for contract administration.

- *Minimize the proportion of the labor component in any subcontract*

The economic model highlighted the importance of the "loss slope" of Equation 8.1 in the region where $q < k$ (see Figure 8.3). Where subcontractors can provide their own materials, their economic vulnerability to variability in the rate of provision of work can be reduced by allowing them to do so. Minimizing the proportion of the labor component in subcontracts, and enlarging the material proportion reduces the slope making them more resilient, and less prone to behave defensively. This recommendation must be approached

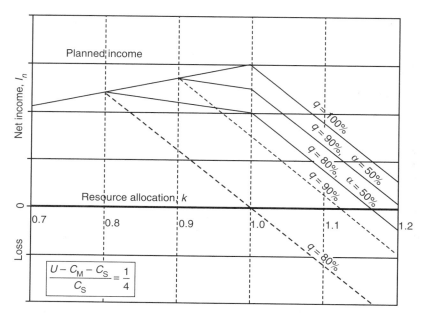

FIGURE 8.7   Adjusted relationship between net income, resource allocation, and plan reliability.

with caution where subcontractors are dependent on suppliers, and may be less able to ensure steady supply than the general contractor.

- *Buffer work where the labor component of a subcontract is high*

A corollary to the previous recommendation is that project managers should consciously apply time and/or work buffers ahead of work to be performed by those subcontractors whose labor proportions are highest.

- *Structure work to reduce the number of handover points between subcontractors*

Resource allocation problems are focused at the handover points between construction activities. Sacks and Goldin (2007) showed how a "complexity index" can be compiled to provide an indication of the number of handovers between subcontractor teams in a project. By consolidating work, and resequencing some activities, the number of handover points can be reduced, thus reducing the potential impact of unreliable resource allocations on the project as a whole.

Naturally, it is not only plan reliability that affects subcontractors' resource allocation decisions. There are additional factors, some of which are beyond the control of any individual project manager. The degree of demand for a subcontractor's resource in the market will impact their reliability, as will the subcontractor's cash flow and liquidity. Knowledge of the factors at work in a subcontractor's economic environment beyond the borders of a project manager's specific project may provide additional leverage for improving reliability. For example, payment terms may be more important to a subcontractor than the price for their work if they function under cash flow constraints.

8.8 Conclusion

There are two sides to the coin of reliability of allocation of resources under construction subcontracts. The perspective of a general contractor's project manager that perceives subcontractors to be inherently unreliable is one-sided; an understanding of the economic imperatives that unit price subcontracts apply would help promote better production management practices with greater chances of achieving win-win outcomes.

The economic motive described by Expression 8.5 and Expression 8.6 shows that subcontractors' net incomes are highly sensitive to the degree of work that will actually become available in any planning period on a project. In order to optimize its income, and avoid the risk of excess capacity, a subcontractor must attempt to allocate resources less than or equal to the amount of work it predicts will become available. The greater the proportion of labor in the unit price, the more acute the sensitivity to plan unreliability. **Given any two otherwise equivalent projects, a subcontractor will prefer to allocate resources to the more reliable of the two.**

A game theory model of the project manager–subcontractor behavior helps explain the relationship between knowledge of plan reliability and resource allocation behavior. In the normal form, the stable equilibrium is one in which the project manager exaggerates demands for resources, and the subcontractor consistently supplies less than demanded. However, as plan reliability improves, and each party's knowledge of the outcome in terms of work availability increases, behavior can shift toward cooperation, and a greater likelihood of achieving mutually optimal results. Lean construction practices such as the Last Planner™ System work to improve plan reliability by involving team leaders directly in making commitments, and by improving transparency of the project among all teams. In this way it can enhance the prospects for cooperative behavior.

Where variation in work provision is high, subcontractors often practice overbooking. They may also attempt to increase reliability by delaying the start of work until sufficient buffers of WIP have accumulated to shield them from upstream variation. Reduction of the wastes of either unutilized resources, or high inventories of WIP requires cooperation between general contractors and subcontractors. However, such cooperation cannot be achieved through partnering or production management alone, as long as the economic motivations remain the same.

Four practical steps are proposed to help project managers achieve more reliable behavior in terms of subcontractor resource allocations:

(a) **Shift part of the risk for lost capacity back to the general contractor**. This can be done by devising a predetermined risk sharing formula for remuneration of the subcontractor, where the cost of waste of underemployed resources is borne in part by the general contractor. This should not be left to be settled through adversarial claims after the fact.

(b) **Minimize the proportion of the labor component in any subcontract.** Work units in which subcontractors provide materials and design, and not only labor and equipment resources, reduce the vulnerability to plan reliability.

(c) **Buffer work where the labor component of a subcontract is high.** Where labor intensive subcontractors are unavoidable, project managers can consciously plan sufficient buffers of WIP to increase the subcontractors' confidence in the work to be made available.

(d) **Structure work to reduce the number of handover points between subcontractors,** by assigning more work to fewer subcontractors, and by reordering activities where possible. It is at these interfaces between subcontractors that resource allocations are least reliable.

The first two should have the effect of making subcontractors more reliable in resource allocations, and more cooperative with production management techniques. The latter two are remedial in nature, helping to ameliorate the deleterious effects where changes to the way labor is subcontracted are difficult.

Acknowledgment

The author gratefully acknowledges the contributions of Michael Harel, a PhD candidate at the Technion, to various aspects of the research on which this chapter is based.

References

Ballard, G. 2000. The Last Planner™ system of production control. PhD diss., University of Birmingham.

Barnes, P. 2006. *The JCT 05 standard building sub-contract.* Boston, MA: Blackwell.

Bennett, J., and Jayes, S. 1995. *Trusting the team.* The University of Reading, Centre for Strategic Studies in Construction.

Bresnen, M., and Marshall, N. 2000. Partnering in construction: A critical review of issues, problems and dilemmas. *Construction Management & Economics,* 18 (2), 229–37.

Costantino, N., and Pietroforte, R. 2002. Subcontracting practices in USA homebuilding: An empirical verification of Eccles's findings 20 years later. *European Journal of Purchasing & Supply Management,* 8, 15–23.

Dainty, A., Briscoe, G., and Millet, S. 2001. New perspectives on construction supply chain integration. *Supply Chain Management,* 6 (4), 163–73.

Edwards, D. J. 2003. Accident trends involving construction plant: An exploratory analysis. *Journal of Construction Research,* 4 (2), 161–73.

Fisher, N., and Green, S. 2001. Partnering and the UK construction industry: The first ten years—A review of the literature. In *Modernizing construction,* ed. J. Bourn, 58–66. London: National Audit Office.

Harel, M., and Sacks, R. 2006. Subcontractor resource allocation in a multi-project environment—Field study. In *Understanding and managing the construction process: Theory and practice, 14th Annual Conference of the International Group for Lean Construction*, eds. R. Sacks and S. Bertelsen, 467–78. Santiago, Chile: School of Engineering, Catholic University of Chile.

Hinze, J., and Tracey, A. 1994. The contractor–subcontractor relationship: The subcontractor's view. *Journal of Construction Engineering and Management*, 120 (2), 274–87.

Howell, G., Ballard, G., and Hall, J. 2001. Capacity utilization and wait time: A primer for construction. *Proceedings of the 9th Annual Meeting of the International Group for Lean Construction*, Faculty of Engineering, National University of Singapore, Singapore.

Hsieh, T. 1998. Impact of subcontracting on site productivity: Lessons learned in Taiwan. *ASCE Journal of Construction Engineering and Management*, 124 (2), 91–100.

Koskela, L. 1992. *Application of the new production philosophy to construction*. Technical Report #72, Center for Integrated Facility Engineering, Stanford University.

———. 2000. An exploration towards a production theory and its application to construction. PhD diss., Helsinki University of Technology, Espoo.

Lazar, F. D. 2000. Project partnering: Improving the likelihood of win/win outcomes. *Journal of Management in Engineering*, 16 (2), 71–83.

Maturana, S., Alarcon, L. F., Gazmuri, P., and Vrsalovic, M. 2007. On-site subcontractor evaluation method based on lean principles and partnering practices. *Journal of Management in Engineering*, 23 (2), 67–74.

O'Brien, W. J. 2000. Multi-project resource allocation: Parametric models and managerial implications. *Proceedings of the 8th Annual Conference of the International Group for Lean Construction*, University of Sussex, Brighton, UK.

Osborne, M. J., and Rubinstein, A. 1994. *A course in game theory*. Cambridge, MA: MIT Press.

Packham, G., Thomas, B., and Miller, C. 2001. Partnering in the Welsh construction industry: A subcontracting perspective. No. 19, WEI Working Paper Series, Welsh Enterprise Institute, University of Glamorgan Business School, Pontypridd. http://www.glam.ac.uk/file_download/487.

Rahman, M. M., and Kumaraswamy, M. M. 2004. Contracting relationship trends and transitions. *Journal of Management in Engineering*, 20 (4), 147–61.

Sacks, R. 2004. Towards a lean understanding of resource allocation in a multi-project sub-contracting environment. In *12th Annual Conference on Lean Construction*, eds. C. T. Formoso and S. Bertelsen, 97–109. Elsinore, Denmark: Lean Construction.

Sacks, R., and Goldin, M. 2007. Lean management model for construction of high-rise apartment buildings. *Journal of Construction Engineering and Management*, 133 (5), 374–84.

Sacks, R., and Harel, M. 2006. An economic game theory model of subcontractor resource allocation behavior. *Construction Management & Economics*, 24 (8): 869–81.

Sakamoto, M., Horman, M. J., and Thomas, H. R. 2002. A study of the relationship between buffers and performance in construction. In *10th Annual Conference of the International Group for Lean Construction*, eds. C. T. Formoso and G. Ballard. NORIE/UFRGS, Porto Alegre, RS Brazil, Gramado, Brazil, 13 p.

Tommelein, I. D., Riley, D. R., and Howell, G. A. 1999. Parade game: Impact of work flow variability on trade performance. *Journal of Construction Engineering and Management*, 125 (5), 304–10.

Womack, J. P., and Jones, D. T. 2003. *Lean thinking: Banish waste and create wealth in your corporation*. New York: Simon & Schuster.

9

Understanding Supply Chain Dynamics Via Simulation

Séverine
Strohhecker (née
Hong-Minh)
Boehringer Ingelheim
Pharma GmbH & Co. KG

Stephen M. Disney
Cardiff University

Mohamed M. Naim
Cardiff University

9.1   Introduction

Simulation is a technique that replicates a real-world situation, usually by means of a simplified model representation, and its dynamic behavior reproduced. The model is used to understand cause and effect relationships leading to insights as to why certain behaviors occur. From such an analysis various alternative scenarios to improve behavior may be tested. The advantage of simulation is that the scenario may be tested at minimum risk to the real-world situation—either due to the fact that the scenario may actually lead to a detriment to performance, or it may be too costly to implement in the real-world without full testing beforehand.

Simulation is a well-established form of analysis in construction engineering and management. Previous research in construction simulation has focussed on the on-site physical processes [e.g., AbouRizk and Halpin 1990; Martinez and Ioannou 1999], and with a strong emphasis on discrete modelling [e.g., González, Alarcón, and Gazmuri 2006], or discrete event simulation [e.g., Nassara, Thabetb, and Beliveauc 2003; Kamat and Martinez 2008]. Less well researched is the role of simulation in modelling construction supply chains, and their associated material, information, and resource flows.

This chapter has two substantive aims. The first is to introduce the reader to the concept of simulation as applied to construction supply chains. The second aim is to show the application of system dynamics simulation in the context of testing supply chain management (SCM) principles. More specifically the chapter investigates material flow management principles as defined in Chapter 7, namely, centralization of supply, the total cycle time reduction, and improved relationships between the trading partners can be tested via simulation modelling. We will analyse if these principles can be used to improve dynamic performance. To achieve this, we will endeavour to "teach by example," giving an overview of the application of a simulation study to a low-value fit-out housing supply chain. We start with an explanation of a generic supply chain modelling methodology, and review the types of dynamic modelling and simulation techniques available. We then describe a low-value fit-out supply chain where a number of specific problem areas or "hot spots" have been identified.

This is followed by a presentation and evaluation of possible SCM solutions to cool these "hot spots." The first solution considered is the centralization of supply by turning away from regional procurement with multiple merchants. The second solution consists of sharing customer information across the supply chain, and finally the third step considers the synchronization of lead-times between trading partners.

The performance is assessed based on the data collected during an in-depth business diagnostic of the supply chain, and simulation is used to assess its dynamic performance. Finally, the results are summarized, and the "best" performing scenario is highlighted.

9.2   Modelling and Simulation

Whilst Chapter 2 provides a construction-specific perspective of modelling, this section, following an overview of the generic principles for modelling supply chains, focuses specifically on simulation as applied to the material, information, and resource flows that constitute a housing supply chain. Modelling and simulation should be considered as part of an extensive structured method to the reengineering of supply chains. Naim and Towill (1994) have proposed a framework to analyse the dynamic properties of supply chains. This complements the work of other researchers and practitioners in the field by integrating "soft," or qualitative [e.g., Checkland and Scholes 1990], and "hard," or quantitative [e.g., Disney and Towill 2003] approaches. The framework, shown in Figure 9.1, aims to develop valid models of business processes that constitute the supply chain so as to optimize the total flow of information, materials, and other resources, such as people, money, and machines, amongst others.

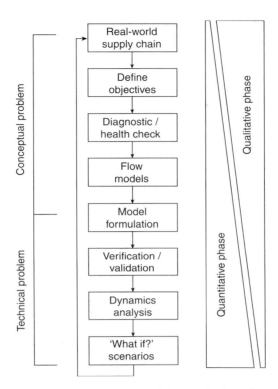

FIGURE 9.1 Modelling and simulation framework for supply chain design [Naim and Towill 1994].

The framework consists of two discrete but overlapping phases. The qualitative phase consists of acquiring sufficient knowledge to understand the current structure and operation of a supply chain. The quantitative phase develops and analyses numerical models with simulation. The two phases overlap in varying proportions as modelling becomes more quantitative. Therefore, the skills required of the team reengineering the supply chain will vary according to which stage of the framework is reached.

The initial conceptual stage of systems analysis aids in defining the boundaries of the problem and interfaces within the system under study. During this stage a flow analysis is undertaken. This flow analysis typically takes the form of value stream maps or process charts, and helps to evaluate time delays and logistics constraints involved in both value adding, and nonvalue adding activities. The flow analysis also provides a means to uncover the cause and effect relationships present in supply chain variables, such as, for example, orders, call-offs, inventory levels, and backlog levels. At this stage of the simulation process, supply chain diagnostic methods that have been applied in many sectors may be used, such as the Quick Scan (QS) in the automotive industry [Lewis et al. 1998; Naim et al. 2002], and a QS adaptation in the construction industry, named the terrain scanning methodology (TSM) [Barker, Hong-Minh, and Naim 2000]. As in the flow analysis, such diagnostic methods may be based on questionnaires, interviews, and archival data in order to obtain a rich picture of the current situation, including the main problems and their possible causes.

The information obtained from the flow analysis can support the development of suitable conceptual models. Two conceptual modelling tools are used in this chapter. These are rich pictures [Checkland 1981; Checkland and Scholes 1990], and causal loop diagrams [Wolstenholme 1990]. Both of them represent the causal relationships between those variables driving the dynamic behavior of the supply chain.

As the nature of the flows becomes known, the conceptual model is converted into a simulation or an analytical form. As we will see later, there are software packages, such as Stella/iThink® and Vensim®, which allow the direct input of a causal loop diagram and the generation of numerical expressions to represent relationships between variables.

The first stage of the quantitative phase is the selection of possible modelling techniques for analysing the supply chain. A model is an abstract representation of reality that is easier to study than the real system. Simulation models may be used for producing new results, verifying results, or demonstrating relationships [Kramer and de Smit 1977]. Naim, Disney, and Towill (2004) identify various types of modelling techniques within the context of SCM:

(a) Management games: these are typically tabletop games that reenact a management decision in the supply chain. The dynamic behavior resulting from the decisions taken by the players can be demonstrated and then discussed. These games may even be incorporated into computer software or embedded into Web pages to facilitate the development of decision-making experience. Games are limited in the sense that, in general, nothing can be rigorously proved from the game in itself. However, they do have the advantage of allowing participants to experience specific learning outcomes for themselves. The most notable SCM game is the Beer Game [Sterman 1989], which has been extended or computerized by various authors including van Ackere, Larsen, and Morecroft (1993), and Lambrecht and Dejonckheere (1999a, 1999b).

(b) Statistical investigations: they usually involve the analysis of difference equations via conditional expectation, and are often highly mathematical. This type of contribution typically provides statistical insights about the impact of demand properties, such as standard deviation and autocorrelation, and supply chain properties, such as lead-times and information paths, on inventory costs and capacity requirements. Statistical methods are often used to quantify performance of real situations. Recent significant contributions of this type include Lee, So, and Tang (2000) and Chen et al. (2000).

(c) Continuous control theory techniques: these have also been used in the SCM field. Typically they have been developed for the analysis of physical hardware systems by electrical and mechanical engineers. However, Simon (1952) described how to use linear deterministic control theory for production and inventory control. Axsäter (1985) presents a useful review paper of early work, summarizing the advantages and limitations of the field. He concludes that control theory "illustrates extremely well dynamical effects and feedback," but cannot incorporate sequencing and lot-sizing issues. Continuous control theory suffers from the fact that some scheduling and ordering scenarios are inherently discrete, and the continuous representation of pure time delays is mathematically complicated but not impossible [Warburton and Disney 2007].

(d) Discrete control theory: this is a digital version of continuous control theory, and it is a very powerful way of investigating sampled data systems (i.e., a scheduling and ordering systems and computer system which are inherently discrete). Vassian (1955), inspired by Simon's work in the continuous domain, studied a production-scheduling algorithm using discrete control theory. Burns and Sivazlian (1978) considered a four-level traditional supply chain using z-transforms. Popplewell and Bonney (1987) have investigated manufacturing resource planning (MRP) systems. Disney (2001) has been using discrete control theory to investigate Vendor Managed Inventory supply chains.

(e) Classical operations research (OR) and management science: at about the same time as the seminal work of Simon (1952) and Vassian (1955), this separate but parallel strand of research established a dynamic programming approach to the inventory control problem. Arrow (2002) provides a review of the search for the optimal inventory policy. While not actually investigating structural dynamics, this approach aims to establish optimal policies for inventory control. For example, the assumptions may be concerned with the cost structure, demand pattern, lead-times, and planning horizons. Towill et al. (2003) have undertaken a detailed comparison of the control theory and OR approach to decision support system (DSS) design for managing supply chain dynamics.

(f) Simulation and system dynamics: it was advocated by Forrester (1958, 1961) as a method of investigating the dynamical effects in large nonlinear systems as a means of avoiding having to resort to complicated mathematical or control theory based models [Edghill and Towill 1989]. Previous work using simulation is very prolific and includes (but it is by no means limited to) Forrester (1958, 1961), and Coyle (1982), who studied traditional supply chain structures, Cachon and Fisher (1997) and Waller, Johnson, and Davis (1999) who studied Vendor Managed Inventory supply chain structures.

It is the latter modelling method that we will concentrate on in this chapter. Naylor et al. (1966) define simulation as "a numerical technique for conducting experiments on a digital computer, which involves certain types of mathematical and logical models that describe the behavior of a system over extended periods of real time."

Hoover and Perry (1989), and Law and Kelton (1991) identify three dimensions of simulation:

(a) Static vs. dynamic: static simulation is a representation of a situation at a particular time (or a representation based on averages), such as the linear programming or Economic Order Quantity techniques, whereas dynamic simulation represents the system as it evolves over time. The main advantages of dynamic simulation can be summarized as:

- It incorporates the impact of time into the performance evaluation [Bowersox, Closs, and Helferich 1986]

- It is flexible, especially in comparison with analytical tools; details can be included which could not be possible with analytical models [Hoover and Perry 1989]
    - The researcher has control over the other variables in comparison with the real system [Johnsson 1992]
(b) Deterministic vs. stochastic: the lack of any random components makes a simulation model deterministic; that is, the outcome is determined once the model relationships and initial stages have been defined, whereas the outcome of stochastic models are random variables.
(c) Continuous vs. discrete: in continuous simulation, time is assumed to pass continuously; that is, events can happen at any point in time. In discrete simulation, the state variables change instantaneously only at separate points in time. This means that the system states can change only at a countable number of points in time. These points in time may be based upon a clock, in which simulation proceeds in uniform steps in time, or upon events, in which simulation has steps that are governed by the events that occur along the simulation, such as those that begin or end.

System dynamics makes use of both deterministic and stochastic changes. Although system dynamics is a clock-driven discrete simulation, it is one that attempts to minimize the uniform steps whilst approximating to a continuous time simulation.

9.3   Example of System Dynamics in a House Building Supply Chain

9.3.1   The Low-Value Fit-Out Supply Chain Current State

This case study description is the outcome of an interventionist action research program with a housing developer in the United Kingdom. The case study follows the framework given in Figure 9.1. The real-world supply chain investigated was a low-value fit-out supply chain. The purpose of the research was to investigate the dynamic performance of several reengineering strategies. The program involved a team of researchers undertaking the TSM diagnostic to ascertain the current state followed by the secondment of one member of the research team to the collaborating organization. The secondee worked with the change management team responsible for developing and implementing the organization's supply chain strategy. The secondee was responsible for the model development involving interviews with key personnel in the organization, with the whole team working together in scenario development.

The main subsystems of a house can be divided into three categories, the house shell, the high-value fit-out, and the low-value fit-out. By house shell we mean the foundations, interior and exterior walls, and roof. High-value fit-out is concerned with the items such as the bathroom suite and kitchen. The low-value fit-out is concerned with items such as doors, skirting, electricity, and plumbing. For ease of explaining the principals of the simulation approach, this chapter will focus on the case of the low-value fit-out.

Typically for bulk or low value items such as skirting boards, steel lintels, door linings, doors, hinges, etc., the supply process is carried out via multiple merchants and manufacturers on a regional basis. Thus, these products are purchased from a wide range of suppliers based on a minimum purchase cost.

Figure 9.2 illustrates a rich picture representation of the flows of a house building supply chain. The focus is placed upon the major issues of planning and control for the supply chain. The products are ordered by the regional buyer and called off by the site manager when required. The products are then delivered to site in a stockyard from which the products are selected, sorted, and moved to the exact construction location.

Seven "hot spots" have been identified during the research and have been reported in Naim and Barlow (2003). Hot spots are the main problem areas from a supply chain perspective, and are represented by the symbol that looks like a "bomb." They are summarized in Table 9.1. These "hot spots" cover a wide range of issues, from the lack of customer information to a high level of waste and a lack of supply chain integration. These "hot spots" can be identified by their symptoms, such as poor supplier delivery performance, poor availability of material on-site, or unsatisfied customers. However, what needs to be tackled are the root causes. These may be such things as the adversarial approach to trading, regional buying arrangements, purchase decisions based solely on price, or even the lack of strategy for utilizing the information available in planning activities.

For simplicity of understanding and analysis, the traditional low-value fit-out supply chain can be simplified, as shown in Figure 9.3. Only two manufacturers have been considered, a door manufacturer (A) and a skirting boards manufacturer (B). The lead-times were established according to data collected in the business diagnostic case study. The site

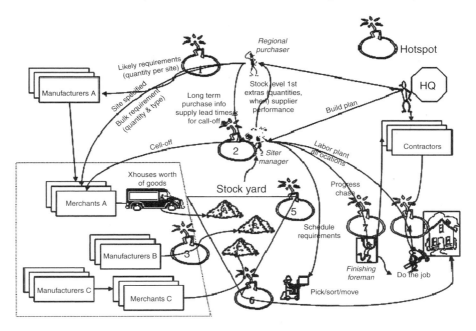

FIGURE 9.2 Rich picture representation of the traditional supply chain for a generic house building supply chain. (From Naim, M.M., and Barlow, J., *Construction Management and Economics*, 21, 593–602, 2003. With permission.)

TABLE 9.1   Summary of Specific "Hot Spots" in the Traditional Supply Chain of Low-Value Fit-Out Material

Hot Spot	Root Causes	Symptoms
Hot spot 1: little use of market knowledge for buying and calling off the material	• Regionally based buying agreements • Purvchase based on price • No time scale guarantee for actual delivery	Suppliers have little visibility of long-term market requirements
Hot spot 2: lack of supply chain integration	• The site manager needs to coordinate a large amount of people and tasks • He holds a considerable amount of information • No clear strategy of how best to utilize the information	Very poor information transfer and use across the supply chain
Hot spot 3: no time compression strategy	• Lack of supplier development and adversarial relationships • Volatile short-term call-off information from the site • Late changes in site requirements	Poor supplier delivery performance
Hot spot 4: inability to rapidly reconfigure	• No medium-term planning horizon given to subcontractors • High work uncertainty pushes subcontractors to commit themselves to several tasks	Poor availability of contractors on-site
Hot spot 5 and 6: excessive muda, or waste	• High level of stock on-site to buffer against uncertainties • No designated stocking area and proper recording mechanism lead to damage, mislaying, or theft of material	Poor availability of material on-site
Hot spot 7: excessive human resource	• Above problems lead to the need for a finishing foreman to chase material, chase labour, and assign rework and snag list	Dissatisfied customer (poor total value)

Note: Based on the generic concept from Naim and Barlow (2003).

manager has to wait a minimum of seven days after the order is placed by the regional buyers before he can call off the material from the merchants. It usually takes 14 days for the merchant to fulfil that order, but only one day is used to transport material to the site. Several different merchants are used by different regional buyers. However, even within the same regions a multiple (generally two to three per site) number of merchants are used.

The replenishment of doors requires a 10-day lead-time, while the replenishment of skirting board requires a seven-day lead-time. Furthermore, as mentioned before, the relationships between the companies tend to be adversarial. Home builders' satisfaction is generally low with regards the availability of material, as highlighted in "hot spots" 5 and 6. Customer satisfaction is also perceived as low, as identified in "hot spot" 7.

Predictably, on-costs and lead-times are also high. "On-costs" are unnecessary costs associated with the highly variable workload. Some materials, believed to be in the stockyard, might have been misplaced, stolen, or damaged, and replacement material will need to be ordered, causing delays in the construction process. Furthermore, the unreliability of deliveries also makes it difficult to predict the finishing date.

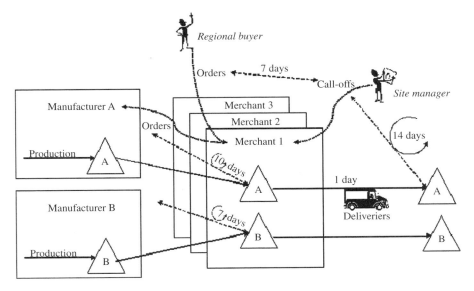

FIGURE 9.3 Traditional low-value fit-out supply chain.

9.3.2 Re-Engineering the Low-Value Fit-Out Supply Chain

This section describes three alternatives developed by the change management team for improving the performance of the low-value fit-out supply chain in relation to the traditional practices. The first alternative is to move away from local supply through multiple merchants, and use one national distributor. The second alternative focuses on improving the information flow across the supply chain. Finally, the third alternative is concerned with the synchronization of lead-times across the supply chain.

9.3.2.1 Merchant's Integration

"Hot spots" 3, 5, and 6 highlighted poor supplier delivery performance and lack of availability of materials on-site. One possible solution is to use only a single merchant, which will be called "the kitter." This strategy is based upon a future supply chain state our case company was just about to achieve. There are different reasons behind this strategy. First, regional buyers usually select merchants based on price. Therefore, many different merchants are used for the procurement of low-value fit-out materials. Instead, one single merchant is used in this strategy. This means that regional buyers do not have to search for the cheapest merchant available, but simply place their orders to the single supplier. This is only possible because the kitter has been vertically integrated, and it is therefore part of home builder. The kitter/home builder believed that the vertical integration would mean lower prices.

Second, only using the kitter to procure all material needed on a nationwide basis increases its buying power. It also allows the kitter to negotiate directly with most manufacturers, as the buying quantities are larger. The approximate turnover for a large merchant is £250 million, £40 million for the kitter, and £30 million for a small to medium-sized merchant. The builder's merchants sector has been consolidating recently as large merchants have greater buying power [Agapiou et al. 1998; Anon 1998]. The kitter is dealing with 26 suppliers, of which 18 are manufacturers. If a third of the materials are still purchased from the merchants, it is due to the reluctance of the manufacturers to deal directly with a rather small customer compared to the large merchants. As mentioned above, the prices are also guaranteed to be lower through the kitter. For example, a specific type of skirting board would be sold at £0.45 per metre, while a standard merchant would sell it for £0.53 per metre, or in other words, the kitter would be 15% cheaper. For a specific architrave, the kitter would sell at £1.13 per metre against £1.30 per metre from a merchant, which is 13% cheaper.

Third, not only are all the low-value fit-out materials bought from the kitter, but they are also sent in packs. All materials for a low-value fit-out are distributed in seven packs for masonry construction and only four packs for timber frame construction. Each pack is specifically aimed at different stages of construction of a house. For example, in one pack, items such as external doors, skirting, architrave, doorstop, internal doors, hinges, and door latches and locks are packed together for a specific house type. The idea behind the use of packs is to reduce waste on-site arising from damaged, mislaid, and stolen material in the stockyard, making possible to reduce stocks on-site. Furthermore, it also reduces on-site deliveries, as one delivery of packs could correspond to up to six deliveries from the individual merchants. Finally, it assures a faster assembly process, as the whole kit is available at once and therefore all the parts needed for one part of the construction process are readily available. This concept of packs is similar to kitting, which has a long history in the electronic industry. Bozer and McGinnis (1984) define a kit as "a specific collection of components and/or subassemblies which, with other kits (if any), support one or more assembly operations for a given product." The use of such a system is appropriate for products with numerous parts, or high-value components, or for the quality assurance reasons [Johansson 1991].

Fourth, using the kitter allows a reduction from 14 to seven days for delivery lead-time from the call off. This seven-day lead-time is made up of four days to prepare the packs, and three days for the delivery of the packs, as shown in Figure 9.4. Furthermore, as home builder has vertically integrated the kitter, the stock control responsibility is kept within home builder. This will improve upon the merchant's poor service level.

9.3.2.2 Information Flow Integration

Improvements in the information flow need to be made in order to address the "hot spots" 1, 2 and 3: little visibility of long-term market requirements by suppliers, very poor information transfer and use across the supply chain, and poor supplier delivery performance. Among the root causes of those three "hot spots" are the insufficient amount and availability of information. One way to share information with several different organizations in a timely and accurate fashion is to use information technology (IT). In our case company, the home builder has decided to use the SAP/R3 system in the near future. The new information system would be accessible by site managers, regional buyers, the kitter, and the manufacturers. The construction schedule will be posted on the system and updated as required. Therefore, all the organizations involved will have

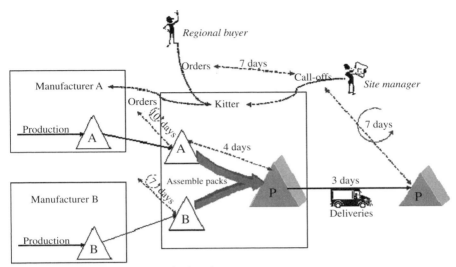

Note: A represents doors and B skirting boards

FIGURE 9.4   Low-value fit-out supply chain using kitter.

access to accurate information on the site progress. Furthermore, the ordering and call-off processes will also be carried out by managers with the support of an information system.

It is envisioned that the total order cycle time will be reduced to five days. One day advance notice before calling off the packs will suffice for the kitter. This is based on the assumptions that the house design is standardized, and that the kitter is in possession of the drawings. Therefore, the packs could be assembled within three days. Finally, as shown in Figure 9.5, the delivery lead-time will be cut down to one day. This is already happening in most cases, the three days presented in the previous section being a buffer rather than a necessity. This planning time buffer is common in the industry [Wegelius-Lehtonen and Pahkala 1998].

9.3.2.3   Synchronization

The last SCM configuration that has been investigated is the synchronization of lead-times in the supply chain [Stevens 1989; Sabath 1995; Towill 2000], whereby the lead-times are established to ensure a continuous flow of material through the supply chain without any queues. This scenario was specifically developed by the change management team with the collaboration of the procurement manager from the kitter. With the kitter working at full capacity and the SAP system fully implemented, lead-times could be further reduced to achieve a total order cycle time of three days, as shown in Figure 9.6. Only one day is required for pack assembly as the personnel have gone through their learning curve and house designs have been standardized. Thus, variations from one pack to another are limited. It will still be necessary to allow one day for the transfer of the packs to site. As the relationship with the manufacturers shifts from

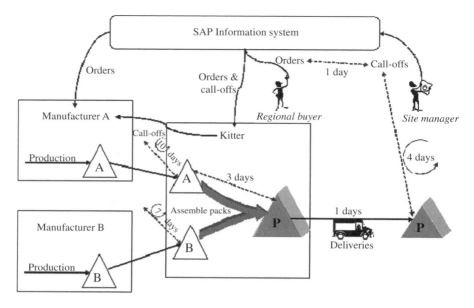

Note: A represent doors and B skirting boards

FIGURE 9.5   Low-value fit-out supply chain with information integration.

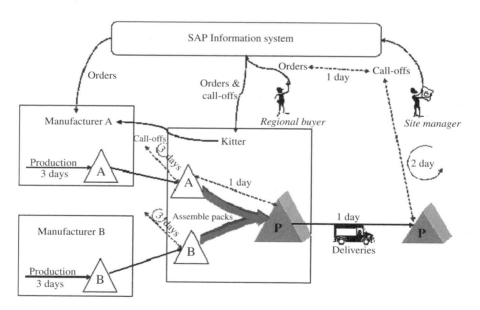

Note: A represent doors and B skirting boards

FIGURE 9.6   Synchronized low-value fit-out supply chain.

TABLE 9.2   Summary of the Reengineering Scenarios

Scenarios	Baseline	Kitter	Integrated Information	Synchronized
Supply chain structure	Developer, multiple merchants, manufacturers	Developer, single merchant, manufacturers		
Role of the developer	Order from merchants	Order from kitter		
Role of merchants/kitter	Order from manufacturers	Order from manufacturers and prepare packs		
Role of manufacturers	Deliver to merchants	Deliver to kitter		
Innovation	–	Use of kitter	Use of an information system across the supply chain	Synchronized lead-times across the supply chain
Total order cycle time (days)	21	14	5	3

being adversarial to being more collaborative, the manufacturers will have access to up-to-date information from the IT system, and the total order cycle time will be reduced to three days. This means that the different organizations in the supply chain will work on the same three-day order cycle time, and will therefore be synchronized.

9.3.3   Summary of the Reengineering Scenarios

Table 9.2 summarizes the four scenarios for the low-value fit-out supply chain in terms of supply chain structure, the involvement of each agent, the type of innovation intro-duced, and the total order cycle time.

9.4   Implications on Supply Chain Dynamics

The simulation model was used to test if one point of control with one stocking point, improved information flow, reduction of lead-times, and synchronization can improve the dynamic performance of the supply chain. This section will present the simulation model, the simulated scenarios, and the simulation results.

9.4.1   Simulation Model

We do not endeavour here to give a detailed description of the model as it is rather lengthy. A consistent lesson from any simulation study is that the effort and resource is skewed initially to the problem definition and "real-world" understanding front end of the simulation process, as given in the previous sections, followed by a considerable effort in understanding the simulation outputs, and the implications for "real-world" implementation. It is the latter that we concentrate on in the remainder of the case study description.

Several different models are used in this simulation, all of which are based on common features and on the same basic model. A structural overview of this basic model is presented in Figure 9.7, and is composed of six subsystems. The exogenous demand is the construction plan defining the rate of units purchased by the end-customers. This demand is inputted into the home builder planning and control system, where the orders and call-offs are generated. Home builder generates orders and call-offs which are transmitted to the merchants. The merchants use this information to generate their own orders, which are sent to the manufacturer. The manufacturer then decides on the appropriate production level to fulfil these orders. A production allocation system has been implemented to determine which merchant should receive which quantities of material in the case of shortages. Then construction on-site can take place using the material delivered by the merchants. Finally, some performance indicators are calculated to assess the model.

The underlying simulation logic is the same for all the organizations included in the structure presented in Figure 9.7. The logic is based on Forrester's production and distribution system [Forrester 1958], and on the Inventory and Order Based Production Control System (IOBPCS) model [Towill 1982]. The IOBPCS model represents the rule for generating order requirements, and is analogous to the classic order-up-to rule often used in inventory control systems [Lalwani, Disney, and Towill 2006]. The IOBPCS model, first analysed by Coyle (1977), has been extensively studied by the Logistics Systems Dynamics Group at Cardiff University since 1982, following the framework outlined by Naim and Towill (1994) in Figure 9.1 [see Ferris and Towill 1993; John, Naim,

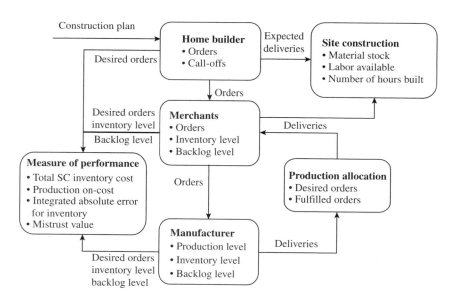

FIGURE 9.7   Model structure overview.

and Towill 1994; Cheema 1994; Towill, Evans, and Cheema 1997; Lewis 1997; Disney, Naim, and Towill 2000].

Figure 9.8 illustrates an IOBPCS-based representation of the merchants' subsystem using Vensim®. The level of production required (the order rate) is based upon the level of demand which has been averaged (over a period of T_a time units), and the level of current inventory in comparison with a target inventory. T_i represents the time to adjust for the differences between the actual and target inventory levels, and T_p, the production delay.

A feature present in Forrester's (1958) original model and not previously represented in the IOBPCS was added to the IOBPCS model. Usually an IOBPCS model is linear and it is assumed that whatever is asked for will come out of the pipeline after a delay. Forrester incorporated a backlog function utilizing a nonlinear representation. In this case, if there are no products available in stock, the products cannot be delivered. This feature takes into account the unfilled orders or order backlog. The order backlog is equal to the previous backlog plus new orders received less any shipments that have been made in the same period of time.

The settings of T_a (time to average consumption), T_i (time to adjust inventory) and T_p (the production delay) are based on John, Naim, and Towill (1994), Towill and Del Vecchio (1994), Mason-Jones, Naim, and Towill (1997) and Mason-Jones (1998). Several studies showed that in order to reduce demand amplification in IOBPCS models, a good setting would be $T_a = 2T_p$ and $T_i = T_p$ [John, Naim, and Towill 1994; Mason-Jones 1998]. However, Towill and Del Vecchio (1994) argue that different settings could be used depending on the purpose of the model and where the company is positioned in the supply chain. Therefore, the settings proposed by John, Naim, and Towill (1994) were used for the manufacturer, and those of Towill and Del Vecchio (1994) for the merchants; that is, $T_a = T_i = 2T_p$.

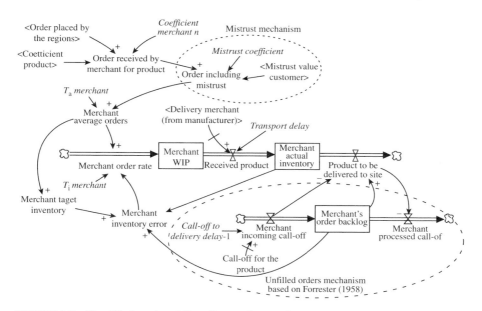

FIGURE 9.8 Simplified stock and flow diagram for merchants.

A mistrust mechanism has been included in the model. The modelling of mistrust is based on real-life observations during the case study, and on Sterman (2000). Mistrust is understood as being the lack of trust between trading partners. Very often this lack of trust is especially tangible when customers do not receive the full quantity of what they have ordered. Instead of trusting that the supplier will deliver the missing products as soon as they become available, customers over-order to make sure that they will receive the real quantities they need. This principle has therefore been reflected in the model as follows: whenever the customers do not receive the full delivery of what they have ordered, the next order they will place will be increased by a percentage of the quantity of product undelivered.

An information enrichment mechanism has also been incorporated into the model, and can easily be switched off. This mechanism is based on Mason-Jones's (1998) work and has been placed within the manufacturer. The information enrichment mechanism allows the manufacturer to utilize the smoothed market demand (merchant average orders), and the smoothed customer orders (manufacturer average orders) to decide on the production level.

For example, when there is 100% information enrichment, the manufacturer is relying solely on the market sales data from home builder to decide how many products to produce, and could thus easily respond incorrectly to customer (merchants) demand. If there is 50% information enrichment, then the manufacturer will base his decision on 50% of the market demand and 50% of its customer demand. Finally, setting the model to 0% information enrichment means the manufacturer relies solely on its' customer demand (the merchants).

9.4.2   Simulation Model Verification/Validation

Model validation and verification are also very important issues for all modellers. However, Sterman (2000) clearly states that, "no model can ever be verified or validated," as by definition they are a simplified representation of the reality, and therefore vary from the real world in many different ways. Forrester and Senge (1980) add that "validation is the process of establishing confidence in the soundness and usefulness of a model. Validation begins as the model builder accumulates confidence that a model behaves plausibly and generates problem symptoms or modes of behavior seen in the system." Furthermore, a model can be considered as realistic "to the extent that it can be adequately interpreted, understood, and accepted by other points of view" [Churchman 1973].

Solberg (1992) emphasizes the benefits of simple models, stating that "the power of a model or modelling technique is a function of validity, credibility, and generality. Usually the simplest model which expresses a valid relation will be the most powerful." Hence, models need, one way or another, to be validated, however there is no single test that would allow the modellers to assert that their models have been validated. Rather, the level of confidence in the model can increase gradually as the model passes more tests [Forrester and Senge 1980].

A wide range of tests to build confidence in the model has been developed [e.g., Forrester and Senge 1980; Barlas 1989; Barlas 1990; Barlas 1996], of which Sterman (2000) presents a summary. These 12 tests are as follows:

The boundary adequacy test is concerned with the appropriateness of the model's boundary. For this test the main question to answer is if the appropriate concepts have been included in the model to address the problem. As stated previously, the aim of the model for this chapter is to compare different supply chain scenarios where a few parameters change, and assess the impact of these changes on the dynamic behavior. The area concerned is the house building supply chain. The main players in the supply chain have been represented (i.e., home builder, merchants, and distributor). The focus of the model is placed upon the material and information flow. Both flows have been repeatedly and successfully studied previously using a member of the IOBPCS model, also used here. Furthermore, although simplified, the model was considered as representative of the real situation by managers from the companies modelled.

The structure assessment test, as its name indicates, is concerned with the consistency of the structure of the model by verifying if the structure of the model represents the real system. As previously presented, the basis of the model is the IOBPCS models. In addition, Coyle (1977), Edghill (1990) and Berry, Evans, and Naim (1998) asserted that the IOBPCS model was representative of production practice, and replicates the dynamic behavior of real-world systems to a reasonable degree of accuracy. Finally, some changes have been made to the basic IOBPCS model to take into account some real life issues, such as unfulfilled orders, and the distinction between orders and call-offs.

The dimensional consistency test examines if the units of measurement used in the model are consistent. This was carried out using the dimensional consistency function available in Vensim®.

The parameter assessment test compares the model parameters to knowledge of the real system to determine if the parameters correspond conceptually and numerically to real life. The parameter values principally on real data collected during interviews.

The extreme condition test analyses the behavior of the model under extreme conditions to verify that the model behaves in a realistic fashion. In the model studied in this chapter, the extreme condition test was carried out for an extremely high demand, the stock level and the amount of labor available then dropped to zero. The number of houses completed reflected the labor capacity. However, as no capacity restriction had been made in the model, the stock level still recovered using an extremely high production level. Capacity restriction was not introduced into the model so as to keep it as simple as possible, and also because the purpose of the model is not to study capacity issues, but broader issues of dynamic behavior.

The integration error test verifies if the time step utilized for the simulation and the integration method are appropriate for the purpose of the model. In the present case, the time step was set at one day, however in order to test the model, the time step has been cut in half, in quarter, and in eights, and the results compared. For the purpose of this model, the differences were marginal. The test was also carried out using a different integration method.

The behavior reproduction test assesses the model's ability to reproduce the behavior of the real system. This test is generally used for a model whose purpose is to reproduce very accurately the real-world system by comparing simulation results and real historical data. Although this test does not apply here, and the model has already been proven to be representative of a production control system (as stated in the structure assessment test), the model was presented to the case company's personnel. They all agreed that it represented their supply chain.

The behavior anomaly test, which examines the importance of specific relationships by deleting or modifying them, was utilized during the model development process. This test helped in analysing the influence of specific variables.

The family member test asks whether a model could be used to represent other more particular models. As stated previously, the IOBPCS used in the model is already part of a model family. Furthermore, the model was used to simulate different scenarios by "switching on or off" the relevant subsystems, and therefore the model proves to be a general one that can be adapted to represent specific members.

The surprise behavior test is concerned with unexpected behavior displayed by the model. The test is passed when the behavior does indeed occur in the real world. This was the case for the build up of stock observed at the merchants and suppliers level.

When using a normal IOBPCS model, the stock first diminishes before regaining a stable state. However, in the model studied here, the stock first increases and then diminishes. This is, however, happening in the real world where companies stock up in advance of a large order.

Sensitivity analysis tests the robustness of the model. As the model is based on an IOB-PCS model, sensitivity analysis have already been carried out and showed the robustness of the IOBPCS model [e.g., Edghill 1990; Disney, Naim, and Towill 2000]. However, a new sensitivity analysis has also been carried out for the overall model. The analysis took into consideration the three parameters influencing the ordering policy; that is, T_a (time to average consumption), T_i (time to adjust inventory), and production delay. The analysis was carried out to study the impact of these parameters on the manufacturer order rate, and the stock of products at the manufacturer level. T_a was tested for a range from 1 to 60, T_i from 2 to 30, and production delay from 1 to 30, which is the maximum range utilized during the simulations.

The validated model is then subjected to extensive dynamic analysis. The objective of this stage is to determine how the supply chain responds dynamically to various test inputs. For example, we may wish to see how the supply chain behaves to a sharp, step change in customer demand. It is at this point that the strength of simulation analysis becomes apparent as various scenarios may be adopted and evaluated relatively easily and quickly. A structured approach to exploiting the supply chain model is presented later in this chapter.

9.4.3   Scenarios Description

The four scenarios taken into consideration are those studied in the previous sections: (a) the baseline scenario represents the traditional low-value fit-out supply chain; (b) in the kitter scenario the merchants are replaced by a single organization; (c) in the integrated information scenario an information system is used to transfer information across the supply chain; and finally (d) in the synchronized scenario the lead-times across the supply chain are synchronized.

In the baseline scenario, the regional buyer places the orders, and the site manager calls off the material. As mentioned before, two types of product have been considered: doors and skirting boards. As several merchants are used across the country, eight have been modelled for each product, making up a total of 16 merchants. Two manufacturers have been represented, Manufacturer A that produces doors, and manufacturer B that produces skirting boards. Finally, mistrust has been modelled between the agents to represent the adversarial relationships. For this scenario, mistrust was set at 100%, which means that whenever the customers do not receive what they have ordered, they will increase the next order they place by 100% of the quantity of the product undelivered, as explained in Sterman (2000).

The kitter scenario only models one distributor instead of 16 different merchants. The kitter assembles packs, which means that both doors and skirting boards need to be available too, before assembling packs and delivering them to the site. As in the previous scenario, two manufacturers have been modelled. Finally, mistrust has been lowered to 75%, as it was agreed with the interviewees that the relationships in this case are not fully adversarial, but a lack of trust is still present as the kitter has only just become operational.

The integrated information scenario is based on the kitter scenario, but uses an information system to transfer information across the supply chain. An information enrichment mechanism, as used by Mason-Jones (1998), was implemented with an information enrichment set at 50%. This means that the manufacturer bases its requirements 50% on the original orders placed by the regional buyer, and 50% on the orders received from the kitter. However, even though information is shared through the supply chain, it was agreed with the interviewees that mistrust should still be set at 50%, as trust is slowly building up between companies, but they are not yet ready to trust each other fully.

The synchronized scenario also uses an information system, but this time it was set at 75% which, according to Mason-Jones (1998), is one of the best settings. The lead-times have been synchronized across the supply chain, which means that at each level, the total order cycle time is set at three days (based on the interviewees' responses). Finally, mistrust has been taken out of the model by setting it at 0%, which is the equivalent of total trust between the partners. Thus, even though customers may not receive what they ordered, they trust their suppliers that missing products will be delivered as soon as possible, and therefore they do not need to over-order.

The lead-times and other parameters for each scenario are summarized in Table 9.3.

TABLE 9.3 Summary of the Parameters for the Four Low-Value Fit-Out Scenarios

Factor	Baseline	Kitter	Integrated Information	Synchronized
Order to call off lead-time (days)	7	7	1	1
Call off to delivery lead-time (days)	14	7	4	2
Order to delivery from Manufacturer A (days)	10	10	10	3
Order to delivery from Manufacturer B (days)	7	7	7	3
Mistrust (%)	100	75	50	0
Information enrichment (%)	0	0	50	75

TABLE 9.4   Ranking of the different scenarios for dynamical performance criteria at the manufacturers for step change in the demand

Scenarios	Peak Value	Peak Time	Order Recovery	Stock Depletion	Trough Time	Stock Recovery	Scenario Performance
Baseline	****	*	**	****	*	****	***
Kitter	**	**	*	***	**	**	*
Integrated information	*	***	****	**	***	*	**
Synchronized	***	****	***	*	****	***	****

Note: **** represents the best performance and * the worst.

9.4.4   Simulation Results

Each scenario has been simulated daily for a step change in demand over a period of three years. The step change in demand increased from 100 to 120 houses at day 20. First of all, it is interesting to analyse the ranking of the scenarios for the step change in demand, taking into consideration the six dynamic criteria presented in Table 9.4, namely:

- Peak value—it is the maximum order or production rate of a system and determines the highest capacity requirements. The peak value should be kept as low as possible.
- Peak time—indicates how quick the peak is reached and it is an indication of the speed of response of the system. The peak time should be as short as possible.
- Order recovery—expresses how well our production system is doing at tracking real demand in steady state. If our system never "catches up" with demand then we are always in backlog, thus smaller order recovery times are desirable.
- Stock depletion—expresses how far stock levels deplete and it is a surrogate for the risk of stocking out. This depletion should be kept as small as possible.
- Trough time—the time to reach the maximum stock depletion, and again indicates how fast we respond to changes in demand. The trough time should be kept as small as possible.
- Stock recovery—indicates how long it takes for the stock level to recover following a step change in demand. Systems that recover quicker are better then systems that take a long time to recover.

Using a linear scale, where four stars is the best and one star the worst, the results presented in Table 9.4 represent the response at manufacturer level. For ease of presentation, only the door manufacturer response is presented. Synchronized scenario achieves the best overall performance, followed by baseline scenario. The integrated information scenario achieves the worst performance for peak value. In contrast, the synchronized scenario registers the worst performance for stock depletion, while achieving the best performance for peak time and throughput time due to the short lead-times.

The dynamic performance, assessed using the six criteria above, can be summarized using only two criteria. These are:

1. Production on-costs—based on the cumulative absolute error between a given behavior and the actual behavior. Hence the order rate is compared with the actual

end-customer orders, and after having integrated the absolute error, the result is cubed so as to give the production on-costs [Wikner, Towill, and Naim 1991].

2. Manufacturer's inventory deviations—calculated for the actual inventory level and is based on the same principle as for the production on-costs. The area considered is the difference between the actual inventory level and the target inventory.

Furthermore, the total supply chain inventory costs can be simply calculated based on the total inventory holdings over the period of simulation. These criteria have been calculated using simulation. Table 9.5 presents the ranking of the scenarios using these three performance criteria.

The production on-costs are minimized in the case of synchronized scenario, which means that synchronized scenario achieves the smallest demand amplification among the four scenarios. The baseline scenario registers the worst results in terms of total supply chain inventory costs. It could be suggested that this is due to the large number of merchants. However, the stock level for each merchant was set to at least four times the average demand, knowing that the total demand placed on the merchants is the same as for kitter.

In order to improve the understanding of the above results, the magnitude of the impact that each scenario has on performance needs to be studied. A comparison of each scenario in relation to the baseline scenario is presented in Table 9.6. It can be seen that the total supply chain inventory costs are reduced for all three scenarios. Furthermore, the synchronized scenario improves the production on-costs by 30% in comparison with baseline scenario. Finally, all three scenarios increase the inventory deviation in comparison with baseline scenario, especially synchronized scenario with a 16% increase.

The results of the four strategies for a step change in demand have been analysed. However, the impact of each SCM principle cannot be fully understood, as more than

TABLE 9.5 Scenarios' Ranking for a Step Change in Demand

Scenarios	Production on-Costs	Manufacturer's Inventory Deviations	Total SC Inventory Costs	Scenario Performance
Baseline	***	****	*	***
Kitter	*	**	***	*
Integrated information	**	***	****	****
Synchronized	****	*	***	***

Note: **** represents the best performance and * the worst.

TABLE 9.6 Impact of the Scenarios on Performance Criteria in Comparison With Baseline Scenario for a Step Change in Demand

Scenarios	Production on-costs	Manufacturer's Inventory Deviations	Total SC Inventory
Kitter (%)	+22	+10	−.8
Integrated information (%)	+21	+6	−1.1
Synchronized (%)	−30	+16	−.8

one parameter has been changed in each scenario. Therefore further simulations have been carried out to analyse the impact of every single change made to move from one scenario to the next.

The first comparison has been carried out between the baseline scenario and the kitter scenario. The SCM principles implemented in this comparison were: centralization of supply, the total cycle time reduction, and improved relationships between the trading partners. This was simulated by:

- "No merchant" strategy: replacing the merchants by the kitter
- "Delay call off" strategy: reducing the lead-time from order to call off from 14 to seven days
- "From stock" strategy: taking material out of stock earlier, in order to assemble packs, from one to eight days
- "Mistrust customer" strategy: reducing the mistrust level between the regional buyer and the merchants from 100 to 75%
- "Mistrust merchants" strategy: reducing the mistrust level between the merchants and the manufacturers from 100 to 75%

Table 9.7 presents the amplitude of impact that each strategy has in comparison with baseline scenario. "No merchant" strategy improves both production on-costs (by 38%) and the manufacturers inventory deviations (by 11%). This means that it improves the dynamic behavior of the supply chain. Therefore, moving away from multiple merchants on a regional basis and choosing one single company on a national basis not only improves the dynamic performance, but also reduces the total supply chain inventory costs. This confirms Charatan's (1999) observation that centralization on a national basis of supply has almost always been beneficial. However, reducing the lead-time between order and call-off has a negative impact on the dynamic behavior. This is understandable as the advance notice of what is going to be called off is shorter, and therefore manufacturers have less time to react to changes in demand.

In a similar manner, "from stock" strategy worsens the dynamic behavior as materials are taken from stock earlier on, and thus the manufacturers do not have much time to build up their stock against the increase in demand. All three strategies ("no merchant," "delay call off," and "from stock") reduce the total supply chain inventory costs.

TABLE 9.7   Impact of Each Strategy from Baseline Scenario to Kitter Scenario

Scenarios	Production on-Costs	Manufacturer's Inventory Deviations	Total SC Inventory Costs
No merchants (%)	−38	−11	−.2
Delay call off (%)	+65	+14	−.2
From stock (%)	+77	+17	−.2
Mistrust customer (%)	(−1)	(−.2)	(+.001)
Mistrust merchants (%)	(−.01)	(+.3)	(−.001)
Kitter (%)	+22	+10	−.8

Finally, the impact of "mistrust" strategies is indicated in brackets as it only has a marginal impact. Furthermore, it does not have the same level of safety stock as other scenarios. However, it gives an interesting insight into the way in which the increase of trust between trading partners affects performance. Interestingly, "mistrust customer" increases the total supply chain inventory costs, while "mistrust merchants" reduces it. Therefore, when the level of mistrust is reduced between the site and the merchants, the total supply chain inventory costs increase, while the dynamic performance at the manufacturer level improves. This can be explained by the fact that there is less disturbance or noise in the demand signal received by the manufacturers.

The reduction of mistrust between merchants and manufacturers improves the production on-costs, but increases the manufacturer's inventory deviations. This can be explained by the fact that as mistrust diminishes, the demand received by the manufacturer is lower (only 75% of the product quantities that have not been received is added to the demand instead of 100%). However, it also means that the manufacturer does not overproduce and its stock level diminishes more rapidly.

The kitter and integrated information scenarios were also compared. The SCM principles implemented in this comparison were: the use of an information system to share end-customer demand, reduction of total cycle time, and improved relationships between trading partners.

These principles were implemented by (again, introducing the strategies one at a time):

- "Information enrichment" strategy: passing on the site demand to the manufacturers. The manufacturers based their requirements 50% on the site demand and 50% on the orders received from kitter.
- "Delay call off" strategy: reducing the lead-time from order to call off from seven days to one.
- "Mistrust customer" strategy: reducing the mistrust level between the regional buyer and kitter from 75 to 50%.
- "Mistrust kitter" strategy: reducing the mistrust level between kitter and the manufacturers from 75 to 50%.

As presented by Mason-Jones (1998), the implementation of an information enrichment mechanism improves the dynamic behavior of the supply chain. It also reduces the total supply chain inventory costs. As seen previously, the "order to call off" strategy has a negative impact on the dynamic behavior (by increasing both production on-costs and the manufacturer's inventory deviations), but improves the total supply chain inventory costs.

Again, by reducing the mistrust level from 75 to 50%, the total supply chain inventory costs are increased (Table 9.8). The marginal increase of the manufacturer's inventory because of "customer mistrust" is due to a greater drop in inventory level in the case of 50% mistrust. However, as there are fewer disturbances in the demand signal, the inventory level recovers more rapidly.

The increase in production on-costs for the "mistrust kitter" strategy is explained by the fact that the production level peaks higher than for 75% mistrust. The marginal reduction of manufacturer's inventory deviations for inventory is due to a smaller trough in inventory levels.

TABLE 9.8 Impact of Each Strategy from Integrated Information Scenario to Synchronized Scenario

Scenarios	Production on-Costs	Manufacturer's Inventory Deviations	Total SC Inventory Costs
Information enrichment (%)	−21	−13	−.1
Order to call off (%)	+13	+8	−.2
Mistrust customer (%)	(−4)	(+.1)	(+.1)
Mistrust kitter (%)	(+.1)	(−2)	(+.01)
Integrated information (%)	−1	−3	−.4

9.4.5 Summary of the Simulation Results

Several important lessons were learnt from this simulation study. First of all, integrated information scenario achieves the best overall performance for a step change in demand. All three scenarios—kitter, integrated information, and synchronized—improve the total supply chain inventory costs in comparison with the baseline scenario.

Using one single national merchant instead of several regional merchants improves all three performance criteria (production on-costs, manufacturers' inventory deviations, and total supply chain inventory costs). This is supported by Charatan's (1999) and Henkoff's (1994) observations on the positive impact of centralization of supply. Reducing the delay between placing an order and the call-off has a negative impact on dynamic performance. This is also the case for the "from stock" strategy, which takes material out of stock several days before delivery.

Information enrichment improves the performance criteria studied, as generically postulated by Mason-Jones (1998), whilst reducing manufacturing lead-times; whilst keeping the parameters in the ordering rule constant has a detrimental effect on both the total supply chain and the manufacturer's inventory costs. However, it has a positive effect on production on-costs. Finally, reducing mistrust either between customer and merchants/kitter, or between merchants/kitter and manufacturers has a positive effect in terms of faster recovery to a stable state. However, in all cases, reduction of the mistrust level between trading partners increases the total supply chain inventory costs.

9.5 Discussion

A low-value fit-out supply chain has been studied in this chapter. First of all, an analysis of the current state was undertaken. Seven "hot spots" were identified as being the major problems of the traditional low-value fit-out supply chain. In order to address these "hot spots," three steps were identified with the ultimate aim of improving performance. The first step is to move away from regional supply using multiple merchants, to a single national supplier, who can also prepare the materials in packages to reduce the number of deliveries and wastage on-site. As seen during the simulations, this scenario reduces total supply chain costs, but it has a negative effect on the dynamic behavior.

The second step is to use customer information across the supply chain through an information enrichment mechanism. This not only reduced total order cycle time, but also improved the dynamics of the supply chain. Finally, the last step is to synchronize lead-times across the supply chain. Here again, total supply chain inventory costs can be reduced and total ordering cycle time reduced.

Simulation has shown us that centralizing the supply (no merchants strategy) generally has a positive effect, improving dynamic behavior and reducing total supply chain inventory costs, as identified by Charatan (1999) in the retail sector. The reduction of the delay between placing the order and calling-off, or in other words, advance notice given to suppliers, has a detrimental effect on the dynamics of the models. It was confirmed that using the information enrichment strategy improves performance, while reducing the levels of mistrust between trading partners helps the system to return faster to a stable state.

TABLE 9.9   Overall Performance for the Four Different Scenarios

Key Performance Indicators (KPIs)	Traditional Low-Value Fit-Out Supply Chain	Kitter's Low-Value Fit-Out Supply Chain	Integrated Information for the Low-Value Fit-Out Supply Chain	Synchronized Low-Value Fit-Out Supply Chain
SCM KPIs				
Stock levels	80*	126*	132*	155*
Total order cycle time (days)	21	14	5	3
Inventory costs	£1.037 M*	£1.029 M*	£1.025 M	£1.029 M
Customer satisfaction	Low (home builder)	Medium (house builder)	Medium-high (house builder)	High (house builder)
House building KPIs				
(Product and service)	Low (customer)	Medium (customer)	High (customer)	High (customer)
Product quality	Low	High	High	High
Predictability costs	Low	Medium	Medium-high	High
Predictability time	Low	Medium	Medium-high	High
Dynamic KPIs				
Peak value	132*	137*	137*	136*
Peak time	70*	60*	58*	52*
Order recovery	302*	313*	280*	285*
Stock depletion	80*	126*	132*	155*
Trough time	59*	54*	53*	46*
Stock recovery	283*	348*	374*	342*

*Indicates relative values from the simulations.

9.6   Conclusions

The summary of the performance results for the four scenarios is presented in Table 9.9. The main benefit of the simulation was to demonstrate to managers the long-term implications of the dynamic behavior of various supply chain designs. With the simulation we were also able to investigate the trade-offs and scenarios in a safe environment.

Finally, we have described, via a case study, a method for undertaking a simulation-based business diagnostic. Notably, we have not focussed on the "software" aspects of the simulation approach, but on the presimulation data collection (including variables, parameters, and cause and effect relationships), simulation analysis, and postsimulation managerial interpretation. We were also able to verify some generic supply chain management principles in a specific supply chain setting, and provide advice on how the company may exploit the newly introduced IT system.

References

AbouRizk, S., and Halpin, D. 1990. Probabilistic simulation studies for repetitive construction processes. *Journal of Construction Engineering and Management*, 116 (4), 575–94.

Agapiou, A., Flanagan, R., Norman, G., and Notman, D. 1998. The changing role of builders merchants in the construction supply chain. *Construction Management and Economics*, 16, 251–361.

Anon. 1998. Merchants of bloom take the lion's share. *Construction News June*, 11, 24–25.

Arrow, K. J. 2002. The genesis of 'optimal inventory policy'. *Operations Research*, 50 (1), 1–2.

Axsäter, S. 1985. Control theory concepts in production and inventory control. *International Journal of Systems Science*, 16 (2), 161–69.

Barker, R., Hong-Minh, S., and Naim, M. M. 2000. The terrain scanning methodology: Assessing and improving construction supply chains. *European Journal of Purchasing and Supply Management: Special Issue on Construction Supply Chains*, 6 (3–5), 179–93.

Barlas, Y. 1989. Multiple test for validation of system dynamics type of simulation models. *European Journal of Operation Research*, 42 (1), 59–87.

———. 1990. An autocorrelation function-test for output validation. *Simulation*, 55 (1), 7–16.

———. 1996. Formal aspects of model validity and validation in system dynamics. *System Dynamics Review*, 12 (3), 183–210.

Berry, D., Evans, G. N., and Naim, M. M. 1998. Pipeline information survey: A UK perspective. *International Journal of Management Science, OMEGA*, 26 (1), 115–32.

Bowersox, D. J., Closs, D. J., and Helferich, O. 1986. *Logistical management—A systems integration of physical distribution, manufacturing support and materials procurement*. New York: Macmillan.

Bozer, Y. A., and McGinnis, L. F. 1984. *Kitting: A generic descriptive model.* MHRC-tr-84-04. Atlanta, GA: Georgia Institute of Technology.

Burns, J. F., and Sivazlian, B. D. 1978. Dynamic analysis of multi-echelon supply systems. *Computers and Industrial Engineering*, 2, 181–93.

Cachon, G., and Fisher, M. 1997. Campbell soup's continuous replenishment program: Evaluation and enhanced inventory decision rules. *Production and Operations Management*, 6 (3), 266–76.

Charatan, A. 1999. Retail supply chain integration. In *Global logistics and distribution planning—Strategies for management*, ed. D. Waters, 155–72. London: CRC Press, Kogan Page.

Checkland, P. 1981. *Systems thinking, systems practice.* Chichester, UK: John Wiley.

Checkland, P., and Scholes, J. 1990. *Soft systems methodology in action.* Chichester, UK: John Wiley & Sons.

Cheema, P. 1994. Dynamic analysis of an inventory and production control system with an a adaptive leadtime estimator. PhD thesis, Cardiff University.

Chen, F., Drezner, Z., Ryan, J. K., and Simchi-Levi, D. 2000. Quantifying the bullwhip effect in a simple supply chain: The impact of forecasting, lead-times and information. *Management Science*, 46 (3), 436–43.

Churchman, C. 1973. Reliability of models in the social sciences. *Interfaces*, 4 (1), 1–12.

Coyle, R. G. 1977. *Management system dynamics.* London: Wiley.

Coyle, R. G. 1982. Assessing the controllability of a production and raw materials system. *IEEE Transactions on Systems, Man and Cybernetics*, SMC-12 (6, November/December), 867–76.

Disney, S. M. 2001. The production and inventory control problem in Vendor Managed Inventory supply chains. PhD thesis, Cardiff Business School, Cardiff University.

Disney, S. M., Naim M. M., and Towill, D. R. 2000. Generic algorithm optimisation of a class of inventory control systems. *International Journal of Production Economics*, 68 (3), 259–78.

Disney, S. M., and Towill, D. R. 2003. On the bullwhip and inventory variance produced by an ordering policy. *OMEGA: The International Journal of Management Science*, 31 (3), 157–67.

Edghill, J. 1990. The application of aggregate industrial dynamics techniques to manufacturing systems. PhD thesis, Cardiff University.

Edghill, J. S., and Towill, D. R. 1989. The use of systems dynamics in manufacturing systems. *Transactions of the Institute of Measurement and Control*, 11 (4), 208–16.

Ferris, J. S., and Towill, D. R. 1993. Benchmarking of a generic family of dynamic manufacturing ordering and control models. *Journal of Systems Engineering*, 3, 170–82.

Forrester, J. 1958. Industrial dynamics: A major breakthrough for decision makers. *Harvard Business Review*, 36 (4), 37–66.

———. 1961. *Industrial dynamics.* Cambridge, MA: MIT Press.

Forrester, J. W., and Senge, P. M. 1980. Test for building confidence in system dynamics models. In *System dynamics*, eds. A. A. Legasto and J. W. Forrester, 209–28. Amsterdam: North-Holland.

González, V., Alarcón, L. F., and Gazmuri, P. 2006. Design of work in process buffers in repetitive building projects: A case study. *Proceedings of IGLC-14*, 25–27 July, Santiago, Chile, 165–76.

Henkoff, R. 1994. Delivering the goods. *Fortune November*, 28, 64–78.

Hoover, S., and Perry, R. 1989. *Simulation—A problem-solving approach.* Boston, MA: Addison-Wesley.

Johnsson, M. 1992. *Discrete event simulation, an evaluation tool for logistic systems.* Lund, Sweden: Lund Institute of Technology.

Johansson, M. I. 1991. *Kitting systems for small parts in manual assembly systems.* London: Taylor & Francis.

John, S., Naim, M., and Towill, D. 1994. Dynamic analysis of a WIP compensated decision support system. *International Journal of Manufacturing System Design*, 1 (4), 283–97.

Kamat, V. R., and Martinez, J. C. 2008. Generic representation of 3D motion paths in dynamic animations of simulated construction processes. *Automation in Construction*, 17 (2), 188–200.

Kramer, N., and de Smit, J. 1977. *Systems thinking—Concepts and notions.* Leiden, The Netherlands: Stenfert Kroese, Martinus Nijhoff Social Sciences Divisions.

Lalwani, C. S., Disney, S. M., and Towill, D. R. 2006. Observable and controllable state space representations of a generalized order-up-to policy. *International Journal of Production Economics*, 101 (1), 173–84.

Lambrecht, M. R., and Dejonckheere, J. 1999a. A bullwhip effect explorer. Research Report 9910, Department of Applied Economics, Katholieke Universiteit, Leuven, Belgium.

———. 1999b. Extending the Beer Game to include real-life supply chain characteristics. *Proceedings of the EUROMA International Conference on Managing Operations Networks*, 237–43.

Law, A., and Kelton, W. 1991. *Simulation modeling & analysis.* New York: McGraw-Hill.

Lee, H. L., So, K. C., and Tang, C. S. 2000. The value of information sharing in a two-level supply chain. *Management Science*, 46 (5), 626–43.

Lewis, J. 1997. An integrated approach to re-engineering material flow within a seamless supply chain. PhD thesis, Cardiff University, UK.

Lewis, J., Naim, M., Wardle, S., and Williams, E. 1998. Quick Scan your way to supply chain improvement. *Institute of Operations Management Control*, 24 (5), 14–16.

Martinez, J. C., and Ioannou, P. G. 1999. General-purpose systems for effective construction simulation. *Journal of Construction Engineering and Management*, 125 (4), 265–76.

Mason-Jones, R. 1998. The holistic strategy of market information enrichment through the supply chain. PhD thesis, Cardiff University.

Mason-Jones, R., Naim, M., and Towill, D. 1997. The impact of pipeline control on supply chain dynamics. *The International Journal of Logistics Management*, 8 (2), 47–62.

Mason-Jones, R., and Towill, D. 1998. Shrinking the supply chain uncertainty circle. *Control*, 24 (7), 17–22.

Naim, M. M., and Barlow, J. 2003. An innovative supply chain strategy for customised housing. *Construction Management and Economics*, 21, 593–602.

Naim, M. M., Childerhouse, P., Disney, S. M., and Towill, D. R. 2002. A supply chain diagnostic methodology: Determining the vector of change. *Computers & Industrial Engineering: An International Journal*, 43 (1–2), 135–57.

Naim, M., Disney, S., and Towill, D. 2004. Supply chain dynamics. In *Understanding supply chains: Concepts, critiques and futures*, eds. Steve New and Roy Westbrook, 109–32. Oxford: Oxford University Press.

Naim, M. M., and Towill, D. R. 1994. Establishing a framework for effective materials logistics management. *International Journal of Logistics Management*, 5 (1), 81–88.

Nassara, K., Thabetb, W., and Beliveauc, Y. 2003. Simulation of asphalt paving operations under lane closure conditions. *Automation in Construction*, 12 (5), 527–41.

Naylor, T., Balintfy, J., Burdick, D., and Kong, C. 1966. *Computer simulation techniques*. New York: John Wiley.

Popplewell, K., and Bonney, M. C. 1987. The application of discrete linear control theory to the analysis of multi-product, multi-level production control systems. *International Journal of Production Research*, 25 (1), 45–56.

Sabath, R. 1995. Volatile demand calls for quick response: The integrated supply chain. *Logistics Information Management*, 8 (2), 49–52.

Simon, H. A. 1952. On the application of servomechanism theory to the study of production control. *Econometrica*, 20, 247–68.

Solberg, J. 1992. The power of simple models in manufacturing. In *Manufacturing systems—Foundations of world-class practice*, eds. J. Hein and W. Compton, 215–23. Washington, DC: National Academy of Engineering Press.

Sterman, J. 1989. Modelling managerial behaviour: Misperceptions of feedback in a dynamic decision making experiment. *Management Science*, 35 (3), 321–39.

———. 2000. *Business dynamics—Systems thinking and modelling for complex world*. Boston, MA: McGraw-Hill Higher Education.

Stevens, G. 1989. Integrating the supply chain. *International Journal of Physical Distribution and Logistics Management*, 19 (8), 3–8.

Towill, D. 2000. A route map for substantially improving supply chain dynamics. *International Journal of Manufacturing Technology and Management*, 1 (1), 94–111.

Towill, D. R. 1982. Dynamic analysis of an inventory and order based production control system. *International Journal of Production Research*, 20 (6), 671–87.

Towill, D., and Del Vecchio, A. 1994. The application of the filter theory the study of supply chain dynamics. *Production Planning & Control*, 5 (1), 82–96.

Towill, D., Evans, G., and Cheema, P. 1997. Analysis and design of an adaptive minimum reasonable inventory control system. *International Journal of Production Planning and Control*, 8 (6), 545–57.

Towill, D. R., Lambrecht, M. R., Disney, S. M., and Dejonckheere, J. 2003. Explicit filters and supply chain design. *Journal of Purchasing & Supply Management*, 31 (2), 73–81.

van Ackere, A., Larsen, E. R., and Morecroft, J. D. W. 1993. Systems thinking and business process redesign: An application to the Beer Game. *European Management Journal*, 11 (4), 412–23.

Vassian, H. J. 1955. Application of discrete variable servo theory to inventory control. *Journal of the Operations Research Society of America*, 3 (3), 272–82.

Waller, M., Johnson, M. E., and Davis, T. 1999. Vendor managed inventory in the retail supply chain. *Journal of Business Logistics*, 20 (1), 183–203.

Warburton, R., and Disney, S. M. 2007. Variance amplification: The equivalence of discrete and continuous. *International Journal of Production Economics*, 110 (1–2), 128–37.

Wegelius-Lehtonen, T., and Pahkala, S. 1998. Developing material delivery processes in cooperation—An application example of the construction industry. *International Journal of Production Economics*, 5657, 689–98.

Wikner, J., Towill, D. R., and Naim, M. M. 1991. Smoothing supply chain dynamics. *International Journal of Production Economics*, 22, 231–48.

Wolstenholme, E. F. 1990. *System enquiry: A system dynamics approach*. Chichester, UK: Wiley.

I

Commentary

Glenn Ballard
University of California

A COMMENTARY MAY BE WRITTEN WITH DIFFERENT PURPOSES and assumptions. The commentator may adopt the position of the expert with the purpose of providing an authoritative interpretation. However, this commentary is intended to help the reader develop their own understanding of what they have read. What should the reader, both practitioner and theorist, take away from these chapters? What questions are provoked? What opportunities are revealed? Each reader has to answer these questions for him or herself. This commentary is one reading and one perception of context and connections. Hopefully it will help others develop their own.

Overview

The eight chapters in section 1 are all concerned with production and operations analysis. They offer a variety of perspectives on supply chain management in construction.

- Methods of modeling and simulation (Chapters 2 and 9)
- Application of supply chain management concepts and methods to the management of design (Chapters 4, 5, and 6)
- Production planning and control (Chapters 3, 7, and 8)
- Two different methodological perspectives on construction supply chain management: value management (Chapter 5) and lean project delivery (Chapter 6)
- Case studies in the management of task specific and nontask specific materials (Chapter 7)
- Managing the supply of services in the construction context; an analytical modeling of the dynamics between general contractors and specialty contractors (Chapter 8)

Contribution to Knowledge

The contribution of these chapters to knowledge can be organized under the functions of production system management; namely, design, operation, and improvement.

Production system design involves:

- Creating capacity to meet specifically characterized demand
- Aligning the interests of the participants
- Structuring the system for value generation and flow

Production system operation involves:

- Setting specific performance objectives
- Steering toward objectives through system transparency and production control
- Correcting objectives or steering in response to experience

Production system improvement involves:

- Developing people
- Tracking performance metrics; especially leading rather than lagging indicators
- Learning from experiments (planned deviations from standard)
- Learning from breakdowns (unplanned deviations from standard)
- Reducing system buffers (funds, inventories, capacities, durations) to reveal and attack sources of variation

It is apparent from the table that most chapters concentrate primarily on the design function. A few chapters focus on system operation. No chapters are devoted specifically to system improvement.

Chapter	Design	Operate	Improve
2	X		
3		X	
4	X	X	
5	X		
6	X		
7	X	X	
8	X		
9	X		

Takeaways

- Modeling and simulation can be used to evaluate possible changes in system design.
- Traditional supply chain management concepts and methods can be usefully applied to construction, especially to the supply of materials to project sites.

- Representing and understanding the complete construction network of suppliers and customers has not yet been done, and poses challenges that may require developments beyond traditional supply chain management methods.
- There have been successful innovations in design management, but their generalization, their distillation into principles with clear rules for application remains to be done. Everything about managing the design phase of construction projects needs to be better understood—from the role of the design team in defining value, to designing the design process, to integrating downstream players. These chapters provide encouraging and informative examples of successful explorations of these issues.
- Historically, individual construction projects have been the focal point for managing the supply of materials and services, with projects taking what the market offers without attempting to structure supply chains.
- Coordination of the activities of multiple, interdependent specialists has been improved through production and people-oriented methods, such as those described by Formoso and Isatto, where work flow control is achieved through the management of handoffs between specialists, as distinct from direct supervision of work processes. However, this approach has not yet been extended far beyond those traditionally understood to constitute the design team, or far beyond those working directly on construction sites. The development of supply chain management in construction requires that extension.
- As construction projects have increased in complexity, both technological and organizational, managers have tended to focus increasingly on contract management; on creating advantage through contract terms, on detecting deviations from performing to contractual requirements, and on collecting evidence in anticipation of claims. This has shifted attention away from managing how things are designed and made. The application of supply chain management concepts and methods to construction can adopt either the contract focus, abstracting away from production, or the production focus, in which contracts are restructured to facilitate the delivery of value through designing and making. These chapters are excellent representatives of the latter approach.
- Construction projects are now dominated by traditional practices, such as shifting risk without regard to the larger system and to system dynamics, but this traditional practice is being challenged by new approaches (value management, lean project delivery), by production-oriented experiments in materials management, and by fact-based analysis of system dynamics.
- It is a time for new ideas and exciting experimentation.

Questions Provoked by the Readings

- Should the project be the unit of analysis for construction supply chain management? For example, might the methods illustrated by Arbulu in chapter 7 be applied to multiple projects, either of a single company (owner or contractor) or of multiple companies?

- Is supply chain management the same as production system management, where production is understood to involve both designing and making goods and services? Might it look different because system boundaries are drawn more broadly, including players and processes formerly excluded, but the management functions are the same; namely designing, operating, and improving production systems?
- Can logistics services companies provide the construction industry with logistics centers, cross-docking, kitting, JIT deliveries, milk runs, etc.? Or will these be developed first by repeat buyers of construction services or by providers of construction services with high and predictable volume in specific geographic regions?
- Where to draw boundaries between competitors and collaborators? Could multiple projects collaborate in the use of shared resources, both materials and services? Could industry agreements, perhaps region by region, exploit standardization as a means for reducing costs across-the-board?
- How to better support production system (supply chain) design? For example, how best represent the design options that result in different supply chain configurations, with different boundaries between what is temporary and what is permanent?

II

Organizational Perspectives

10

Review of Organizational Approaches to the Construction Supply Chain

Ruben Vrijhoef
Delft University of Technology

Kerry A. London
The University of Newcastle

10.1 Introduction

Supply chain management (SCM) for an individual organization emerged in the late 1990s as a distinct field of research in the construction management discipline [London and Kenley 2001]. Lean construction has been universally supported in the literature and has had uptake by selected large clients and/or contractors. SCM has been ad hoc in its diffusion by practitioners in the property and construction industry; yet at the heart of the lean production concept is the supporting system of interfirm relationships along the supply chain [Hines 1994; Nischiguchi 1994; Lamming and Cox 1995]. Lamming's empirical research, where he defined the concept of lean supply, identified the importance of the strategic arrangement of the firms that would support such a system. Lamming's (1993) concept of lean supply is important to the understanding of strategic procurement in the supply chain.

Inherent in the research related to organization and SCM is the perspective that organizations who take this approach are developing longer-term relationships with each other outside the boundaries of an individual project. In much of the construction management research to address better relationships between firms there has been a propensity to focus on landmark type projects or lighthouse projects, which have used particular relational contracting strategies with the assumption that this will then diffuse change throughout the industry. For example, relational contracting including such strategies as partnering or alliance contracts cannot simply be translated to another project which has a completely different set of conditions, and more than likely a completely different set of supply chains. There are also various processes and practices which support projects, but may be outside a specific project environment and the responsibilities of those involved in a project, and need top level support and engagement by senior management of firms to come to fruition.

Research regarding organizational approaches towards construction supply chains can be categorized at four main levels and increasing complexity:

- Intraorganizational: within organizations within an individual supply chain
- Interorganizational: between organizations within projects and supply chains
- Cross-organizational structure: across clusters of many supply chains and many suppliers with many clients

10.2 Intraorganizational

Intraorganizational supply chain research typically involves research conducted on the internal environment of an organization to support SCM. It is nearly impossible to achieve improved integration between firms linked in a supply chain if internally organizations are not focused on improving their customer and supplier interfaces.

SCM is simultaneously a philosophy, concept, system, process, strategy and method, and a state of mind [Ross 1997]. Particularly interesting is Ross' discussion of SCM as a state of mind. He claims that "… as a major business philosophy, the effective application of the principles of SCM requires broad acceptance by the staffs of all supply channel members. In fact, leveraging the capabilities and creative vision of the people who comprise the supply channel is, without a doubt, the essential key to realizing superior performance through SCM. Achieving 'buyin' however requires companies to tackle the difficult challenges to the traditional concepts of management and work occurring as a result of enterprise reengineering, the growth of 'virtual' organizations, application of quality management techniques, and team-based organizational styles. Activation of organizations ready to implement SCM requires environments that foster the generation of ideas which will continually improve quality and productivity, management policies and practices that support individual initiative and entrepreneurialism, and new workplace values enabling the convergence of specialized resources to be found among channel members to exploit new opportunities or solve critical problems that would be impossible if attempted from the standpoint of a single company" [Ross 1997, 335].

10.3 Interorganizational

10.3.1 Lean Supply Organization

Lean construction is perhaps one of the most significant examples in the last decade of interorganizational SCM; however, strategic procurement and management of the supply chain is much wider than the lean movement and other factors will be discussed after a consideration of lean construction. The lean construction movement has, from 1993, led much discussion on supply chains through the International Group for Lean Construction annual conferences. It is a concept applicable to all firms and not simply those involved in production and manufacturing. A significant part of strategic procurement is concerned with business alliances. Cooperation among firms has grown rapidly since the early 1980s as alliances have proliferated in one industry after another [Gomes-Casseres 1996]. Alliances are, however, only one method available for strategic procurement.

Lean construction evolved from lean production also referred to as Toyota Production System, and is a movement that is centered primarily upon a production philosophy for construction. In so doing, key protagonists have explored workflow and conversion processes, waste reduction and efficient use of resources, through lean project management, lean supply, lean design, lean partnering, and cooperative SCM [Alarcon 1997]. The central themes have been eliminating waste and improving workflow in the construction.

Those researchers in the wider lean production field have understood the importance of strategic management of supply chains supports lean thinking [Lamming 1993; Nischiguchi 1994; Hines 1994; Sugimoto 1997]. The contextualization of lean production that supports lean thinking has been provided through organizational and industrial organizational economic descriptions of the automotive and electrical industry supply chains. This understanding of the organization of the supply chain is an important part of the philosophy of lean thinking [London 2008].

Lean construction has not been without its detractors [Green 1999]. In his paper, "The Dark Side of Lean Construction: Exploitation and Ideology," Green challenged the narrowly defined instrumental rationalist approach currently undertaken in the movement. He aimed at introducing literature to the lean community that provided evidence of the human cost of lean methods in Japanese industry [Sugimoto 1997], including repression of independent trade unionism, societal costs (pollution and congestion), and regressive models of human resource management [Kamata 1982; Sugimote 1997]. He referred the lean construction community to Kamata (1982) who describes another perspective of the Toyota Production System—one where "workers were often required to live in guarded camps hundreds of miles from their families and suffered high levels of stress at the workplace as they struggled to meet company work targets." Green argued that "whilst the *lean* rhetoric of flexibility, quality and teamwork is persuasive, critical observers claim that it translates in practice to control, exploitation and surveillance."

According to London and Kenley (2001), Green's argument is reminiscent of the dualist theory of subcontracting, of which the core suggests that economic agents located in different segments of the economy are treated unequally, regardless of their objective worth. Traditionally, it has been argued that behind the prosperity of Japanese industries, particularly in the automotive and electronics sectors, lies the sacrifice of many subcontractors. They are characterized as sweatshops with cheap labor and labor-intensive technologies.

Lean proponents then defended the lean movement with the argument that it is based upon a long history of production management thinking, particularly the physics of production [Ballard and Howell 1999], and that lean thinking simply offers a new way to organize production. Both protagonists are partially correct; lean is dependent upon a long history of production management thinking, however, it is also dependent upon a history of economic, technological, political, industrial relations, and industrial organizational influences. Green, although using emotive language, was accurate in highlighting political, social, moral, and industrial relations evidence that contextualizes the lean movement.

Lean production implementation by large producers would not have been possible had it not been supported by highly organized governance structures in the supply chain. Supply chains were organized into hierarchical clusters of tightly tiered structures of subcontracting firms known as *keiretsu* [Nischiguchi 1994; Hines 1994]. Lean construction researchers, in their quest for production efficiency, in many cases have forgotten that organizing and controlling the market on a very wide and deep scale was instrumental in lean implementation. Lean construction proponents typically borrowed the concept and developed normative models for project integration with little regard for the situational context [London and Kenley 2001].

Japanese economists have typically debated the nature of Japan's subcontracting small enterprises from two perspectives. The first position relies upon the dualist theory, which holds that "big businesses accumulate their capital by exploiting and controlling small businesses which have little choice but to offer workers low pay under inferior working conditions" [Sugimoto 1997]. The prosperity of Japanese industries, particularly in the automotive and electronics sectors, lies with the sacrifice of many subcontractors. The core dualist theory suggests that economic agents, either workers or firms located in different segments of the economy, are treated unequally, regardless of their objective worth.

The second position emphasizes the "vitality, dynamism and innovativeness of small businesses that have responded flexibly to the needs of clients and markets" [Sugimoto 1997]. Nischiguchi (1994), along with Sugimoto (1997), attempts to reject the dualist theory claiming that Japanese business is more complex. Nischiguchi presented empirical evidence of sustainable growth and high asset specificity of the small to medium-sized subcontractor firms within the lean system. He also showed that union membership in Japan has remained the same, and interscale wage differentials between large and small firms are not as marked as some suggest.

Both positions exist and the variation in value orientations and lifestyle of workers is dependent upon the extent of control of the small businesses by the larger companies at the top of the hierarchy. Those who tended to diversify their connections were less controlled and more innovative, participatory and openly entrepreneurial [Sugimoto 1997].

Lean supply, which is supplier procurement, coordination, and development is supported through allied business partners and strategic collaborative partnerships to enable lean production to take place [Lamming and Cox 1995]. Lamming's perspective is in contrast to Porter's (1985) view of the value chain and suppliers, which is often the accepted view that control of suppliers is key to a lean supply chain. Lamming suggested that achieving lean supply is a complex matter because of the nature of competition in markets, as the suppliers are involved simultaneously in several other chains. Jealous guarding of expertise cannot be maintained in the lean enterprise as it requires trust between firms.

Porter's value chain (1985) is a product of understanding the relationships between supply chain actors for the individual organization's gain. Although his work originates from considering industrial organization concepts, it is firmly located within a strategic management perspective from the perspective of an individual market leader. Typically it involves positioning a firm competitively in the marketplace by developing appropriate sourcing and management strategies for suppliers. The strategic management of the supply chain is reinterpreted by Porter (1985) as the management of the "value chain" to achieve competitive advantage. Porter (1985) developed the concept of the value chain as a tool for firms to improve competitive advantage in an industry. He tends towards understanding the supply chain as the relative distribution of power between an individual organization and its suppliers [Porter 1985]. The value chain arises from identifying the discrete activities a firm performs, and then developing appropriate strategies to optimize these activities to position the firm to achieve competitive advantage. Integral to the concept is that it is desirable to purchase from suppliers who will maintain or improve the firm's competitive position in terms of their own products and/or services. The question is how to purchase so as to create the best structural bargaining position.

10.3.2 Portfolio and Project SCM

Construction SCM is not only the management of the construction materials supplier during a project, but it embraces a broader perspective and involves working with organizations outside the boundaries of projects; it will include strategic procurement, project SCM and supplier development, and coordination after projects are completed.

The client is the more likely proponent and beneficiary for the management of the supply chain [London and Kenley 1999; London, Kenley, and Agapiou 1998]. Further to this, therefore, as initiator, the client has a greater stake in effective SCM, whether they directly manage the chain or abrogate the responsibility to first-tier suppliers. Of course large organizations in any industry can equally take a wider and more long-term perspective of various supply chains which they depend upon—just as we have seen in various industries such as automanufacturing. A repetition of project activity, longevity, and a strategic perspective of procurement in the construction industry provides an ability for selected customers/clients to go the next level in an industry, and develop a range of portfolio and project-based activities related to their supply chain in response to a deep consideration of supply markets, an analysis of their own demand, a risk and expenditure analysis, and an organizational audit towards developing a supply strategy [London 2008].

Many different organizational arrangements have evolved including strategic alliancing, network outsourcing, partnering, and joint ventures. These differing arrangements have developed primarily for one or a combination of the following reasons: reduced costs and financial risk; improved innovation (product/process) and reduced risk; entry into new markets, trust and reciprocity in volatile markets.

Early research into such concepts as strategic alliances, serial contracting [Green and Lenard 1999], multiple project delivery [Miller 1999], organizational design [Murray et al. 1999], vertical integration [Clausen 1995], and supply chain procurement [Cox and Townsend 1998], supply chain clusters [Nicolini, Holti, and Smalley 2001], supply chain alliances [Dainty, Briscoe, and Millett 2001], supply chain constellations [London 2001], and supply chain transaction cost economics [Winch 2001] are indications of the awareness of strategic organizational management in construction supply chain research. Much of this research has as the basic premise that to develop better supply chain relationships and sustain an effective supply chain relationship, the interorganizational relationships need to be developed for a portfolio of projects.

Strategic procurement management is the development of an external sourcing and supply strategy designed to maintain a sustainable position for that organization in the total value chain [Cox and Townsend 1998]. Cox and Townsend (1998) proposed the critical asset and relational competence approach to construction SCM which relied upon clients controlling the supply chain. The authors advocated for clients to understand the underlying structural market characteristics of their own construction supply chains, and to develop contingent approaches to procurement based upon this understanding. They considered the UK construction research, based upon lean and agile manufacturing inappropriate because it lacked contextual understanding of the construction industry. They even suggested that:

"It is our view that if the Latham report, and the somewhat naive research industry into automotive partnerships and lean and agile manufacturing processes that it has spawned, had devoted more time to analyzing and understanding the properties of the unique supply chains which make up the complex reality of the UK construction industry a greater service might have been done to value improvement in construction" [Cox and Townsend 1998].

This view was largely based upon their findings reported in the seminal text *Strategic Procurement in Construction: Towards Better Practice in the Management of Construction Supply Chains* [Cox and Townsend 1998]. In this text they reported on the findings of their study where they analyzed the management methods taken by six client organizations. The profiles of these organizations differed and included an international Japanese "design and build" management contractor, one multinational restaurant chain client, two UK international transportation clients, one UK national property developer, and a US multinational client organization that is involved in the development of innovative products.

This critical asset and relational competence model relied upon observing the strategic and operational approaches of six organizations that appeared to indicate better practice construction procurement strategies and methods. The six firms were client organizations and were from the private sector, two whose core business was in the construction industry. The criteria for choosing these six were that they exhibited better practice and that they were private sector clients [Cox and Townsend 1998].

Other researchers have conducted similar case study research on strategic procurement and SCM. Olsson (2000), through a qualitative case study on SCM of a Swedish housing project driven by Skanska and Ikea, highlighted that a conventional construction approach was found to be too expensive to meet particular client demands. Similar to the work of Cox and Townsend, the conclusion of the research was that for SCM to be effective, the construction industry rearrangement of existing structures may be necessary. Both studies focused upon strategic supply relationships of project team members.

Consistently, researchers have concentrated upon a small group of firms and the SCM concept related to an individual project. For example, Clausen's Danish (1995) study also focused upon the key firms in the main construction contract as they evaluated a government program where the government acted as a large client, intended on improving productivity and international competitiveness in the construction industry. The central argument to the program was the "need for vertical integration of the different actors and their functions in the construction process," with the premise that key actors in the process should be involved in strategic decisions at the outset. Therefore, through a tendering process, four consortia were selected to carry out experimental building projects. The core of the consortia include a contractor, an architect, a consulting engineer, and "manufacturing firms and suppliers of materials and components [who] were associated on a more or less permanent basis."

Clausen (1995) determined that the program was much less successful than anticipated, because there was discontinuity in the supply of projects to the consortia; firms were concerned about the financial risk of committing their resources to a single client. The conclusions suggest that the degree of uncertainty in supply of projects and the inherent risk for firms involved is a very important factor in SCM.

10.3.3 Supply Chain Constellation

The interplay between supply and demand, the balance of power or control and incentive, have been considered by others in the form of serial contracting [Green and Lenard 1999], and multiple project delivery [Miller 1999]. In 2001, the supply chain field developed further with strategic procurement management concepts such as work clusters and supply chain integration [Nicolini, Holti, and Smalley 2001], firm constellation of network of alliances [London 2001b], subcontractors and barriers to supply chain integration [Dainty, Briscoe, and Millett 2001], and transaction cost economics and supply chains [Winch 2001] being explored in greater detail.

Nicolini, Holti, and Smalley (2001) reported on two UK pilot construction projects which were organized on a work cluster basis aiming at supply chain integration. Members of the cluster included engineers, architects, subcontractors, suppliers, and the contractor (depending upon which work packages were identified as a work cluster). Traditional roles and hierarchies were challenged as a cluster leader was appointed. An action research approach was undertaken and the benefits of the cluster approach were reported through comments and observations made by participants in the process, and researchers involved in the process in both formal and informal data collection situations.

Although not as deeply entrenched in the ethnographic tradition of research, an equally qualitatively based study on a supply chain constellation indicated the leveraging power that is possible with small to medium-size enterprises when firm to firm procurement relationships are considered outside the boundaries of a single project

[London 2008]. A strategic approach to procuring and managing the supply chain was identified in London's study of very small Australian construction firms that had formed a network of alliances to penetrate international markets. In construction, alliances have been suggested as a form of governance structure to solve procurement issues, as there is a need for improved firm to firm relationships.

Project alliancing has found favor in the construction industry in recent years. In the project alliance the contractual relationships between key firms form for a project; then, when the project is completed, the temporary organization is disbanded as there is no contractual relationship binding the parties. The project alliance has a contractual relationship tied to an individual project, and is largely short term in focus, and includes the client in the relationship.

The network of alliances identified in that case study differs from the project alliance perspective. The perspective of the network of alliances is that the relationships were formed for more than one project and had a longer-term perspective. The case study indicated that there were alliances related to the procurement of individual projects, however they are simply one small part of what was termed a constellation of firms and alliances.

In the constellation, the governance structure is much broader than the usual individual construction project approach. The case study indicated the existence of different types of alliances—learning, positioning, and supply [Gomes-Casseres 1996]. In some cases, all three elements occur within the one alliance. The focus of the case study is not specifically on the type of alliance or a single alliance; rather, it is on describing the context of the constellation of alliances, that is, the relationship between the network of alliances and the strategy for the group.

The network of firms joined by various types of alliances evolved over some 20 years. Over the two decades, the key decision makers in the core of the constellation learnt and reacted to the market environment, and gradually clustered together more and more firms with a variety of contractual relationships. At the time of the study approximately 20 firms were involved in the constellation. The firms were typically classed as small or even micro, employing in some cases two or three people. The constellation was structured according to the strategy to penetrate the international market for affordable housing through using an innovative building product system of prefabrication. As the case study revealed, the affordable housing market is closely allied to the search for markets for innovative building products using waste material. The case study demonstrated some of the conflicts, constraints, and issues that concerned the actors in small companies that are involved in the process. The importance of this study was the focus on describing the variety of firm–firm relationship types, the connection of the relationship types to the business strategy or market, and the description of the cluster of firms that are contractually linked outside individual projects.

10.3.4   Barriers to Supply Chain Integration

Regardless of the government initiatives and debate and discussion on the benefits of SCM, there are significant barriers to supply chain integration by smaller subcontractors and suppliers. Largely, "there remains a general mistrust within the SME companies that make up the construction supply chain and a general lack of belief that there are mutual benefits in supply chain integration practices" [Dainty, Briscoe, and Millett 2001]. The researchers suggested that leading clients should take responsibility for engendering the necessary attitudinal change throughout the supplier networks if further performance is to be realized within the sector.

Much of the strategic organizational-based SCM research is related to governance of contractual relationships. Winch (2001) explored the application of *transaction cost economics* to construction project and supply chains. Transaction cost economics is an economic theory derived to develop the theory of the firm, and is concerned with how the boundary of firm is governed by the attempt to reduce the cost of transacting with other firms. This has connections to industrial organization economics and is explored in greater detail in the following chapter. It is discussed now as it was presented by Winch as a conceptual framework for understanding the governance of construction project processes. As noted by Winch (2001), there were previous attempts to apply this theory over the years [Eccles 1981; Lai 2000; Masten, Meehan, and Snyder 1991; Pietroforte 1997; Reve 1990; Walker and Lim 1999; Winch 1989; Winch 1995]. Previously, transaction cost economics had only been applied to the principal contractor procurement, whereas Winch's contribution was to consider its application across all project procurement relationships and further down the supply chain. The paper was theoretical and proposed that empirical work was required to explore this further.

In all these models and/or empirical observations, construction supply chains are largely viewed from the perspective of the individual firm, situation, or project perspective—perhaps the London small case study on the constellation of firms was beginning to move away from the individually focused organizational perspective. However, this is quite a limited case study.

The type of relationships between organizations is central to supply chain strategic procurement, management, and coordination, however the type of relationships cannot be fully comprehended without an understanding of the markets. Many authors have moved the debate regarding supply chains with respect to: the need for the development of appropriate relationships, the problems of unreliable supply, the different degrees of control between firms, and the difficulties due to the temporary nature of a project-based industry. However, these are all characteristics of the real-world construction industry, and although considered to be barriers to supply chain integration, are still constraints that must be dealt with. These are factors which are typically outside the control of an individual organization regardless of how powerful and significant they are within the industry. Few individual organizations have the capability to make fundamental structural and behavioral change on their own—although "governments" have some degree of capability in this area. Therefore, perhaps what would be more instructive is to take a wider perspective across a sector and consider the supply chains across the sector. This type of approach is considered in the following Section 10.4 Cross-Organizational, and it accepts the characteristics of the real world in the first instance and the inherent structural characteristics of a project-based industry as opposed to a process-based industry, and rather than attempting unachievable, inappropriate, and unrealistic changes to an idealistic model of a supply chain [Cox and Townsend 1998], it seeks to understand how the supply chains "work."

10.4 Cross-Organizational

The industrial organizational (IO) structure is concerned with understanding all the various supply chains that arise within a given sector, and the associated structural and behavioral characteristics which give rise to these different organizational structures. This is perhaps the most wide ranging, and therefore complex approach to studying SCM. The lines blur between the previous research and this research, but the fundamental difference is that the analysis is not taken from the perspective of an individual organization and its supply chains, but also includes the chains of their competitors; largely it is a "helicopter" view of all the various supply chains within the sector with the view to improving the sector performance. If any "organization" is to be considered to have a stake in this type of research, it would largely fall to government agencies, who have a series of mandates to develop construction industry policy, to support and improve and intervene where necessary towards improving the performance of the industry.

The industrial organization methodology is a large body of work and a complete field within itself, and therefore it is not the intent to completely describe the field. It is also noted that it is more far-reaching than simply an attention on the supply chain concept. In summary, it deals with the performance of business enterprises, and the effects of market structures on market conduct (pricing policy, restrictive practices, and innovation), and how firms are organized, owned, and managed. The most important elements of market structure in these models refer to the nature of the demand (buyer concentration, number and size of buyers), existing distribution of power among rival firms (seller concentration, number and size of sellers), entry/exit barriers, government intervention, and physical structuring of relationships (horizontal and vertical integration).

As discussed in greater detail by London (2008), selected supply chain research, published in mainstream management literature, has studied the complex system of supply chains through interorganizational structure. These are important models that merge the field of IO and supply chain theory [Ellram 1991; Harland 1994; Hines 1994; Hobbs 1996; Lambert, Pugh, and Cooper 1998; Nischiguchi 1994]. Ellram (1991) took an industrial organizational perspective which was cognizant of the market concept, albeit from a single organization's ability to manage the supply chain. She suggested types of competitive relationships that firms undertake from transactional, to short-term contract, long-term contract, joint venture, equity interest, and acquisition. These involve increasing commitment on the part of the firms. She described the key conditions under which SCM relationships are attractive according to an industrial organization perspective. The main thrust was that SCM is "simply a different way of competing in the market" that falls between transactional type relationships and acquisition and assumes a variety of economic organizational forms [Ellram 1991]. Her paper analyzed the advantages and disadvantages of "obligational contracting" and vertical integration. She continued by describing the key conditions under which SCM relationships are attractive according to the industrial organization literature. Situations conducive to SCM included:

1. Recurrent transactions requiring moderately specialized assets
2. Recurrent transactions requiring highly specialized assets
3. Operating under moderately high to high uncertainty

What is perhaps more instructive for the industrial organization domain of research is that this work was one of the first discussions to explore the implications of Williamson's transaction cost economic (TCE) theory and industrial organization economics related to SCM. TCE has close ties with industrial organization economics, but it is not the one and same—which it often is mistakenly assumed.

As well as this, transaction cost economics theory has just as many critics as supporters. One of the main criticisms is that it has tended to assume a market and hierarchy dichotomy [Ellram 1991]. Theorists have found it difficult to explain contractual relationships between firms where clearly the transaction costs were high and yet firms did not vertically integrate. There are a variety of institutional arrangements between the two extremes of market versus hierarchy which do not fall neatly into the transaction cost model and clearly demonstrate that markets are not the only way prices are coordinated [Alter and Hage 1993]. However, there is potential for future research relating transaction cost economics to the supply chain movement for the construction industry. Transaction cost economics tends to focus upon individual contractual relationships, whereas supply chain theory aims to understand many interdependent relationships as the unit of analysis.

10.4.1 Network Sourcing

Hines (1994) and Nischiguchi (1994), who are clearly advocates of the lean system of supply, merging the supply chain concept and industrial organization theory, explored the nature of sourcing in the Japanese manufacturing industry and found it was an example of network sourcing. Some of the more significant contributions of their research were the descriptions of the historical, organizational, and economical structure of the Japanese system of supply across automotive and electronics industries. In many ways this has provided a *richer* picture of lean production and supply chains than other writings which portray an apocalyptic posturing of the field's success.

Suppliers are categorized and organized into either specialized subcontractors or standardized suppliers, based upon the degree of complexity of the supply item. It is within the specialized subcontractors that the pyramidal Japanese subcontracting system or the concept of clustered control lies. As Hines and Nischiguchi used the terms networks and clusters interchangeably for the same industrial sourcing scenario in Japan, for the remainder of this discussion, networks and clusters are considered the same.

This system has traditionally been described as a pyramid with an individual assembler corporation at the top and successive tiers of highly specialized subcontractors along the chain, increasing in number and decreasing in organizational size at each progressive stage. More importantly, though, is that each tier would procure, coordinate, and develop the lower tier through supplier associations. Hines (1994), in his seminal text, *Creating World Class Suppliers*, described this procurement as network sourcing. In his comprehensive research on suppliers, Hines (1994) considered network sourcing as one example of a type of buyer–supplier relationship with its origins in what he termed the Japanese School. He suggested that the buyer–supplier relationships can be categorized into three schools based upon the origins of the relationship:

1. Trust school: relationships are primarily due to a complex mix of social and moral norms with technological, economic, and government policies also of some importance. There is, therefore, some suggestion that such approaches may be more difficult, or even impossible, given the set of external and internal factors in the Western world.
2. Partnership school: relationships are developed on the basis of a one-to-one partnership with individual strategic suppliers—with the emphasis on the formal creation of the partnership. This primarily UK model plays down the potential problems of cultural specificity in following an approach designed to form relationships of the type exhibited in Japan [Ellram 1991].
3. Japanese school: which appears to take the middle ground of the above schools, the Japanese school suggests that although conditions are different in the West, a somewhat modified or developed Japanese-style approach can be translated to other cultures. A number of authors describe the route to developing the desired supplier–buyer relationship in this mode [Nischiguchi 1994; Burt and Doyle 1993; Lamming 1993; Hines 1994] and it involves supplier grading, supplier coordination, and development.

The network sourcing model was developed within the Japanese school context [Hines 1996]. A study using data collected on 40 Japanese companies within the automotive, electronics, and capital equipment industries, through semistructured interviews and questionnaires, illustrated the relationship between the 10 causal factors:

1. A tiered supply structure with a heavy reliance on small firms.
2. A small number of direct suppliers, but within a competitive dual sourcing environment.
3. High degrees of asset specificity among suppliers, and risk sharing between customer and supplier alike.
4. A maximum buy strategy by each company within the semipermanent supplier network, but a maximum make strategy within these trusted networks.
5. A high degree of bilateral design, employing the skills and knowledge of both customer and supplier alike.
6. A high degree of supplier innovation in both new products and processes.
7. Close, long-term relations between network members, involving a high level of trust, openness, and profit sharing.
8. The use of rigorous supplier grading systems increasingly giving way to supplier self-certification.
9. A high level of supplier coordination by the customer company at each level of the tiered supply structure.
10. A significant effort made by customers at each of these levels to develop their suppliers.

Nischiguchi (1994) explored strategic industrial sourcing through his exhaustive empirical investigation as part of the internationally acclaimed MIT Motor Vehicle Research Program (MVRP). He suggested the concept of clustered control. This represents a single company network encompassing all the relevant tiers necessary to produce the end product, suggesting a closed system. The common conception of the Japanese subcontracting system has been that it was a simple, tiered, pyramid structure. This system has traditionally been described as a pyramid with an individual assembler corporation at the top, and successive tiers of highly specialized subcontractors, along the chain, increasing in number and decreasing in organizational size at each progressive stage. However, this was challenged. The Japanese system of subcontracting is no longer the closed, highly integrated, pyramidal, and hierarchical structure it used to be. The type of relationship and the forms of collaboration have diversified.

In the early 1990s it was identified that the Japanese structure of subcontracting had become a network system, and it was noted that before long it will become quite an extensive network system [Mitusi 1991]. Hines (1994) enlarged the industry-specific view to look at the wider economy and has suggested even further that, rather than this closed rigid system, the Japanese subcontracting system had moved more towards a structure of interlocking supplier networks. In this system, many firms supply more than one industry sector and potentially operate in different tiers.

Nischiguchi (1987) aggregated the pyramids that were described previously and suggested that the structural formation was similar to a series of mountain peaks. He called this the Alps structure. The Alps structure of supply chains represents a series of overlapping pyramids resembling mountain Alps across an industry where each mountain represented a large assembler [Nischiguchi 1994]. First-tier suppliers supply to many assemblers across the whole industry. This enlarged the industry-specific view to look at the wider economy and suggested that rather than this closed rigid system, the Japanese subcontracting system had moved towards a structure of interlocking supplier networks. In this system, many firms supply more than one industry sector and potentially operate in different tiers; for example, the electronics suppliers operate in a number of sectors. The most significant finding in this assessment was that the structure was not as simple as previously claimed, and that far from the Japanese system of subcontracting being a closed *keiretsu*, as some Western researchers had observed, it was more open. Subcontractors supply to many final assemblers and at higher levels of independence compared to Western automotive firm counterparts.

Much of the supply and industrial network literature builds upon the industrial networks movement of the 1980s [Piore and Sabel 1984]. This body of research has tended to suggest that close-knit interorganizational networks produce superior economic performance and quality, and that there should be a move away from the large, vertically integrated firms [Alter and Hage 1993].

The AEGIS (1999) study in Australia was perhaps the first step towards understanding the construction industry supply chain using an industrial organization economic perspective. AEGIS (1999) through the AEGIS model for the Building and Construction Industry Cluster described the industry as a chain of production, and conceptualized the industry through five main sectors: onsite services; client services; building and construction suppliers; product manufacturers; and fasteners, tools, machinery equipment. Existing statistics were used to describe the sectors in terms of industry income. However, the researchers themselves note that this is contrived, as sufficiently detailed data is not available. The major firms are organized and listed according to groups: suppliers, project firms, and project clients. The discussion focused upon general information about size and turnover, and addressed a market view of some key markets and the major players in the key markets, but does not seem to address the firm or project level of supply chains. It certainly didn't explore in practice the realities of the construction supply chain of contracts, rather tending to describe the industry as clusters of firms.

There have however been various methods developed to describe and classify chain organization at the sector level. According to Hoek (1998), the study of supply chain channel configurations goes back to the 1960s, and there are five general approaches used to study and describe supply chain include: descriptive institutional, channel arrangement classification, graphic, commodity groupings, and functional treatments. This method for categorizing the approaches is derived from a logistics perspective. According to London (2008), in the 1990s two other methods for describing chain organizations were developed including the structural mapping of Lambert, Pugh, and Cooper (1998) and Nischiguchi's (1994) tiers of pyramids. The five approaches noted above are now discussed, followed by the two developed in the 1990s:

1. Descriptive institutional
2. Channel arrangement classification
3. Channel structure: graphic
4. Commodity groupings
5. Functional treatments
6. Structural mapping
7. Pyramid tiers

10.4.2 Construction Supply Chain Economics

London (2004) used the structural mapping approach described by Lambert, Pugh, and Cooper (1998) for individual supply chains towards developing a series of aggregated channel structures in the manner described as the channel structure: graphic [Bowersox and Closs 1996], in an endeavor to describe the industrial organization of supply chains in nine different sectors (glazing, aluminum, steel, concrete, bricks, formwork, fire products, tiles, and mechanical services). This mapping extended typically from client through the various tiers of the chain to the suppliers, to the manufacturers, or materials suppliers. These two approaches are now briefly described.

The structural mapping is described as a network of members and the links between members of the supply chain, and suggests that there are three critical dimensions to the overall organization of the supply chain:

1. Horizontal structure refers to the number of tiers across the supply chain, which is, in effect, the number of different functions that occur along the supply chain, and indicates the degree of specialization; chains may be long with numerous tiers or short with few tiers.

2. Vertical structure refers to the number of suppliers and customers represented within each tier. This reflects the degree of competition amongst suppliers. A company can have a narrow vertical structure with few companies at each tier level, or a wide vertical structure with many suppliers and/or customers at each tier level.
3. Horizontal position is the relative position of the focal company within the end points of the supply chain.

A channel structure: graphic approach is a useful technique to identify the flow of commodities between the various ranges of alternatives in firm types as they are grouped by the similar service they provide in the marketplace. This approach indicates general patterns or the general channel structure. It begins with the producer or materials suppliers/manufacturer and tracks all the different functional types of firms at each tier and the various contractual connections through to consumer or client. For example, a producer may supply directly to a consumer in a particular sector, or it may supply to a retailer who then supplies to the consumer, or it may supply to a wholesaler who then supplies to a retailer who then supplies to the consumer, thus this sector would have three different channel arrangements. This approach maps all the different permutations of supply chains evident in a particular sector, and thus describes the channel structure alternatives. The most useful aspect to this approach is that at an industry or sector level the structure is presented, although at times the simplicity of the aggregation might belie the multiplicity of different channel structures that can occur at the firm level. The advantage of this approach is the ability to show, in a logical sequence, the variety and positioning of firms that participate in ownership transfer in general patterns. The disadvantage is that it tends to understate the complexities of chains for individual firms and their immediate competitors.

This mapping was only a part of the research by London (2004) towards understanding the underlying structure of the various supply chains within a sector—it is coupled with descriptions of the economic characteristics of the markets. The structural and behavioral characteristics of the markets at each tier along the supply chain provide the background to how the chains are organized in the manner in which they are. Procurement decisions at each level in the various tiers provide a key ingredient as to the nature of the economics of the construction supply chain.

"Construction supply chain economics is the manner in which the economic market structural characteristics and the firm and sector behavioral characteristics interact to produce attributes which describe types (classes) of supplier firms, procurement relationships, supply chain industrial organization and supply chain performance. Structural and behavioral characteristics produce:

1. Supplier firm classes which rely upon two key related attributes of commodity significance and countervailing power in the customer–supplier relationship.
2. Procurement relationship classes which rely upon three key interconnecting attributes of formation based upon risk and expenditure, transaction significance based upon control requirements, and then negotiation strategies based upon strategies.
3. Supply chain classes which are reliant upon four key attributes of uniqueness, sector type, internationalization, and fragmentation." [London 2008].

Construction supply chain economics is an important part to understanding the effectiveness of SCM by individual organizations.

10.5   Conclusion

Construction supply chains are characterized by a large number of firms involved in designing, procuring, and assembling construction. In order to improve construction it is needed to address the arrangements and relationships between organizations, and within organizations. This chapter has observed various approaches to understanding and improving organizational issues in the construction supply chains. These observations influence the way in which clients and firms in construction organize production and the supply chain, procure products and materials, innovate technology and processes, and improve products and processes.

In the following chapters of this section, these issues are further discussed and exemplified. In Chapter 11, the uniqueness of construction is discussed, and the application of concepts from other sectors to improve construction. The examples presented demonstrate the benefits for construction to apply alien concepts if carefully managed. In Chapter 12, the management of procurement by construction clients is discussed, and the need for construction clients to have a clear understanding of the market structure to be good procurers. The project-based nature of construction projects supports capital expenditure (CAPEX) decisions, rather than operational expenditure (OPEX) decisions in repetitive relationships. Depending on these two types of decision, construction clients need to consider different relationships for different types of procurement routes. In Chapter 13, an interdisciplinary model and an empirical study is presented focusing on supplier procurement based on merging industrial organization economic theory and object-oriented modeling. Complex core commodity supply chains, such as façades supply chains, require specialist design and construction solutions, and a high level of integration of suppliers. Finally, in Chapter 14, the application of a series of innovation management tools in the construction supply chain is observed on the same three organizational levels presented in this Chapter. The use of these tools and the approach to innovation appear to vary considerably on all three levels. However, in all cases, the aim is to improve the quality of cooperative ties between organizations in the construction supply chain.

References

Alarcon, L. ed. 1997. *Lean construction: Compilation of 1993–1995 IGLC Proceedings.* Rotterdam: Balkema.

Alter, C., and Hage, J. 1993. *Organisations working together,* Vol. 1. London: Sage Publications.

Australian Expert Group for Industry Studies (AEGIS). 1999. *Mapping the building and construction product system in Australia.* Department of Industry, Science and Technology, Canberra, Australia.

Ballard, G., and Howell, G. 1999. Bringing light to the dark side of lean construction: A response to Stuart Green. *Proceedings of the 6th Annual International Group for Lean Construction Conference*, Berkley, CA, 26–28 July, 1999.

Bowersox, D., and Closs, D. 1996. *Logistics management: The integrated supply chain process.* New York: McGraw-Hill.

Burt, D., and Doyle, M. 1993. *The American keiretsu: Strategic weapon for global competitiveness.* Homewood, IL: BusinessOne Unwin.

Clausen, L. 1995. *Report 256: Building logistics.* Danish Building Research Institute, Copenhagen, Denmark.

Cox, A., and Townsend, M. 1998. *Strategic procurement in construction: Towards better practice in the management of construction supply chains*, Vol. 1, 1st ed. London: Thomas Telford Publishing.

Dainty, A., Briscoe, G., and Millett, S. 2001. Subcontractor perspectives on supply chain alliances. *Journal of Construction Management and Economics*, 19, 841–48.

Eccles, R. 1981. The quasifirm in the construction industry. *Economic Behaviour and Organisation*, 2, 335–57.

Ellram, L. 1991. Supply chain management: The industrial organisation perspective. *International Journal of Physical Distribution and Logistics Management*, 21 (1), 13–22.

Gomes-Casseres, B. 1996. *The alliance revolution: The new shape of business rivalry.* Cambridge, MA: Harvard University Press.

Green, S. 1999. The dark side of lean constriction: Exploitation and ideology. *Proceedings of the 7th Conference of the International Group for Lean Construction (IGLC-7)*, (ed.). I. D. Tommelein, 21–32. University of California, Berkeley, CA.

Green, S., and Lenard, D. 1999. Organising the project procurement process. In *Procurement systems: A guide to best practice in construction*, eds. S. Rowlinson and P. McDermott. London: E&FN Spon.

Harland, C. M. 1994. Supply chain management: Perceptions of requirements and performance in European automotive aftermarket supply chains. PhD thesis, Warwick Business School.

Hines, P. 1994. *Creating world class suppliers: Unlocking mutual competitive advantage.* London: Pitman Publishing.

———. 1996. Network sourcing: A discussion of causality within the buyer–supplier relationship. *European Journal of Purchasing and Supply Management*, 2 (1), 7–20.

Hobbs, J. 1996. A transaction cost approach to supply chain management. *Supply Chain Management*, 1 (2), 15–27.

Hoek, R. 1998. Reconfiguring the supply chain to implement postponed manufacturing. *International Journal of Logistics Management*, 9 (1), 95–103.

Kamata, S. 1982. *Japan in the passing lane: An insider's account of life in a Japanese auto factory.* London: Unwin Paperbacks.

Lai, L. 2000. The Coasian market-firm dichotomy and subcontracting in the construction industry. *Journal of Construction Management and Economics*, 18, 355–62.

Lambert, D. M., Pugh, M., and Cooper, J. 1998. Supply chain management. *The International Journal of Logistics Management*, 9 (2), 1–19.

Lamming, R. 1993. *Beyond partnership: Strategic for innovation and lean supply.* Hemel Hempstead, UK: Prentice-Hall.

Lamming, R., and Cox, A. 1995. *Strategic procurement management in the 1990s: Concepts and cases.* London: Earlsgate Press.

London, K. 2004. Construction supply chain procurement modelling. PhD diss. thesis, University of Melbourne.

———. 2008. *Construction supply chain economics.* London: Taylor and Francis.

London, K., and Chen, J. 2006. Construction supply chain economic policy implementation for sectoral change: Moving beyond the rhetoric. *The Construction Research Conference of the Royal Institute for Chartered Surveyors*, London, 6–8 September.

London, K., and Kenley, R. 1999. *Client's role in construction supply chains — a theoretical discussion.* CIB Triennial World Symposium W92, Cape Town, South Africa.

———. 2001. An industrial organisation economic supply chain approach for the construction industry: A review. *Journal of Construction Management and Economics*, 19 (8), 777–88.

London, K., Kenley, R., and Agapiou, A. 1998. Theoretical supply chain network modelling in the building industry. *Proceedings of the 13th ARCOM Annual Conference*, University of Reading, UK.

Masten, S., Meehan, J., and Snyder, E. 1991. The costs of organisation. *Journal of Law, Economics and Organisation*, 7 (1), 1–25.

Miller, J. 1999. Applying multiple project procurement methods to a portfolio of infrastructure projects. In *Procurement systems: A guide to best practice in construction*, eds. S. Rowlinson and P. McDermott. London: E&FN Spon.

Mitusi, I. 1991. *A unique Japanese subcontracting system from a global point of view.* Customer-Supplier Relationship Study Tour, EC-Japan Centre for Industrial Cooperation.

Murray, M., Langford, D., Hardcastle, C., and Tookey, J. 1999. Organisational design. In *Procurement systems: A guide to best practice in construction*, (eds.). S. Rowlinson and P. McDermott. London: E&FN Spon.

Nicolini, D., Holti, R., and Smalley, M. 2001. Integrating project activities: The theory and practice of managing the supply chain through clusters. *Journal of Construction Management and Economics*, 19, 37–47.

Nischiguchi, T. 1994. *Strategic industrial sourcing: The Japanese advantage*, Vol. 1, 1st ed. New York: Oxford University Press.

Olsson, F. 2000. *Supply chain management in the construction industry: Opportunity or utopia? Unpublished licentiate thesis*, Department of Engineering Logistics, Lund, Sweden.

Pietroforte, R. 1997. *Building international construction alliances: Successful partnering for construction firms.* London: E&FN Spon.

Piore, M., and Sabel, C. 1984. *The second industrial divide: Possibilities for prosperity.* New York: Basic Books.

Porter, M. 1985. *Competitive strategy: Creating and sustaining competitive advantage*, Vol. 1. New York: Free Press.

Reve, T. 1990. The firm as a Nexus of internal and external contracts. In *The firm as a Nexus of treaties*, (eds.) Aoki, M., Gustafsson, B., and Williamson, O. London: Sage Publications.

Ross, D. R. 1997. *Competing through supply chain management*, Vol. 1, 1st ed. New York: Chapman & Hall.

Sugimoto, Y. 1997. *An introduction to Japanese society.* Melbourne: Cambridge University Press.

Walker, A., and Lim, C. 1999. The relationship between project management theory and transaction cost economics. *Journal of Engineering, Construction and Architectural Management,* 6, 166–76.

Winch G. 1989. The construction firm and the construction project: A transaction cost approach. *Journal of Construction Management and Economics,* 13, 3–14.

Winch, G. 1995. Project management in construction: Towards a transaction cost approach. Le Groupe Bagnolet Working Paper No. 1, University College, London, UK.

Winch, G. 2001. Governing the project process: A conceptual framework. *Journal of Construction Management and Economics,* 19, 799–808.

11

Construction Supply Chain and the Time Compression Paradigm

Denis R. Towill
Cardiff University

11.1 Introduction

The construction sector is often perceived as "unique" in the range of problems encountered. However, closer examination shows that there is no single problem which is special to the sector: what is certainly true is that there is a unique combination of problems which face the industry. But the fact that every problem is to some extent shared elsewhere does mean that new practices proven in (say) the manufacturing sector may well find profitable application in construction (Evans, Naim, and Towill, 1998). In this paper the theme advanced is "experience transfer" into construction via the total cycle time (TCT) compression paradigm. Here the emphasis is on the reengineering of business processes to reduce the time between customer need identified and customer need satisfied [Towill 1997].[1]

---

[1] A glossary of terms appears in the appendix of this paper.

Some observed characteristics of construction are:

- Tendency to make money out of other "players" alleged mistakes
- Material suppliers cutting capacity leads to higher process costs and longer delivery times
- Uncertainties arise from many sources
- Tendering has become expensive and laborious
- "Players" aim to bid low and then make a profit from penalty payments
- Clients often end up paying up to as much as 50% extra for projects completed late

Fortunately, as George (1996) emphasizes, there are exceptions to the above somewhat gloomy picture, and construction sector market leaders are identified by their ability to:

- Complete projects ahead of time
- Complete projects within budget
- Resist compromises on safety or quality standards

The Innovative Manufacturing Initiative (IMI) is well illustrated in the comparison with housebuilding undertaken by Gann (1996). There is an increasing realization that innovation rather than invention is what gives a company a competitive edge. In this respect, the definition by Arthur Koestler (in his *The Act of Creation*) is highly relevant [Sherwood 1998]. He argues that "Innovation is borne out of the rearrangement of existing component patterns and is not creative in the Old Testament sense. It does not create something out of nothing. Instead it uncovers, selects, reshuffles, combines, synthesizes already existing facts, ideas, faculties, and skills. The more familiar the parts, the more striking the new whole."

Thus, as Jack Trout (1999) has emphasized, something borrowed is simpler to exploit than a new invention. It is therefore likely to be much easier to implement such an analogue. There is also with a corresponding increase in the chance of success. An idea therefore needs only to be original in its adaptation to the "new" problem. Hence, in the search for transferable ideas the benchmarking concept has proved particularly helpful. So in searching for "best practice" it is perfectly acceptable (and indeed actively encouraged) to look at other market sectors. Thus, in the IMI program referred to above, one specific aim is to promulgate such best practice as widely as possible irrespective of market sector of origin. It is the purpose of this paper to demonstrate that the TCT compression paradigm represents such an innovations opportunity for best practice transfer between industries, countries, and market sectors.

11.2　Construction Targets for the New Millennium

It is important to emphasize that in seeking to attain international competitiveness, improvement in one business performance metric must not be sought at the expense of another, otherwise there will be no overall gain to the company. Fortunately, the industry appears to be well aware of this problem. For example, Table 11.1 lists US construction

TABLE 11.1   US Construction Sector Performance Improvement Targets to be Achieved by Year 2003

Construction Sector Performance Metric	Year 2003 Target Improvement	Ranking (*) in Importance
Total project delivery time	Reduce by 50%	First
Operation, maintenance, and energy lifetime costs	Reduce by 50%	Second
Productivity and comfort levels of occupants	Increase by 50%	Fifth=
Occupant health and safety costs	Reduce by 50%	Seventh
Waste and pollution costs	Reduce by 50%	Fifth=
Durability and flexibility in use over lifetime	Increase by 50%	Third
Construction worker health and safety costs	Reduce by 50%	Fourth

Source: From Wright, R.N., Rosenfield, A.H., and Fowell, A.J. in National Science and Technology Council report on federal research in support of the US construction industry, Washington, DC, 1995. With permission.

(*) Ranking in industry importance obtained from White House Construction Industry Report workshop participants representing the residential, commercial, institutional, industrial, and public works construction sectors.

sector improvements targeted for the year 2003. Note the emphasis on delivery time, operating costs, lifetime durability/flexibility, and safe working, and the quest to enable these improvements in parallel. A practical innovative example of what has already been achieved in the United Kingdom is the reduction in construction time for similar supermarkets of 50%. Perhaps even more important is the demonstration that for identical building projects is the parallel reduction in "snags on handover" from 1800 down to 3 achieved via the "partnering" way of working [Johnston et al. 1997]. This demonstrates that substantial reductions in TCT can be accompanied by impressive gains in quality and starts moving construction away from the scenario painted in Section 11.1.

In seeking transfer of relevant best practice between market sectors, it is crucial to critically examine available evidence from a wide variety of sources to ensure that we are dealing with a true paradigm and not a mere transitory "fad" [Towill 1999]. Hence, this paper provides the substantial evidence available on the successful application of the TCT paradigm across other market sectors. This material is introduced ahead of a review of an impressive case study comparing 300 building projects analyzed via a regression model. The results therein suggest that properly reengineered via the TCT paradigm, construction projects deliver business benefits every bit as impressive as previously obtained in the manufacturing and service sectors. Also, quality and safety are *not* prejudiced by proper application of the TCT paradigm, as demonstrated in *Lean Thinking* [Womack and Jones 1996] to which time compression techniques are closely linked.

11.3   Paradigms Old and Paradigms New

According to business process reengineering (BPR) experts Morris and Brandon (1993), the word "paradigm" has received more publicity than understanding, so we need to consider carefully exactly what it means. They quote the dictionary definition of a paradigm "as an example or pattern, especially in outstandingly clear or archetypal one." However, at a practical working level, they expand this definition so that a paradigm:

- Is a set of rules which establish boundaries and describes how to solve problems within these boundaries
- Influences our perception and help us organize and classify the way we look at the world
- Is a model which helps us comprehend what we see and hear
- Can disable objective thinking based on this information
- Operates at the subconscious level
- May be seen as sets of unquestioned, subconscious business assumptions

Morris and Brandon (1993) then describe a paradigm shift as:

- A significant change in the rules, assumptions, and attitudes related to an established way of doing something
- Has the effect of a new beginning; that is, is a new way of doing things

This is the "paradigm found." But often to achieve success an existing paradigm which stands in the way of progress may also need to be paradigms lost. This is because:

- An old paradigm may be held so strongly that it prevents detection and acceptance of the need for change
- Tensions between old and new paradigms can lead to erratic and underperforming experience curves
- There are real dangers in working methods associated with the new paradigm regressing back to the old familiar routines

Morris and Brandon (1993) have also concluded that in BPR terms: a new paradigm is essential if a significant change in business is sought because the future cannot be viewed through old paradigms.

A well-known case of "old" and "new" paradigms concerns the move in the automotive industry from Just-In-Case (large stock levels) towards Just-In-Time (small stock levels). The former paradigm accepts losses both in time and material throughout the delivery system, whereas the latter paradigm seeks to eliminate waste at source, hence reducing uncertainty throughout the system. It is manifest that it is just not feasible to have these two paradigms operating in parallel. The consequence of continuing to operate with conflicting paradigms is likely to result in temporary improvement followed by regression to the original status quo as typified by the classic work in progress (WIP) reduction experiments by Gomersall (1964). Thus, when encouraged to reduce backlog via better working methods, the operators "beat the system" to protect what they considered to be their WIP "comfort level."

The message here is clear. If the TCT compression paradigm is to deliver benefits into the construction sector, it must be understood and actively supported at all levels in the company, from CEO to individual operators.

11.4   What is the TCT Compression Paradigm?

The TCT compression paradigm may be simply expressed as "the principle of reducing the time taken to execute a business process from perception of customer need to the satisfying of that need." Although the benefits of doing things quickly (and well) have been understood for some time, the TCT compression paradigm as a formal mechanism for increasing company performance is of recent origin. Industrialists in the UK, such as Jack Burbidge (1983) and John Parnaby, (1995) were early advocates of the paradigm, but it was left to management consultants, such as Stalk and Hout (1990), to widely publicize the approach. Early successes were reported in electronic products, mechanical engineering, banking, and insurance sectors.

Table 11.2 is an outstanding example of how TCT compression was marketed by the management consultants as a set of slick, easily remembered rules. The associated Stalk and Hout (1990) text is certainly thought provoking (as was intended). It also leaves many questions unanswered for the busy executive seeking to influence change. The contribution by Thomas (1990), and also a consultant, is complimentary to Stalk and Hout, as can be seen from the bar chart of Figure 11.1. Here the order-of-magnitude improvements resulting from a sample of TCT programs are displayed and confirm that subject to proper reengineering, all normal performance criteria are bettered. Thus, the proposition may be advanced that performance improvement via the TCT paradigm is not in dispute. However, the leverage exerted may well be sector-dependent, which is a powerful reason for ongoing research in construction.

Thomas (1990) also established two very important key points associated with TCT compression programs. The first is that the only worthwhile goal is to reduce TCT from customer need right through to customer need satisfied. In other words, it is counter-productive to reduce process times that do not significantly affect TCT (and which also might require as much effort to be expended as doing something really worthwhile). The second key point is that as TCT is progressively reduced, so does the process variance

TABLE 11.2   Time Compression Paradigm—An Outstanding Early Example of the Management Consultant Perspective

No.	Meaning
Rule 1: the .05 to 5 rule	"Most products and many services are actually receiving value for only .05 to 5% of the time they spend in their value delivery systems"
Rule 2: the 3/3 rule	"Lost time is spent in three roughly equal categories. These are waiting for our batch to be completed; queuing for the next process; and management decision making delays"
Rule 3: the ¼-2-20 rule	"Every quartering of the cycle time doubles labor and working capital productivity with costs reductions at up to 20%"
Rule 4: the 3×2 rule	"Companies that cut cycle times in their value delivery systems achieve growth rates three times the national average with profit margins twice the industry average"

Source: From Stalk, G.H. Jr, and Hout, T.M. in *Competing against time: How time based competition is re-shaping global markets*, Free Press, New York, 1990. With permission.

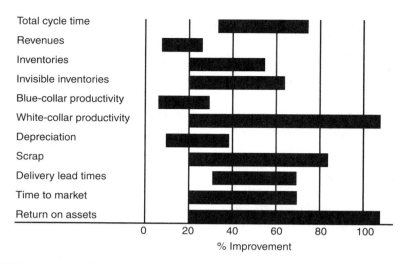

FIGURE 11.1   How reducing TCT paradigm leverages the company "bottom line:" a range of industrials results reports by Thomas (1990). (From Thomas, P.R., *Competitiveness through total cycle time*, McGraw-Hill, New York, 1990. With permission.)

reduce. In other words, not only does a TCT compression program reduce the expected cycle time, but achievement on target is also better guaranteed.

11.5   TCT Results From the Manufacturing Sector

Whilst the opinions of management consultants are clearly of importance in shaping industrial change strategy, to move from "fad" to "paradigm" is in itself an importantt process in which corroborating evidence must be forthcoming from a variety of sources [Towill 1999]. Detailed industrial case studies have a very important part to play, despite the reluctance of companies to release the information or to spare their executives time to write up a meaningful account of the change program. One industrial example is shown in Table 11.3. This summarizes results obtained for a UK Aerospace Actuator Company. They confirm that good reengineering of the product delivery process (PDP) is rewarded by improved performance measured by every business metric; that is, no tradeoff or engineering compromise is required.

A further step in the provision of evidence to support the transition from fad to paradigm is publication of the results of carefully designed industrial surveys. The aim here is to perform independent analyses to establish causal relationships of high statistical significance. Table 11.4 summarizes the results obtained by Schmenner (1988) in a large-scale (of several hundred companies) experiment by surveying three different market sectors, and then testing the results for correlation between cause and effect. Of the 10 factors tested in Table 11.4, only TCT reduction was found to have a significant impact on productivity.

Once evidence such as that portrayed in Table 11.2 through Table 11.4 becomes widely available, it would be expected that take-up of the TCT compression paradigm would

TABLE 11.3 The Manufacturing Industry Case Study Perspective: Typical Results Quoted by Parnaby (1995) on Time Compression Paradigm Applied to an Aerospace Actuator Company

Benchmark	Improvement
Leadtime	Down 75%
Manufacturing costs	Down 75%
Material movements	Down 90%
Inventories	Down 75%
Work in progress	Down 75%
Adherence to schedule	Up 30%
Product "ownership"	Much improved

TABLE 11.4 Industrial Survey Perspective on the Time Compression Paradigm: Results of Large-Scale Review Establishing that Cycle Time Reduction is a Significant Productivity Driver

Factors Tested for Statistical Significance	Significant?
Investment in high technology	No
Setting up gain sharing plans	No
Investment in Class A MRP II systems	No
Operator focused industrial engineering	No
Age of plant	No
Size of plant	No
Global location of plant	No
Degree of union activity	No
Process/nonprocess industries	No
Total cycle time reduction	Yes

Source: From Schmenner, R.W. *Sloan Management Review*, Fall, 1988, 11–17. With permission.

become widespread. This is indeed the case, at least for top management consultants McKinsey and Co., Arthur Andersen, Boston Consultancy Group, Ernst and Young, and ABB. For example, Werr, Stjernberg, and Docherty (1997) demonstrated convergence between these consultants' business process reengineering practices by establishing four shared operational principles. These extremely important shared principles are:

1. A holistic view of organizations leading to a systems model of the enterprise
2. Time as the explicit improvement target reflected in the performance metrics used during the diagnosis of the organization undergoing change
3. A focus on learning especially during the change process followed by competence transfer within the client organization
4. Highly structured methods which seek to identify and support the many steps in the change process

All flow of these principles relate to the TCT compression paradigm. This is because any change program which links customer need to customer satisfaction must take an end-to-end; that is, systems approach. Time as an explicit target is self-explanatory as it is a performance metric which travels unambiguously across company and national boundaries. Also, focus on learning is a necessary prelude to continuous improvement in performance [Thomas 1990]. Finally, having established the need to change the "what" of how we do things, then there is a need for a structured methodology for doing this, including published checklists, software packages, and questionnaires. Good descriptions of suitable methodologies include those provided by Meyer (1993) and Rummler and Brache (1995). The latter is particularly detailed and provides the basis of certification of individual consultants within ABB.

11.6   Importance of Process Mapping

It must be emphasized that attention is focused on a *business process* which spans the range of activities between customer need and that need being satisfied. An example is the single integrated design and construction business process known as T_{40}, and which will be described in Section 11.8. Such a business process may contain many work processes, each of which will contain many tasks (Harrington, 1991; Watson, 1994). In the light of the shared principles listed in Section 11.5, TCT compression programs take a holistic view of organizations. The reengineering is preceded by the creation of a total systems model (usually as a process map) of the business. Full details of a suitable, highly structured mapping approach are given in Evans, Naim, and Towill (1997) together with a construction example, and need not be repeated here. What must be emphasized is that we are concerned with reengineering *how* we do things to the highest possible standard (a procedure usually associated only with *what* we do) using TCT as an explicit performance metric against which alternative designs may be compared. According to Meyer (1993), process analysis involves answering all the following questions:

- From whom is the process input received?
- What is the final deliverable from this process?
- What is needed before the process can start?
- How do we know when the process is finished?
- How long does it take to complete the process?
- To whom is the process output delivered?

Regrettably, many executives want to skip the mapping part of reengineering, wrongly believing that the business process is already fully understood. The same executives then wonder why their reengineering programs fail to deliver the predicted benefits. The truth of the matter is "one fact is worth a dozen opinions" and "one process flowchart contains a dozen facts," so flow charting is an essential and major step in reliable business process modeling [Towill 1997].

For example, a detailed study of an electronics assembly process showed that, at the shop floor level "real work" involving doing things right first time accounted for only 30% of the time available. The traumatic shortfall in this assembly process was largely attributable to unrecognized complexity due to machine and human errors, and rework occurring within the actual process. Unfortunately, these actions within the work processes were not identified in the existing documentation, which wrongly assumed zero defects on the part of both products and people. It therefore represented a false and extremely idealistic view of the assembly process. The message here is that an erroneous model of a business process is potentially every bit as damaging to a company as an erroneous model of an artifact.

11.7   Detail of Time Compression

Once a reliable process flowchart of the business process is available, various practical ideas for reengineering to reduce TCT may be explored. As part of the process analysis, the time taken for each task and each delay is logged. Some process charts are indeed laid out such that the activity may be immediately recognized as essential or otherwise [Scott and Westbrook 1991]. In construction there are three classifications of elapsed time in any process. These are:

1. Value-added time (VAT) during which the product/project is enhanced in value in the direction required by the customer
2. Nonvalue-added time (NVAT) which is wholly waste since it contributes nothing to the product, project, safety, or quality
3. Essential nonvalue-added time (ENVAT) which is regarded as an essential part of the process, such as safety/quality audits

In a nutshell, TCT compression is based on reengineering a business process to eliminate NVAT, and to systematically (and safely) reduce VAT and ENVAT [Towill 1996]. Table 11.5 lists the basic principles by which these ideas may be implemented via elimination, compression, integration, and concurrency. Note that the integration route is concerned with better engineering of business interfaces across all the organizations involved in a project. In particular, this is the opportunity to be seized to banish forever the "functional silo" mentality, which is easier said than done. It should instead be replaced by seamless operation in which all the actors think and act as one in delivering the project on time, on cost, and on quality. In other words, adversarial attitudes need to be eliminated and replaced by a systems approach to total project management [Muir Wood 1996; Lewis 1998].

TABLE 11.5   Basic Reengineering Principles for Taking Time Out of the Value Stream

Time Compression Principles	Correspondence Engineering Procedure
Elimination	Totally remove an unnecessary work process from the value stream
Compression	Eliminate NVA time and/or streamline essential NVA/VA time within the work process
Integration	Eliminate interfaces between successive work processes to streamline material and information flows
Concurrency	Develop ways of executing work processes in parallel, not sequentially

Many of the individual reengineering tools commonly used to effect TCT are already familiar to the construction sector. What is innovative is the particular way they are brought together for the sole purpose of effectively and efficiently removing time from the business process [Towill 1996]. It is convenient to group the tools under the four headings of industrial engineering, IT, product engineering, and operations engineering. Table 11.6 gives simple construction sector examples under each heading. They have all been widely applied within the industry; all we ask here is that as an integrated package they be specifically directed at TCT reduction, since we have seen from the manufacturing sector that *all* performance metrics will then be bettered.

11.8 T_{40} TCT Program

This case study was devised to enable a new industry-wide process for construction. As shown in Figure 11.2, this new process is capable of reducing time to completion of construction by 40% resulting in a consequential costs saving of 25% of the capital value [Ireland 1996]. The T_{40} study concentrated on reengineering and integrating the design and construction business process for a capital facility, such as a building using principles that can be adapted to civil engineering or process plants. An important outcome of T_{40} was substantive evidence that the TCT compression paradigm applies to project-based companies and project-based value streams.

The starting point for the T_{40} project was the knowledge that managing the process of producing a building, a civil engineering structure such as a road or bridge, or an oil refinery involves a similar set of processes. These include the definition of needs, feasibility study, design and detailed specification, and then construction. But while there is similarity between projects, essentially every project is also different and is a prototype, because the site is different, and hence the design is different. The situation is further complicated in the "traditional" way of working with temporary teams assembled specifically for the project, right across the spectrum from architect to the smallest subcontractor. So in theory there is limited opportunity to learn from repeated cycles. Nevertheless, the T_{40} project is aimed at exploiting the practical similarities between projects as the basis for innovation in construction.

TABLE 11.6   Examples of Use of Standard Reengineering Tools in Time Construction Sector Compression Programs

Source of Performance Improvement	Simple Construction Sector Example of Beneficial Change
Industrial engineering	Line balancing of excavation teams for maximum earth removal rate
Information technology	Order transfer via EDI enabling quick and unambiguous information
Product engineering	Standardization for easier assembly both off site and on site
Operations engineering	Just-In-Time, not Just-In-Case deliveries to site

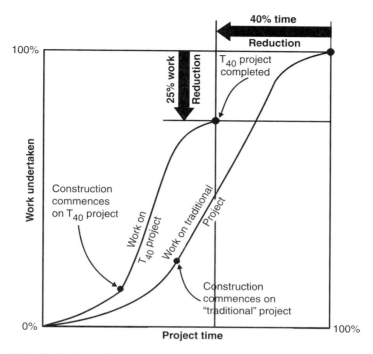

FIGURE 11.2 Work–time relationship as influenced by one-stage design and construction process. (From Ireland, V., *Business Change and Re-Engineering*, 3 (1), 28–38, 1996. With permission.)

A fundamental part of the T_{40} project involved process analysis using the tools already encountered in Section 11.6. These included mapping the "as is" flowchart, highlighting issues for change, identifying the value-added items as judged by the customer, innovating a "should be" flowchart for the redesigned process, analyzing the ability of the proposed changes to meet the program goals, highlighting input–activity–output relationships for all key parts of the process, providing a route map for implementing change, and finally documenting and promulgating the results to get "buy-in" from all the participants. In our experience over a range of industrial sectors obtaining such "buy-in" is probably the critical issue in process reengineering.

According to Ireland (1996), the final integrated design and construction process requires that the "whole solutions team" be involved from the point of determining the customer needs to those needs being satisfied. Specifically this requires the following modus operandi:

- A clear specification of customers' needs, preferably in performance terms
- Acceptance of responsibility by the whole team for the design and construction phases, a significant change from current practice
- Negotiated cost on a particular project, including reference to a third party audit if necessary

- Significantly reduced time performance as established by reference to industry time records
- Single point accountability in the solutions team by the lead contractor

A critical issue is the definition and sharing of risk and rewards so that there is an adequate incentive for members of the solutions group to act as a coherent team. The lead contractor clearly should not have to share rewards without being able to share risks.

11.9 T_{40} Statistical Model

It is the aim of the T_{40} case study to provide a selection system that identifies the best contractor on the basis of the following criteria:

- Past record on time performance, by reference to an industry time database
- Agreed cost based on an adequately developed design
- Third party endorsement of cost as being close to the best price
- Agreed TCT which is significantly better than the industry average for a building of the type (e.g., 40% better time performance which statistically is found to correspond to a 25% reduction in costs)

The approach is based on the fact that time performance on a range of projects can be ranked by reference to a set of performance indicators which are independent of the physical dimensions, building type, and complexity of the building. The industry average performance can then be expressed in terms of a limited number of key variables, such as cost, building type, area, height, and complexity.

The new approach to contractor selection requires an adequate database to be available plus the ability to fit a regression equation linking TCT with other important variables. This was achieved by accessing the 300 building database of the Royal Commission into the Building Industry in New South Wales. The regression equation for industry TCT was based on the following 10 variables:

$1 Cost	$6 Document quality
$2 Building type	$7 Public/private client
$3 Location	$8 Familiarity with technology
$4 Project delivery system	$9 Design complexity
$5 Planning time	$10 Safety index

The result is a performance predictor for proposed projects which is independent of building type and size, and which is claimed can be used to replace the conventional, expensive, and wasteful tendering stage. Table 11.7 lists the most important conclusions to be reached from the T_{40} program. There is thus clear evidence that TCT compression paradigm is applicable to construction. However, the customary warning must be sounded. It is essential to reengineer the process by adopting the understand, document, simplify, optimize (UDSO) routine [Watson 1994]. Jumping in and starting time compression, irrespective of the value added to the customer by the process, is definitely not the way to consistently reduce costs and improve the bottom-line business performance.

TABLE 11.7　Major Conclusions from T_{40} Construction Study

A.	The model
$1	TCT performance can be benchmarked using a 10 variables predictive model
B.	Time cost relationship
$2	Project cycle times can be reduced by 40%
$3	Total costs are thereby reduced by 25%
C.	TCT range
$4	"Best" contractor has TCT 50% less than industry average
$5	"Worst" contractor has TCT twice the industry average
$6	For a given type of building the "worst" contractor takes four times as long as the "best" contractor
D.	About the contractors
$7	Contractors performance consistent from project to project: "best" are always best, "worst" are always worst
$8	"Best" contractors use one stage design and construction process approach

Source: From Ireland, V., *Business Change and Re-Engineering*, 3 (1), 28–38, 1996. With permission.

11.10　Barriers to Change

In an earlier section, we discussed old paradigms inhibiting change. This is usually the result of people problems arising from fear or ignorance. Such difficulties are met irrespective of the market sector, as instanced by Andraski (1994) in the retail chain, and by Belk and Steels (1998) in pharmaceuticals. The latter change program which resulted in a TCT reduction of 77%, and consequently much better customer service, was initially opposed by a staff disbelief that change was necessary, coupled with a rearguard action by "players" to try and protect functional empires; that is, maintain the silo mentality already met in Section 11.7.

T_{40} also met problems in implementation which may be regarded as either unique to construction or of greater significance in construction than elsewhere. These include [Ireland 1996]:

- The need to reduce the cost of a service the first time the process is reengineered
- The challenge of implementing change in projects which typically combine the resources and hence the collaboration of at least 40 "players" to deliver the completed facility
- Craft traditions opposing new ways of working

In the view of the present author, the first of these is in fact the most inhibiting. We have already seen in manufacturing that a real impact on business profitability may take several years before a reliable trend is established [Towill and McCullen 1999]. Furthermore, winning at an international level of competitiveness may require a succession of change programs as new opportunities (especially in information technology)

TABLE 11.8   Seven Rules for Successful Change Management into the Time Compression Paradigm

Rule	Action Required
1: Strategy	Relate process improvement to business strategy
2: Involvement	Involve the right people in the right way
3: Accountability	Give task forces a clear brief and the necessary accountability to achieve it
4: Effectiveness	Do not confuse endless reorganization with effective reengineering
5: People	Understand how process changes affects people
6: Implementation	Always focus on successful implementation
7: MOPs	Ensure task force leave effective monitoring systems in place

Source: From Rummler, G.E., and Brache, A.P. in *Improving performance: How to manage the white space on the organisation chart*, 2nd ed, Jossey-Bass, San Francisco, 1995. With permission.

present themselves. For example, in the reengineering of an electronics products supply chain, four discrete programs may be identified. Although there is limited overlap, these programs are largely engineered sequentially over an eight year period (i.e., approximately two years/program). The result is a level of performance improvement so great that it would be beyond the comprehension of various "players" as forecast at the start of the initial program [Berry, Naim, and Towill 1995]. No doubt this feeling is reciprocated in the poor performers in the T_{40} project.

Given that any change program requires determined and dedicated management if it is to deliver expected benefits to the business, what are the precautions that need to be taken during reengineering? Table 11.8 is a summary of the seven rules devised by Rummer and Brache (1995), who are very experienced and successful business process engineers. They are much expanded in the original reference and mirror the experiences of the present author. Although initially developed for reengineering in manufacturing and related sectors, the rules can be recommended with confidence for transfer to construction.

Basically this is a list of key "enabling" actions which must involve all "players" in the program. Thus, process mapping and process redesign are seen as merely necessary, and far from sufficient conditions for change. In particular, the focus has to be on implementation and not on just the earlier phases of process mapping and documentation. Essential though the latter are as an enabling mechanism, they are but one step on the path to change. However, the company should start seeing some benefits once simplification is enabled prior to full-blown process enhancement. But early payback must not be interpreted as a signal to ease off the change program: the best companies accept the long-haul nature of effective change reengineering mechanisms. It is working hard to implement these rules that puts the business in a position to win, thus minimizing the downside potential well described in Green (1998).

As suggested by Rummler and Brache (1995), the T_{40} case study has also provided suitable measures of performance (MOPs) to monitor the reengineering business process. Those shown in Table 11.9 refer to the internal fitting-out stage of the building [Ireland 1996], and are transparent, unambiguous, and easy to measure. They may therefore be regarded as a good construction sector example of a monitoring system set in place

TABLE 11.9 Simple MOPs to be Monitored as Part of Change Program Towards One Stage Design and Construction

Project Type (MOPS)	Traditional Way of Working	One Stage Design and Construction
Number of discrete activities	32	9
Number of subcontractors	11	5
Number of site visits required	25	9

Source: From Ireland, V., *Business Change and Re-Engineering*, 3 (1), 28–38, 1996. With permission.

by the task force responsible for reengineering the business process. Manifestly, MOPs must be seen as an integral part of managing the change program, which in this case moves from "traditional" to "partnering" mode of operation.

11.11   Conclusions

It has taken a decade for the TCT compression paradigm to graduate from "fad" to a widely accepted principle upon which to build highly successful business process reengineering programs. At the same time it is clear that such programs need the same high standards of engineering to be applied to the "how" we do things; that is, the business process, as we have traditionally applied to the "what" of the business. TCT compression requires good analysis of process requirements, innovation at the process design stage, good vision and planning, and finally sound execution, and effective monitoring.

The initial reported success of TCT compression has been in the manufacturing and similar sectors over a time span of at least a decade. The available evidence is strong, forms part of the process of moving from "fad" to "paradigm," and confirms that the TCT compression paradigm is a valid part of management theory. That is, it is transferable between companies, between industries, and between countries [Micklethwait and Woolridge 1996]. Also, when properly engineered, the paradigm delivers improvements across all the bottom-line performance metrics. It therefore avoids the difficult world of trading off one area improvement against some downside consequence elsewhere in the business. This was illustrated via a reported huge reduction on faults at handover accompanying the 50% TCT compression in supermarket construction.

Finally, the T_{40} project has shown that the paradigm can be applied with equal success in the project-based industries. The same ground rules apply as in manufacture. Unless the business process is reengineered and implemented to a high standard, then the effort is likely to prove abortive. Far from the application of the TCT paradigm producing watered-down benefits in construction, the T_{40} results suggest equality, and possibly even superiority in potential gains to the "players" involved. It may well be the case that "process" concepts fit very naturally into the construction sector. Thus, once the "buy-in" and "functional silo" problems have been overcome, we can expect to see the TCT paradigm deliver innovative change programs in construction which can be used as exemplars of best practice for other sectors to follow. These include such important areas as project management, in which some market sectors have much scope for improvement and would welcome a lead from construction.

Acknowledgment

This chapter appears by kind permission of Construction Management and Economics, and was first published in Vol. 21, pp. 581–91, 2003. The Journal Website is http://www.tandf.co.uk/journals.

Appendix

Glossary of terms useful in business systems engineering applications in the construction sector:

Benchmarking
The continuous process of measuring products, services, and practices against companies renowned as industry leaders. It is essential to benchmark business processes against "best in class" irrespective of the market sector in which the leader operates.

Business process
A linked and natural group of skills and competencies which start from a set of customer requirements and delivers total product or service. It is a key concept in achieving internationally competitive performance.

Business process improvement (BPI)
The means by which an organization can achieve continuous change in performance as measured by cost, delivery time, service, and quality. It is usually driven by empowered work process teams as part of the normal brief to improve productivity as part of their everyday duties.

Business process reengineering (BPR)
The means by which an organization can achieve radical change in performance as measured by cost, delivery time, service, and quality via the application of the systems approach which focuses on a business as a set of customer-related core business processes rather than as a set of organizational functions. It is usually driven by a multidisciplinary task force, seconded for this purpose, and charged with analysis, design, and implementation of the change program.

Business systems engineering (BSE)
BSE is the systems approach to designing new business processes and redesigning existing business processes. It provides a structured way of maximizing both customer value and the performance of the individual business, hence exemplifying a win–win scenario. BSE has available a proven tool kit for business process analysis, which is an essential step in both BPI and BPR.

Empowerment
A means of managing a process whereby the personnel working on the process are also charged with continuously improving its performance. It is an essential part of any learning organization.

Learning organization

The learning organization is an environment where people continually extend their capacity to create new ways of thinking processes that lie behind decision making throughout the business. It uses its ability to learn faster than its competitors in sustaining competitive advantage. This is coupled with a determination to regard every completed task as an opportunity to identify and expand those elements that constitute "best practice."

Partnership

A generic term which is used to describe various ways in which businesses within the supply chain work together as part of an extended organization or family with the objective of greatly improving the competitive advantage of the whole chain.

Process

Any activity or group of activities that takes an input, adds value to it, and provides an output to an internal or external customer. Processes use an organization's resources to provide definitive results on behalf of the business.

Supply chain

A system whose constituent parts include material suppliers, design agencies, production facilities, distribution services, commissioning teams, and customers linked together via the feedforward flow of materials and products, and the feedback flow of information.

Systems approach

A system is the grouping together of parts that operate for a common purpose. In the systems approach the focus is on the behavior of the total enterprise, and the design, operation, and interfacing of the constituent parts so as to achieve best total competitive performance. It is particularly concerned with the streamlining and illumination of functional interfaces.

TCT

A major philosophy adopted in many BPR programs, which focuses on the total time required by the existing business process to satisfy a customer need from the initial enquiry. Because cycle times are relatively easy to predict and measure, they are powerful metrics in planning and achieving change. Consequently, TCT is alternatively known as time-based management (TBM).

References

Andraski, J. C. 1994. Foundations for a successful continuous replenishment programme. *International Journal of Logistics Management*, 5 (1), 1–8.

Belk, K., and Steels, W. 1998. Case study ATS BERK – From arbitration to agility. *Logistics Information Management*, II (2), 128–33.

Berry, D., Naim, M. M., and Towill, D. R. 1995. Business process re-engineering an electronics supply chain. *Proceedings of IEE Science, Measurement and Technology*, 142 (5), 395–403.

Burbidge, J. L. 1983. Five golden rules to avoid bankruptcy. *Production Engineer*, 62 (10), 13–14.

Evans, G. N., Naim, M. M., and Towill, D. R. 1997. Process costing—The route to construction re-engineering. *Proceedings of the Mouchel Centenary Conference*, Civil-Comp Press, Edinburgh, 153–62.

Gann, D. 1996. Construction as a manufacturing process; similarities and differences between industrialised housing and car production. *Construction Management and Economics*, 14, 437–50.

George, B. V. 1996. *A statement on the construction industry*. London: Royal Academy of Engineering.

Gomersall, E. R. 1964. The backlog syndrome. *Harvard Business Review*, September–October, 105–15.

Green, S. D. 1998. The technocratic totalitarianism of construction process improvement: A critical perspective. *Engineering Construction and Architectural Management*, 5 (4), 376–86.

Harrington, H. J. 1991. *Business process improvement*. New York: McGraw-Hill.

Ireland, V. 1996. Case study; T40 radical time production in construction projects. *Business Change and Re-Engineering*, 3 (1), 28–38.

Johnston, C., Lorraine R., Giddings, A., and Robinson, C. 1997. Partnering the team—Changing the culture and practices of construction. CPN Workshop Report,—733L, UK.

Lewis, J. P. 1998. *Mastering project management—Applying advanced concepts of systems thinking, control and evaluation, and resource allocation*. New York: McGraw-Hill.

Meyer, C. 1993. *Fast cycle time: How to align purpose, strategy and structure for speed*. New York: Free Press.

Micklethwait, J., and Woolridge, A. 1996. *The witch doctors—What the management gurus are saying, why it matters and how to make sense of it*. London: Mandarin Books.

Morris, D. C., and Brandon, J. S. 1993. *Re-engineering your business*. New York: McGraw-Hill.

Muir Wood, A. 1996. Systems engineering and tunnelling—A new look. *Gallerie*, 50, 16–24.

Parnaby, J. 1995. Systems engineering for better engineering. *IEE Engineering Management Journal*, 5 (6), 256–66.

Rummler, G. E., and Brache, A. P. 1995. *Improving performance: How to manage the white space on the organisation chart*, 2nd ed. San Francisco: Jossey-Bass.

Schmenner, R. W. 1988. The merit of making things fast. *Sloan Management Review*, Fall, 11–17.

Scott, C., and Westbrook, R. 1991. New strategic tools for supply chain management. *International Journal of Physical Distribution and Logistics Management*, 21 (1), 23–33.

Sherwood, D. 1998. *Unlock your mind: A practical guide to deliberate and systematic innovation*. New York: Gower Press.

Stalk, G. H. Jr, and Hout, T. M. 1990. *Competing against time: How time based competition is re-shaping global markets.* New York: Free Press.

Thomas, P. R. 1990. *Competitiveness through total cycle time.* New York: McGraw-Hill.

Towill, D. R. 1996. Time compression and supply chain management. *Supply Chain Management,* 1 (1), 15–27.

Towill, D. R. 1997. Successful business systems engineering. *IEE Engineering Management Journal,* 7 (1 and 2), 55–64 and 89–96.

Towill, D. R. 1999. Management theory—Is it of any practical use? *IEE Engineering Management Journal,* 9 (3, June), 111–21.

Towill, D. R., and McCullen, P. 1999. The impact of an agile manufacturing programme on supply chain dynamics. *International Journal of Logistics Management,* 10 (1), 83–96.

Trout, J. 1999. *The power of simplicity—A management guide to cutting through the nonsense and doing things right.* New York: McGraw-Hill.

Watson, G. H. 1994. *Business systems engineering.* New York: John Wiley & Sons.

Werr, A., Stjernberg, T., and Docherty, P. 1997. The functions of methods of change in management consultancy. *Journal of Organisational Management Change,* 10 (4), 288–307.

Womack, J. P., and Jones, D. T. 1996. *Lean thinking.* New York: Simon and Schuster.

Wright, R. N., Rosenfield, A. H., and Fowell, A. J. 1995. National Science and Technology Council report on federal research in support of the US construction industry. Washington, DC.

12

Strategic Management of Construction Procurement

Andrew Cox
University of Birmingham

12.1 Managing Repeat and One-Off Games in Construction: The Capital Expenditure (CAPEX) and Operational Expenditure (OPEX) Procurement Context

There has been a great deal of advice for mangers in the construction industry about what constitutes best practice in buying, selling, and relationship management. In essence, most of this advice has been predicated on the view that relationships are optimized if buyers and sellers seek win–win outcomes [CII 1989, 1991, 1994; Bennett and Jayes 1995;

1998; Hellard 1995; Godfrey 1996; Hinks, Allen, and Cooper 1996; Holti and standing 1996; Koskela 1997]. This advice has been particularly marked in reports sponsored by governments that have been focused on improving the competitiveness of the industry [Latham 1994; DETR 1998]. This chapter demonstrates that such advice is, at best, only partially accurate, and at worst, a recipe for disaster—and especially for naive buyers untutored in the realities of construction relationship management.

The first requirement for anyone seeking to provide advice on the most appropriate way in which to buy construction goods and services, is that they understand the demand and supply circumstances within which any exchange relationship between a buyer and supplier is likely to operate—both pre- and postcontractually. This is another way of saying that all buying and selling in construction is in fact a series of games, in which buyers and sellers seek to optimize their returns. If this is so—and if we recognize that games can be short term and never (or rarely) repeated, as well as long term and with degrees of repeatability—then advice to buyers of construction goods and services must explain what are the most appropriate ways of managing in these one-off and repeat games.

This implies that to think strategically about construction procurement requires buyers to understand how to manage appropriately under very different pre- and postcontractual demand and supply—and therefore power and leverage—circumstances. What may be appropriate for a buyer in one scenario may therefore not be appropriate for the same buyer in another. To think—as the authors of government reports seem to do—that there can only be one best way of managing in construction (and that is to adopt long-term collaboration based on win–win outcomes) is patently absurd. This is because construction normally involves many more one-off than it does repeat games; and it is only in repeat games that long-term collaborative approaches to relationship management are appropriate and effective.

The truth of this distinction between one-off and repeat games is clearly well understood by companies that buy construction goods and services on a regular basis. Most major organizations buying construction goods and services make distinctions between CAPEX and OPEX requirements. CAPEX requirements refer to the initial project management side of construction sourcing; while OPEX requirements refer to the continuous operational expenditure after a project has been completed. It does not take much reflection to realize that most—although for some buying organizations not all—CAPEX expenditure tends to be one-off expenditure; while OPEX expenditure tends—although again not always—to be more of a repeat game. It follows, therefore, that a strategic approach to construction procurement requires that organizations understand the most appropriate ways to manage, not only in repeat and one-off games in general, but also in repeat and one-off games with different types of frequency, scale, and scope.

Buyers in construction have to source a wide range of goods and services, either for a major capital project, or for the maintenance of the operational activities of their organization. In making choices about the most appropriate ways to manage these relationships, buyers have to think not only about the demand and supply and power and leverage circumstances that exist pre and post-contractually, but also about the right balance between insourcing (make) or outsourcing (buy) for particular aspects of their CAPEX and OPEX requirements. If they decide to outsource, then they will also have to consider which of the six major approaches to outsourced relationship management is the most conducive given their CAPEX and OPEX requirements [Cox and Townsend 1998; Cox et al. 2004; Cox, Ireland, and Townsend 2006]. These six sourcing and six relationship management options are discussed in more detail in the next section.

12.2 Six Sourcing Options for Buyers in Construction

In what follows, the basic sourcing options available to organizations as they source supply requirements are discussed theoretically. The theoretical discussion is relevant to buying decisions for all types of organizations and is not specific to the construction industry. The relative appropriateness of these six sourcing and six relationship management options, in the context of the demand and supply and power and leverage circumstances that exist in construction, is discussed later.

The discussion that follows focuses first on the fact that an organization with a high volume and frequent demand for CAPEX and OPEX requirements must always consider whether insourcing is a more attractive proposition than outsourcing. This means that sourcing supply internally or through partial degrees of joint ownership is a perfectly acceptable sourcing arrangement if the frequency and scale of what is required makes outsourcing a less attractive value for money proposition than insourcing [Lonsdale and Cox 1998]. If, on the other hand, insourcing is a less attractive value for money proposition than outsourcing, then organizations have to think about the relative attractiveness of four theoretically different ways of undertaking external sourcing. The four basic approaches to sourcing that can be defined theoretically are differentiated on the basis of the level of involvement that buyers and suppliers can have with one another (reactive and arm's-length or proactive and collaborative), as well as the nature and degree of the buyer's involvement in developing the supplier and their suppliers own competencies (at the first tier and/or within the supply chain(s) as a whole). Each of the six broad sourcing options that must be considered as part of a strategic approach to procurement management is discussed below.

12.2.1 Insourcing

It is often overlooked by those developing their approach to strategic procurement management that insourcing or internal supply is often a viable and sensible option. In today's world, where there is something of a fad in favor of outsourcing and concentrating on the core competencies of the firm, many organizations fail to understand that some assets (resources) are critical to business success and should be retained in-house [Cox 1997]. This is normally critical for two categories of requirement—those activities that provide a competitive advantage/differentiation against competitors, and those requirements (which may not provide for a competitive advantage/differentiation) that cannot be safely managed externally without a serious risk of postcontractual lock-in and dependency, leading to a power and leverage situation of supplier dominance.

12.2.2   Joint Ventures

Sometimes an asset may be deemed to be critical to the competitive advantage/differentiation of an organization, but the organization does not possess all of the capabilities to insource on their own. In such circumstances, organizations may seek to create a competitive advantage/differentiation through the signing of a joint venture agreement with an organization that possesses the capabilities it does not. This is a hybrid form of insourcing/outsourcing, since the buying organization is sourcing externally a unique resource/asset in order to close the market for this valuable capability by jointly owning and controlling an external supply resource. Obviously, under such an arrangement both parties are on the same side operationally and commercially because they both have an ownership stake in the joint venture organization. The relationship will necessarily be close and highly collaborative.

If neither insourcing nor joint venture arrangements provide an attractive value for money proposition, it is likely that an organization will opt for the outsourcing of their requirements from external suppliers. As Figure 12.1 demonstrates, there are four basic theoretical sourcing approaches available to buyers when they outsource [Cox et al. 2003].

12.2.3   Supplier Selection

The most common form of external sourcing for all types of organization is supplier selection. This form of external sourcing is based on relatively short term and arm's-length contracting relationships, with the buyer selecting from amongst relatively competent suppliers in the market, who individually decide what will be their supply offering

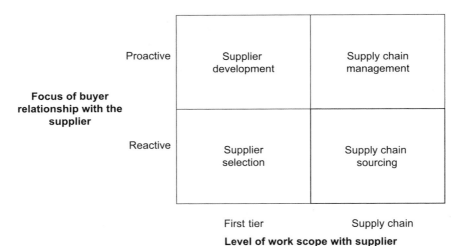

FIGURE 12.1   The four external sourcing options for buyers. (From Cox, A. et al., *Supply chain management: A guide to best practice*, Financial Times/Prentice Hall, London, 2003. With permission.)

to the buyer. The role of the buyer is to select from the available suppliers on the basis of the currently perceived best value for money offering in the market. The relationship is short term and arm's length because there is no basis for the buyer or the supplier to collaborate. This is the standard approach adopted by construction clients going to the supply market infrequently and/or on a one-off basis.

There is nothing wrong with this approach—if it is undertaken effectively—because regular market competition can stimulate innovation. In such circumstances, short-term and arm's-length relationship management approaches based on regular supplier selection techniques can be used continuously to test the market to assess whether incumbent or alternative supplier performance has improved.

Under supplier selection a competent buyer seeks to operate in a commercially opportunistic, but operationally reactive (arm's-length) way with suppliers. The buyer passes basic product or service specification, volume, and timing information to the supplier. The supplier develops its own competence without significant buyer involvement in the process of supply innovation. In this way of working, great emphasis will normally be placed by both the buyer and the supplier on comprehensive clause contracting (either lump sum or reimbursable), with terms and conditions rigorously delineated and enforced pre- and postcontractually.

12.2.4   Supply Chain Sourcing

This reactive way of working can be extended beyond the first-tier relationship with the supplier into the tiers of the supply chains that service the first-tier supplier's offering to the buyer. The buyer uses exactly the same reactive sourcing techniques as outlined above, but now selects from amongst suppliers in as many of the tiers in the supply chain as possible.

At each tier in the chain, the buyer selects from the currently competent suppliers on the basis of the best available value for money offering. The buyer's role is still essentially reactive and at arm's length operationally, providing basic contractual information, and relying on suppliers in the chain to provide innovation through competition. This option is more time and resource-intensive for buyers than supplier selection because they have to incur transaction costs for search, selection, and negotiation with suppliers throughout the chain rather than just at the first tier.

12.2.5   Supplier Development

A buying organization can adopt a far more operationally proactive (collaborative) approach in which the relationship, rather than being arm's length, becomes more long term and both parties collaborate extensively. Under supplier development approaches, both the buyer and supplier make dedicated investments in the relationship, create technical bonds, develop cultural norms to guide the way they work together, and also make relationship-specific adaptations in order to improve the value for money offering for the buyer by working towards the achievement of "stretch" (i.e., beyond current supply market capability) improvement targets [Cannon and Perreault 1999].

Under this sourcing option, the buyer offers the supplier a long-term relationship in return for a commitment from the supplier to provide greater transparency over their input costs, margins, and production techniques (and vice versa), in order to create innovation beyond that which the general supply market can provide. This approach is extremely resource-intensive for both the buyer and supplier when compared with more reactive sourcing approaches.

12.2.6 Supply Chain Management

Supply chain management involves the buying organization working proactively and undertaking the supplier development activities described above, not only at the first tier of the supply chain, but also at all of the stages in the chain, from first tier through to raw material supply. This sourcing option is the most resource-intensive and time-consuming of all for buyers and suppliers, but if it can be implemented successfully, it is capable of providing significant improvements in value for money throughout a supply chain.

There is little doubt that this extended network of proactively focused and highly collaborative buyer and supply relationships has been implemented quite success-fully in significant parts of the automotive, retail, and in some other manufacturing supply chains [Womack and Jones 1996]. The problem is, however, that just because some organizations can achieve this, does not mean that everyone else can. In practice, research in construction shows that very few organizations are, or ever will be, in a posi-tion to undertake either the supplier development or supply chain management options described here [Cox and Townsend 1998; Cox et al. 2004a, 2004b; Ireland 2004]. There are two primary reasons for this.

Supply chain management is by far the most resource-intensive sourcing option for both buyers and suppliers in the chain. It requires the continuous acceptance of high levels of operational and commercial collaboration, and also extensive transaction costs for supplier development work throughout the chain. This undertaking is not possible for very many organizations because they lack the internal capabilities to undertake the work, and the frequency and volume of demand to be able to make the necessary long-term commitments to suppliers—this is a particular problem in the construction industry. Research has also shown that supplier development and supply chain manage-ment souring options tend to work best in power and leverage circumstances of *buyer dominance* and/or *interdependence* [Cox 1999a; Cox, Sanderson, and Watson 2000; Cox et al. 2002, 2003, 2004a].

The organizations that have been able to adopt these types of proactive sourcing approaches tend therefore to have regular and very high levels of volume for fairly stand-ard demand requirements, and they operate in supply markets characterized by high levels of market contestation with relatively low switching costs. This is not a place in which many buyers in construction markets and supply chains find themselves very often. This means that recommendations by the authors of government-sponsored reports for all buyers and suppliers to embrace the supply chain management practices endemic in the automotive sector are likely to have only limited strategic value for most (if not all) participants in the construction industry [Latham 1994; DETR 1998]. This is because the underlying demand and supply, and power and leverage circumstances that support proactive sourcing options (like supplier development and supply chain management) only rarely exist in the industry.

In construction there appear, therefore, to be very few public or private sector organizations that have the internal competence, and a sufficiently conducive external environment, to develop proactive sourcing approaches. This is not to say that no one can adopt these approaches, but that the number is likely to be small. This is because, even when a sufficiently conducive external environment exists (regular standardized and high volume demand with willing suppliant suppliers), there is often a lack of internal competence to develop the effective operational and commercial leverage to make this approach work effectively. This seems to be particularly true in the public sector, where many of the preconditions for effective proactive sourcing exist. There are, of course, private sector organizations—notably in the financial services, house building, oil and gas, utilities, and property management sectors—where proactive sourcing approaches are feasible in construction.

Given this, competence in strategic construction procurement cannot be judged solely on whether or not organizations have adopted proactive sourcing approaches. On the contrary, competence requires that organizations know about all of the six potential sourcing approaches available to them, and also understand which of these is the most appropriate (i.e., provides the best value for money option) given the circumstances they are in. To achieve this level of competence it is necessary to know what are the six souring options, but this must also be linked with knowledge of both power and leverage circumstances, and of the relationship management styles that can be used to manage particular sourcing approaches effectively. Finally, all of these elements must be brought together in order to align a particular sourcing approach with a specific power and leverage circumstance, using the appropriate relationship management style. As we shall see, by doing this, six relationship management approaches can be delineated, each of which may be appropriate as part of a strategic approach to construction procurement.

12.3 Understanding Power and Leverage in Buyer and Supplier Exchange

The only way in which a buyer or a supplier can understand whether or not it is possible to undertake reactive or proactive sourcing is through an understanding of the power regime that exists within a particular supply chain network [Cox, Sanderson, and Watson 2000, 2001; Cox et al. 2002, 2003]. Figure 12.2 and Figure 12.3 show how, based on the use of the buyer and supplier power matrix, the power and leverage situation in a supply chain power regime can be understood. If supplier development and supply chain management sourcing approaches are only really effective in situations of buyer dominance (>) and interdependence (=), it will be obvious from the hypothetical power regime presented in Figure 12.3 that supply chain management is not really possible in this power regime as a whole. While supplier development activities may be possible in certain tiers of the supply chain (those marked with the symbols > or =), for the majority

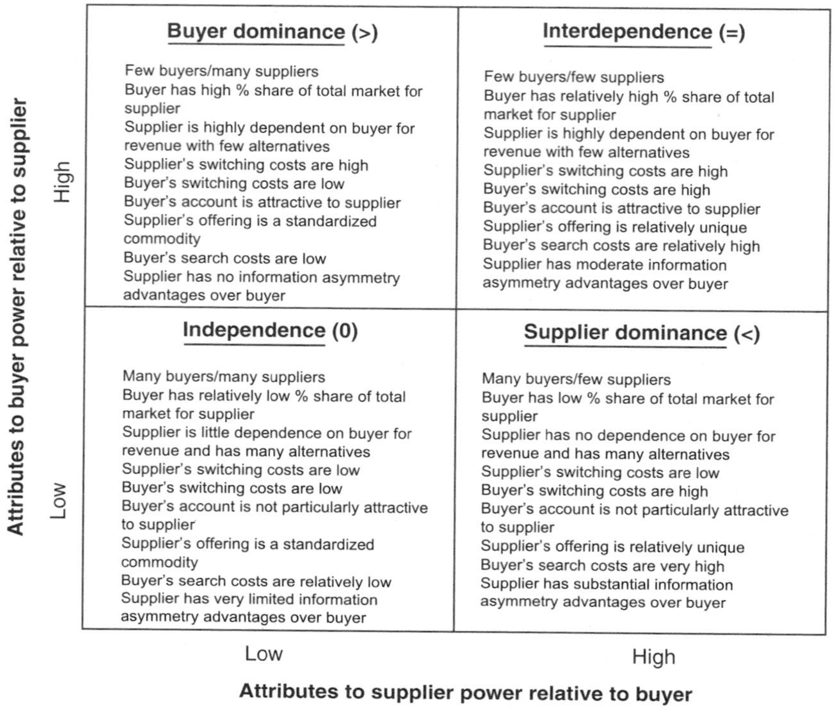

Attributes to buyer power relative to supplier		Buyer dominance (>)	Interdependence (=)
	High	Few buyers/many suppliers Buyer has high % share of total market for supplier Supplier is highly dependent on buyer for revenue with few alternatives Supplier's switching costs are high Buyer's switching costs are low Buyer's account is attractive to supplier Supplier's offering is a standardized commodity Buyer's search costs are low Supplier has no information asymmetry advantages over buyer	Few buyers/few suppliers Buyer has relatively high % share of total market for supplier Supplier is highly dependent on buyer for revenue with few alternatives Supplier's switching costs are high Buyer's switching costs are high Buyer's account is attractive to supplier Supplier's offering is relatively unique Buyer's search costs are relatively high Supplier has moderate information asymmetry advantages over buyer
		Independence (0)	**Supplier dominance (<)**
	Low	Many buyers/many suppliers Buyer has relatively low % share of total market for supplier Supplier is little dependence on buyer for revenue and has many alternatives Supplier's switching costs are low Buyer's switching costs are low Buyer's account is not particularly attractive to supplier Supplier's offering is a standardized commodity Buyer's search costs are relatively low Supplier has very limited information asymmetry advantages over buyer	Many buyers/few suppliers Buyer has low % share of total market for supplier Supplier has no dependence on buyer for revenue and has many alternatives Supplier's switching costs are low Buyer's switching costs are high Buyer's account is not particularly attractive to supplier Supplier's offering is relatively unique Buyer's search costs are very high Supplier has substantial information asymmetry advantages over buyer

Low   High

Attributes to supplier power relative to buyer

FIGURE 12.2   The power matrix: the attributes of buyer and supplier power. (From Cox, A. et al., *Supply chain management: A guide to best practice*, Financial Times/Prentice Hall, London, 2003. With permission.)

of the relationships presented proactive sourcing approaches are not feasible. In such circumstances, buyers are forced to adopt reactive sourcing approaches based on supplier selection or supply chain sourcing principles.

12.4   Six Relationship Management Options and the Problem of Relationship Alignment

Whatever the power circumstance all buyer and supplier exchange transactions have to be managed appropriately. This means that there must be alignment between the operational and commercial strategies being pursued by both the buyer and the supplier in the relationship [Cox et al. 2004a, 2004b]. To understand how these can be aligned it is necessary to understand theoretically the way in which buyers and suppliers can conduct relationships with one another. The basic theoretical choices are outlined in Figure 12.4.

As Figure 12.4 demonstrates, when a buyer and supplier interact there are at least two fundamental aspects to the relationship. The first is the way of working, which

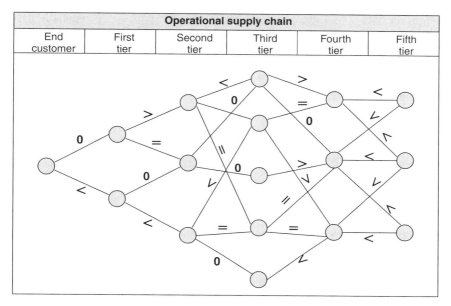

FIGURE 12.3 Hypothetical power regime. (Adapted from Cox, A., *Journal of Supply Chain Management*, 37 (2), 42–47, 2001. With permission.)

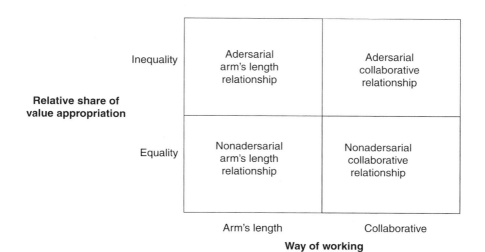

FIGURE 12.4 Relationship portfolio analysis. (Adapted from Cox, A., *Strategic procurement management: Concepts and cases*, Earlsgate Press, Stratford-upon-Avon, UK, 1999b. With permission.)

refers to the level of operational linkage between the two parties. Operationally, buyers and suppliers can choose to make few dedicated investments in their relationship and operate on a fairly short-term contractual basis. This arm's-length way of working requires the buyer only to provide basic specification, volume, and timing information to the supplier, and the supplier need only provide the buyer with limited specification, timing, and pricing information.

There is, of course, an alternative way of working. If collaboration occurs, the buyer and supplier make extensive dedicated investments in the relationship. On top of basic specification, timing, price, and volume information, both parties will normally make relationship-specific adaptations to their operational processes and provide detailed information about future product and service road maps, as well as creating technical linkages in their respective operations [Cannon and Perreault 1999]. These types of relationship tend to be long term in duration.

The second aspect of any relationship is the commercial intent of the two parties when they enter into a transaction. If the buyer or supplier is primarily interested in maximizing their share of value from the relationship at the expense of the other, this is referred to as adversarial value appropriation. If the intention of the buyer or supplier is to provide open and transparent commercial information about profit margins and the costs of operations, such that any improvements can be shared relatively equally, then this is referred to as a nonadversarial value appropriation.

Linking these two aspects creates four relationship management styles that buyers and suppliers can choose from. These are: adversarial arm's-length (one party seeks to maximize commercially using short-term market-testing techniques); nonadversarial arm's-length (both parties accept the current market price without recourse to aggressive bargaining, but test the market actively); adversarial collaboration (both parties collaborate over the long term, but one seeks to maximize commercially); and nonadversarial collaboration (both parties collaborate, but share commercial value relatively equally).

If buyers and suppliers can choose from these four broad options it is not obvious which is the most appropriate. Obviously some commentator argue that win–win outcomes based on nonadversarial collaborative ways of working are always the most appropriate—but as we shall see later, this is an illusion because win–win outcomes are not objectively feasible in buyer and supplier exchange [Cox 2004a, 2004b]. In adopting a strategic approach to procurement management, managers cannot rely on simplistic assumptions about appropriateness. On the contrary, managers have to understand how each of the four relationship management styles described here link with particular sourcing approaches, under specific power and leverage circumstances, to create aligned business relationships that optimize performance for themselves, first, and only secondly for their exchange partners. In order to align business relationships appropriately, buyers and suppliers have to adopt the power and relationship linkages outlined in Figure 12.5.

As Figure 12.5 demonstrates, in some power and leverage circumstances buyer dominance occurs; in others supplier dominance will occur. It is unlikely that relationship and performance outcomes for a buyer or a supplier can be optimized in either of these power circumstances unless the dominant party has a willing suppliant as an exchange

The way of working	Buyer dominant arm's length relationship	Buyer-supplier reciprocal arm's length relationship	Supplier dominant arm's length relationship
Arm's length	• Short-term operational relationship, with limited close working between buyer and supplier • Buyer adversarially appropriates most of the commercial value created and sets price and quality trade-offs • Supplier is nonadversarial commercially and a willing supplicant, accepting work rather than high margins/profitability from the relationship • *Buyer dominance power situation (>)*	• Short-term operational relationship, with limited close working between buyer and supplier • Buyer accepts current market price and quality trade-offs • Supplier accepts normal (low) market returns • Both buyer and supplier operate adversarially commercially whenever possible, but normally have few leverage opportunities • *Independence power situation (o)*	• Short-term operational relationship, with limited close working between buyer and supplier • Supplier adversarially appropriates most of the commercial value created and sets price and quality trade-offs • Buyer is a nonadversarial commercially and a willing supplicant, paying whatever is required to receive given quality standards • *Supplier dominance power situation (<)*
Collaborative	**Buyer dominant collaborative relationship** • Long-term operational relationship, with extensive and close working between buyer and supplier • Buyer adversarially appropriates most of the commercial value created and sets price and quality trade-offs • Supplier is a nonadversarial supplicant commercially, and accepts work rather than high margins/ profitability from the relationship • *Buyer dominance power situation (>)*	**Buyer-supplier reciprocal collaborative relationship** • Long-term operational relationship, with extensive and close working between buyer and supplier • Buyer and supplier share relatively equally the commercial value created • Buyer and supplier agree price and quality trade-offs, with supplier making more than normal returns • Both buyer and supplier operate nonadversarially commercially • *Interdependence power situation (=)*	**Supplier dominant collaborative relationship** • Long-term operational relationship, with extensive and close working between buyer and supplier • Supplier adversarially appropriates most of the commercial value created and sets price and quality trade-offs • Buyer is a nonadversarial supplicant and commercially, and pays whatever is required to receive given quality standards • *Supplier dominance power situation (<)*
	Buyer dominance	Buyer-supplier reciprocity	Supplier dominance

Who appropriates value from the relationship?

FIGURE 12.5 Value appropriation, power and relationship management styles. (From Cox, A., *Supply Chain Management: An International Journal*, 9 (5), 346–56, 2004c. With permission.)

partner. This means that when a buyer is dominant, they can adopt either reactive or proactive sourcing approaches, but they will expect their suppliers to accept either market testing under reactive sourcing, or collaboration under proactive sourcing. Whichever sourcing approach is utilized, the dominant buyer will expect most of the commercial value from the exchange to flow to them.

In some circumstances an equivalence of power and leverage resources exists. In such circumstances a power situation of interdependence exists and this is likely to encourage nonadversarial ways of working commercially, with relative equity in the sharing of commercial value and close and collaborative operational ways of working. Conversely, in a situation of independence (low power resources for both parties), arm's-length reactive relationship approaches are normally the most appropriate for both parties to adopt.

Rather than assuming, therefore, that there is always one sourcing and relationship management approach that is "best practice"—like partnering or long-term collaboration—both parties to an exchange have a range of options from which to choose. Any of these options may be more or less appropriate under different power and leverage circumstances. Thus, an adversarial and arm's-length approach to relationship management may be the most appropriate way to manage particular construction relationships; on the other hand, a more collaborative and nonadversarial approach may be appropriate in other circumstances. Understanding when one is more appropriate than another is clearly the mark of a competent buyer and/or supplier in construction—just as it is in all other forms of buyer and supplier exchange.

This implies that there must be a correlation between power and leverage circumstances and appropriate relationship management styles and sourcing options [Cox 2004c]. Thus, supplier selection and supply chain sourcing approaches, which focus on arm's length rather than collaborative ways of working, may be appropriate for buyers and suppliers to engage in under all power and leverage circumstances. But, for these relationships to be aligned both the buyer and supplier must adopt the appropriate relationship management approaches. As Figure 12.6 demonstrates, if the buyer is dominant and adversarial commercially, the relationship is only aligned if the supplier is nonadversarial. Conversely, if the supplier is dominant then the reverse is true. If the power structure is one of interdependence then both parties should be nonadversarial commercially, and if the power circumstance is of independence then both parties should be adversarial if they can.

If buyers or suppliers wish to adopt a proactive approach (supplier development or supply chain management), this is only likely to be feasible in power situations of buyer dominance, supplier dominance, or interdependence. This is because, if a power situation of independence exists, there is no basis for collaboration to occur for both parties. If buyer dominance or supplier dominance occurs, however, the dominant party will have the opportunity to be commercially adversarial and force the supplicant party to operate nonadversarially and pass the bulk of any value created in the exchange to them. If interdependence occurs then the exchange will be aligned if both parties operate nonadversarially and agree to share the value created equally.

Relationships can, therefore, be aligned or they can be misaligned under both reactive and proactive approaches [Cox et al. 2004a, 2004b]. Furthermore, there is

Sourcing approach	Power and leverage circumstance	Appropriate relationship management styles
Supplier selection	Buyer dominance (>)	Buyer adversarial arm's length / supplier nonadversarial arm's length
	Independence (0)	Buyer and supplier adversarial arm's length
	Interdependence (=)	Buyer and supplier nonadversarial arm's length
	Supplier dominance (<)	Buyer nonadversarial arm's length / supplier adversarial arm's length
Supply chain sourcing	Buyer dominance (>)	Buyer adversarial arm's length / supplier nonadversarial arm's length
	Independence (0)	Buyer and supplier adversarial arm's length
	Interdependence (=)	Buyer and supplier nonadversarial arm's length
	Supplier dominance (<)	Buyer nonadversarial arm's length supplier nonadversarial arm's length
Supplier development	Buyer dominance (>)	Buyer adversarial collaboration / supplier nonadversarial collaboration
	Independence (0)	Not applicable
	Interdependence (=)	Buyer and supplier nonadversarial collaboration
	Supplier dominance (<)	Buyer nonadversarial collaboration / supplier adversarial collaboration
Supply chain management	Buyer dominance (>)	Buyer adversarial collaboration / supplier nonadversarial collaboration
	Independence (0)	Not applicable
	Interdependence (=)	Buyer and supplier nonadversarial collaboration
	Supplier dominance (<)	Buyer nonadversarial collaboration / supplier adversarial collaboration

FIGURE 12.6 Appropriateness in sourcing strategies, power circumstances, and relationship management. (From Cox, A., *Supply Chain Management: An International Journal*, 9 (5), 346–56, 2004c. With permission.)

always the possibility that what may be in the interests of one party to an exchange may not always be in the interests of the other. Even in an interdependence power situation where reciprocal (or nonadversarial) collaboration works well for both parties to the exchange, it can be argued that there may still be superior relationship and performance outcomes for both parties. If this is true then it raises the possibility that commercial exchange between buyers and suppliers will be permanently contested. This issue—which challenges much recent writing about the benefits of mutuality and win–win relationship management in construction—is discussed in the next section.

12.5. Understanding the Problem of Win–Win in Construction

The recent fad in favor of partnering (or long-term operational collaboration) in construction assumes that this is a superior approach because it creates win–win commercial outcomes for both parties. This may sometimes be true, but in other circumstances this may not be the case, and more traditional and reactive approaches may provide superior commercial and operational outcomes for one or both parties than a partnering approach.

There are two primary reasons for this conclusion. First, buyers and sellers often enter into relationships in which they have no desire for long-term relationships and have no qualms about win–lose outcomes. Second, even when buyers and sellers need long term and sustainable relationships, to argue that both sides must gain something from the relationship does not tell us exactly what the commercial outcome is, or should be, in any relationship. It is perfectly conceivable that both sides can gain something commercially from a relationship, even though one party gains a disproportionate share of the value compared with the other. Furthermore, relationships can be sustained even with continuous inequity of commercial outcome. To argue that the needs of both parties must be met (or that they must make mutual gains) tells us nothing about what these needs or gains are, or how much of them must be met for either party to find the exchange acceptable.

The fundamental problem for the analysis of business transactions between buyers and suppliers is that they have dissimilar operational and commercial goals. When entering into exchange relationships all buyers wish to exchange money for goods and/or services, and suppliers wish to receive money in return [Cox et al. 2004a]. The buyer is normally concerned operationally with the functionality (performance, quality, on-time delivery, etc.) of the goods and/or services provided, and commercially with their total costs of ownership. The buyer seeks to maximize the value for money received from any supplier by increasing functionality and/or reducing the total costs of ownership.

The supplier is, normally, most interested in making profits, but has to deliver goods and/or services operationally to receive revenue from a buyer. The supplier is concerned commercially with maximizing the price and margins that can be obtained from the delivery of a particular product and/or service. The ideal for the supplier is, therefore, to increase revenue and to increase returns. Given these different drivers there is an

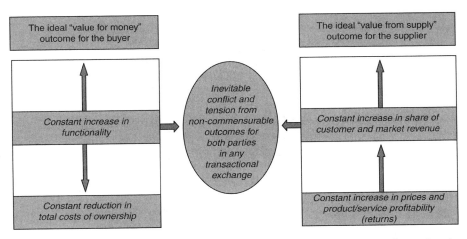

FIGURE 12.7. The conflict and tension in buyer-supplier transactional exchange. (From Cox, A., *Win–win? The paradox of value and interests in business relationships*, Earlsgate Press, Stratford-upon-Avon, UK, 2004a. With permission.)

inevitable and irresolvable commercial tension (outlined in fig. 12.7) in buyer and supplier exchange.

The ideal outcome for a buyer is to achieve increased operational performance with reduced total costs of ownership. If the buyer achieves this outcome, unfortunately, it is not feasible for the supplier to achieve their ideal outcome of constantly increasing revenue and above normal returns. This is because above normal returns must be paid for by the buyer—whose ideal outcome is only that the supplier makes breakeven or normal returns for any given level of operational performance. Some writers argue that both parties should concentrate not on the share of the spoils, but more on growing "the size of the cake" [Carlisle and Parker 1989]. There are two problems with this argument. First, how any "cake" should be divided is still an issue even if it grows. Second, the ideal outcomes for both the buyer and the supplier remain the same (not fully commensurable) irrespective of "the size of the cake."

It is clear, therefore, that win–win (positive-sum) outcomes (in which both parties simultaneously achieve their ideals) are not feasible in buyer and supplier exchange. Despite this, more limited forms of mutuality based on nonzero-sum outcomes are feasible and sometimes outcomes that are negative-sum (lose–lose) and zero-sum (win–lose) can occur.

There are, as Figure 12.8 confirms, eight possible outcomes that are feasible in buyer and supplier exchange. One outcome—win–win (positive-sum) in Cell C in the matrix—is not feasible in practice due to the objective conflict between the buyer's search for cost reductions and the supplier's search for above normal returns. Mutuality can still occur. This arises when one party achieves their ideal but the other party only partially achieves their ideal, or when neither party achieves their ideal but both parties partially achieve their objectively ideal goals. These nonzero-sum outcomes involving win–partial win, partial win–partial win, and partial win–win—Cells B, E and F in the matrix—represent what is actually meant in practice by mutuality in buyer and supplier relationships.

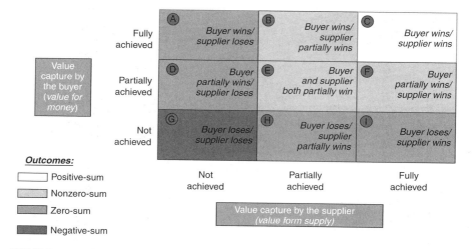

FIGURE 12.8. Possible outcomes for buyers and sellers from transactional exchange (Adapted from Cox, A., *Win–win? The paradox of value and interests in business relationships*, Earlsgate Press, Stratford-upon-Avon, UK, p. 95, 2004a. With permission.)

Interestingly enough, all three of these outcomes provide the basis for long-term sustainable relationships under certain power and leverage circumstances. The problem is, however, that none of the three options provides for an outcome under which both of the parties to the exchange can simultaneously achieve their ideal performance outcome. This is because, even in Cell E, where both parties only partially achieve their performance goals, each has an alternative outcome that they can be expected to prefer if it is available to them (Cells B and F).

This reasoning creates fundamental problems for those who argue that conflict can be eradicated from buyer and supplier exchange through trust [Sako 1992], or by creating order-generating governance mechanisms [Williamson 1996]. This is because there is always conflict in buyer and supplier exchange due to the fact that the rational rank order preference of the supplier for the three nonzero-sum outcomes is the opposite of the buyer's preference. The supplier's objective rank order preference is for Cell F first, because this outcome maximizes their revenue and profits. Suppliers will then normally prefer Cell E, and Cell B is the least desirable of the three nonzero-sum outcomes. This is the opposite of the rank order preference of the buyer—which is for Cell B, then Cell E, and finally Cell F. Thus, it is unlikely that nonzero-sum outcomes can ever eradicate the fundamental commercial tensions and conflict that is inherent in buyer and supplier exchange.

There is, however, another important conclusion that one can make from this understanding of feasible outcomes in buyer and supplier exchange. Despite the bad press that adversarial relationship management has received in construction specifically, and in buyer and supplier exchange more generally, it is inevitable—if conflict is inherent in buyer and supplier transactions—that zero-sum outcomes may actually be appropriate relationship management strategies under some power and leverage circumstances.

Indeed, in certain circumstances zero-sum outcomes may be preferable to nonzero-sum outcomes for at least one party to some exchanges.

Thus, for buyers, zero-sum outcomes in Cell A may be preferable in some circumstances to nonzero-sum outcomes in Cell B. This is because in Cell A the buyer fully achieves the ideal performance outcome, but the supplier does not achieve any of their performance goals. In such circumstances the costs of ownership would be reduced because the supplier was unable to make a profit. One can argue that if this situation were sustainable for the supplier, it would be a superior outcome for the buyer than Cell B (or any other of the Cells available). The problem with this outcome for the buyer is, of course, that it is unlikely to be achievable on a sustainable basis unless the supplier is prepared to undertake a "loss leadership" approach. This does not detract, however, from the argument that there must be circumstances under which Cell A (win–lose) is the most desirable option for the buyer when compared with all other possible options. The same logic applies to Cell D, where the buyer partially achieves their performance goals, but the supplier does not. Clearly this approach may also be preferable to the options available to the buyer in Cells E and F. Once again, however, the problem for the buyer is the sustainability of the relationship if the supplier is not capable of sustaining a "loss leadership" approach over time.

The supplier faces similar choices when considering zero-sum options. A supplier might find the zero-sum option in Cell I a better leverage position than the nonzero-sum outcome in Cell F. In Cell I the supplier is normally dominant in power terms and able to impose quality and price trade-offs in the market. In this situation, the buyer will be a price and quality receiver, and the supplier can be expected to make above normal profits. The major problem for the supplier is also whether this is sustainable. If the supplier believes that they have sufficiently robust isolating mechanisms to sustain their power in the relationship, then this may be a preferable option to Cell F, where the supplier normally has either to increase functionality and/or reduce pricing and profitability demands on the buyer. Similarly, the supplier might be expected to prefer Cell H to Cells E and B, because the buyer in both these options is able to reduce the leverage of the supplier. The same issues arise here for the supplier about sustainability.

One can conclude, therefore, that some zero-sum outcomes (especially those in Cell A for the buyer and Cell I for the supplier) may be preferable to each of three nonzero-sum outcomes in Cells B, E, and F. The only major problem for the zero-sum options is that, other things being equal, they may be less sustainable transactions over time than the three nonzero-sum outcomes. The major reason for this is because one side loses while the other either partially or fully wins. This does not mean, however, that zero-sum outcomes (Cells A, D, H, and I) cannot be sustained for a considerable period of time.

Despite what some writers believe about the need for mutuality in exchange, buyers have no great difficulty operating in Cells A or D if a supplier is willing to operate with them in this situation. This might occur when a supplier wishes to provide a "loss leader" to drive other competitors out of the market; or wishes to develop brand association with a major customer and attempt to recoup profits from premium pricing with other buyers in the market. It can also be possible for buyers to transact with suppliers, operating in highly competitive markets, who must "loss lead" to win the business, but then have to withdraw.

However, the contested nature of the supply market means that there are sufficient suppliers available to allow the buyer to continuously behave opportunistically against them individually. Similarly, suppliers do not have a problem operating in Cells I or H, if they can impose prices and functionality standards without fear that buyers can source elsewhere.

If this is true then the best that a buyer or a supplier can achieve may in fact be to find exchange partners who are willing to operate in zero-sum rather than in nonzero-sum outcomes. It all depends on what the risks are to the buyer or supplier of the power circumstances moving against them if they continue to operate in this fashion. For the buyer it could eventually mean that the supplier goes out of business and they find themselves with a very restricted supply market to source from. The supplier, on the other hand, might drive away customers or force them to seek substitutes, consider insourcing, or the direct development of competitors. Despite these risks from the excessive use of buyer and supplier power, it can be argued that a great deal of business activity occurs in these circumstances and that, if the downside risks can be avoided, these may be highly desirable strategies for buyers and suppliers to pursue.

12.6 Best Practice in the Strategic Management of Construction Procurement

If there is logic in the argument presented above it would seen sensible to suggest that there cannot be one "best practice" approach to construction procurement—or any other type of procurement for that matter. On the contrary, "best practice" must involve an organization and its buyers understanding all the factors outlined here in order to select wisely and appropriately from amongst the complex range of sourcing and relationship management approaches that are available for either party to select from in a specific exchange transaction.

To argue in construction management that there is always one sourcing or procurement approach—partnering or alliancing—that constitutes "best practice" is clearly nonsense. There are a range of sourcing and relationship management approaches that can be adopted by buyers to procure all types of construction goods and services. The task for managers in construction is not to seek only one solution or option, but to understand the short and long-term opportunities that exist to leverage increased value for money from all of their relationships with suppliers. To achieve this, buyers of construction cannot look for a shortcut by labeling some activities as "strategic" and then adopt half-backed partnering or alliancing strategies with a few suppliers based on win–win and trust-based principles.

On the contrary, competence requires that organizations and their buyers undertake five simple—although necessarily time-consuming and resource-intensive—tasks [Cox and Ireland 2002; Ireland 2004; Cox, Ireland, and Townsend 2006]:

1. First, a construction buyer must properly segment their activities and spend profile to understand the nature of the demand profile (regular/serial/irregular/one-off) they possess, and the types of power resources this provides them with to leverage value from the supply market.

2. Having understood their demand profile (and its relationship to the alternative sources of demand available to suppliers in the market), a buyer of construction goods and services must then understand what the nature of the supply market is—this implies understanding the relative power and leverage of suppliers in the market.
3. Having understood the power and leverage situation for the particular projects and/or construction activities/spend to be sourced, a buyer has to understand the opportunity costs and risks of pursuing any of the six sourcing and six relationship management options available to them.
4. Having understood the opportunity costs and risks of these options, the buyer must then decide which of the available commercial outcome is feasible, and seek to test in the market to what extent each of the available options is the most desirable given current market circumstances.
5. Having selected the options for market testing, a buyer must then select the most appropriate approach through a rigorous and robust approach to market testing and contractual negotiation and award.

What all of this implies is that there can be no panacea for construction procurement, and that a strategic approach to construction procurement management involves a commitment to understanding the appropriateness of a range of complex sourcing choices under very different power and leverage circumstances impacting on different types of CAPEX and OPEX activity. Only by understanding from first principles what these options are under particular power and leverage circumstances, can an appropriate approach to construction procurement be developed.

This means that all of the relationship management options available may be appropriate for construction buyers to utilize. Despite this, in the construction industry the majority of organizations in the industry—and especially those involved in CAPEX activities—are not often in a position of dominance over their supply base. This is normally because their demand profile is irregular and one-off. Actors possessing a high level of regular CAPEX and/or OPEX construction spend, and who can standardize their requirements, will normally be in a better position to leverage their suppliers than those who cannot. They will be able to consider proactive sourcing options in repeat games rather more often than the majority of actors in the industry who have to manage in one-off games [Cox, Ireland, and Townsend 2006].

This does not mean, however, that only these types of organization are able to develop a strategic approach to construction procurement; rather it implies that the construction procurement strategy for some organizations will have to be very different to that of other organizations, who possess very different demand and supply and, therefore, power and leverage circumstances within which to operate. It is interesting to note that many of those who recommend partnering and alliancing (repeat game strategies) normally draw on examples from clients or contractors who have regular demand profiles—unfortunately these demand profiles are not available to the majority of those operating either as clients or as contractors or subcontractors in the industry.

This is because the majority of construction procurement is focused on projects that are unique, requiring novel processes, and that are transient, often without a clear start and with an uncertain completion date. This means that what a strategic approach will look like for an organization with this type of demand profile is likely to be very different to that of an organization with a regular demand for standardized construction requirements. This implies that when an organization develops its strategy for construction procurement, the scope for different sourcing and relationship management approaches will be a function of the power and leverage variables that arise from the types of demand and supply profiles that exist in particular sourcing relationships.

Next, some of the strategic sourcing and relationship management options that arise from these different types of demand scenarios are described:

- If demand is similar because of standardized design and specification, and projects and activities can be undertaken either in a consecutive or overlapping manner, then a number of preferred suppliers may be selected to deliver requirements through long-term collaborative relationships.
- If projects and activities have a unique aspect and are irregular or one-off, then it is unlikely that a buying organization will be able to standardize design and specification, and as a result, short term and relatively ad hoc and adversarial arm's-length relationship and sourcing options are likely to be adopted.

Whether the design and specification is standardized or not, it is evident that the nature of regularity involves two separate variables: the regularity of workload and the regularity of relationship. Given this distinction, it is clear that the appropriate management of construction spend and supplier relationships may vary based on the nature of the regularity of the demand and supply variables that have to be managed:

1. For one-off projects with different suppliers it is likely that reactive sourcing approaches (such as supplier selection) based on adversarial and opportunistic short-term relationships are likely to be appropriate.
2. When there are one-off projects with preferred suppliers it is possible for opportunistic and adversarial reactive sourcing approaches to be adopted that include supplier selection and supply chain sourcing.
3. For those with an ongoing portfolio with different suppliers it is possible to be both reactive and proactive, but the proactive sourcing approach would probably only extend to supplier development activities.
4. When there is an ongoing portfolio with preferred suppliers it is also possible to pursue both reactive and proactive approaches, but there may be scope for both supplier development and supply chain management approaches.

Given this, it is evident that simplistic generalizations about what is the appropriate way to manage construction procurement, and what a strategic approach implies, must be avoided at all costs. Organizations within the construction industry face very different demand and supply, and power and leverage, circumstances. As a result, competence in the development of procurement strategies requires that buyers and suppliers understand what is appropriate given the circumstances they are in.

References

Bennett, J., and Jayes, S. 1995. *Trusting the team: The best practice guide to partnering in construction.* Reading: Reading Construction Forum.

―――. 1998. *The seven pillars of partnering.* London: Thomas Telford Publishing, London.

Cannon, J. P., and Perreault, W. D. 1999. Buyer and seller relationships in business markets. *Journal of Marketing Research,* 36 (4), 439–60.

Carlisle, J. A., and Parker, R. C. 1989. *Beyond negotiation.* Chichester, UK: John Wiley.

CII. 1989. *Partnering: Meeting the challenges of the future.* Austin, TX: Construction Industry Institute, Special Publications.

―――. 1991. *In search of partnering excellence.* Austin, TX: Construction Industry Institute, Special Publications.

―――. 1994. *Benchmarking implementation results, teambuilding and project partnering.* Austin, TX: Construction Industry Institute.

Cox, A. 1997. *Business success: A way of thinking about strategy, critical supply chain assets and operational best practice.* Stratford-upon-Avon, UK: Earlsgate Press.

―――. 1999a. Power, value and supply chain management. *Supply Chain Management: An International Journal,* 4 (4), 167–75.

―――. 1999b. Improving procurement and supply competence: On the appropriate use of reactive and proactive tools and techniques in the public and private sectors. In *Strategic procurement management: Concepts and cases,* eds. R. Lamming and A. Cox. Stratford-upon-Avon, UK: Earlsgate Press.

―――. 2001. Managing with power: Strategies for improving value appropriation from supply relationships. *Journal of Supply Chain Management,* 37 (2), 42–47.

―――. 2004a. *Win–win? The paradox of value and interests in business relationships.* Stratford-upon-Avon, UK: Earlsgate Press.

―――. 2004b. Business relationship alignment: On the commensurability of value capture and mutuality in buyer and supplier exchange. *Supply Chain Management: An International Journal,* 9 (5), 410–20.

―――. 2004c. The art of the possible: Relationship management in power regimes and supply chains. *Supply Chain Management: An International Journal,* 9 (5), 346–56.

Cox, A., and Ireland P. 2002. Managing construction supply chains. The common sense approach. *Engineering Construction and Architectural Management.* 9 (516), 409–18.

Cox, A., Ireland, P., Lonsdale, C., Sanderson, J., and Watson, G. 2002. *Supply chains, markets and power: Mapping buyer and supplier power regimes.* London: Routledge.

―――. 2003. *Supply chain management: A guide to best practice.* London: Financial Times/Prentice Hall.

Cox, A., Ireland, P., and Townsend, M. 2006. *Managing in construction supply chains and markets.* London: Thomas Telford.

Cox, A., Lonsdale, C., Sanderson, J., and Watson, G. 2004. *Business relationships for competitive advantage: Managing alignment and misalignment in buyer and supplier transactions.* Basingstoke, UK: Palgrave Macmillan.

Cox, A., Sanderson, J., and Watson, G. 2000. *Power regimes: Mapping the DNA of business and supply chain relationships.* Stratford-upon-Avon, UK: Earlsgate Press.

——. 2001. Supply chains and power regimes: Towards an analytic framework for managing extended networks of buyer and supplier relationships. *Journal of Supply Chain Management*, 37 (2), 28–35.

Cox, A., and Townsend, M. 1998. *Strategic procurement in construction.* London: Thomas Telford.

Cox, A., Watson, G., Lonsdale, C., and Sanderson, J. 2004b. Managing appropriately in power regimes: Relationship and performance management in 12 supply chain cases. *Supply Chain Management: An International Journal*, 9 (5), 357–71.

DETR. 1998. *Rethinking construction.* London: DETR.

Godfrey, K. A. ed. 1996. *Partnering in design and construction.* New York: McGraw Hill.

Hellard, R. B. 1995. *Project partnering; principles and practice.* London: Thomas Telford.

Hinks, A. J., Allen, S., and Cooper, R. D. 1996. Adversaries or partners? Developing best practice for construction industry relationships. In *The organisation and management of construction industry: Shaping theory and practice*, eds. D. A. Langford and A. Retik. London: E & FN Spon.

Holti, R., and Standing, H. 1996. *Partnering as inter-related technical and organisational change.* London: Tavistock.

Ireland, P. 2004. Managing appropriately in construction power regimes: Understanding the impact of regularity in the project environment. *Supply Chain Management: An International Journal*, 9 (5), 372–82.

Koskela, L. 1997. Lean production in construction. In *Lean construction*, ed. L. F. Alarcon. Rotterdam: A. A. Balkema.

Latham, M. 1994. *Constructing the team.* London: HMSO.

Lonsdale, C., and Cox, A. 1998. *Outsourcing: A business guide to risk management tools and techniques.* Stratford-upon-Avon, UK: Earlsgate Press.

Sako, M. 1992. *Prices, quality and trust.* Cambridge: Cambridge University Press.

Williamson, O. E. 1996. *The mechanisms of governance.* Oxford: Oxford University Press.

Womack, J. P., and Jones, D. T. 1996. *Lean thinking.* New York: Simon Schuster.

13

Industrial Organization Object-Oriented Project Model of the Facade Supply Chain Cluster

Kerry A. London
The University of Newcastle

13.1　Introduction

This chapter presents a cluster of chains in relation to the building facade, which supports a model merging industrial organization economic theory and object-oriented modeling. The results are from the interviews with various project managers, firm executives, production, and procurement managers involved in the supply chains for commodities that are clustered around the facade specialist subcontractor on a major building project in Melbourne, Australia. Many of the suppliers are international in that they either directly work on projects in other countries or are within companies that are international.

Researchers have tended to develop normative models to improve industry performance through supply chain integration. Such models are based upon the assumption of a homogenous industry, which is fragmented and composed of numerous small to medium-sized enterprises. Organizations, either governments or corporations, are seeking positive descriptive economic models which are fundamentally informational models to improve firm performance and/or industry performance [London 2008].

The model presented is a summary of the key components, and a more complete discussion can be found in the text *Construction Supply Chain Economics* [London 2008]. This model integrates supply chain economics and supply chain information for project-based industries for decision support in the construction sector. The model presented accepts that the industry is specialized and heterogeneous with varied structural and behavioral characteristics across individual markets, and simply attempts to capture, represent, and display underlying structural and behavioral supply chain procurement characteristics.

A past difficulty with supply chain management in terms of construction research theory and practical application is that, currently, too little is known about these characteristics and how to describe them in a holistic framework. We typically discuss supply chain management and various strategies, but upon closer inspection the strategies often represent supplier management and do not address strategies for the chain of suppliers; particularly the firm interdependencies which are apparent in "good" supply chain management practices at various linkages at numerous tiers. It is difficult to achieve this, as information beyond the tier an organization contracts with is often hidden. The emergence of more sophisticated building information modeling could eventually trigger more sophisticated model market information at each tier, and ultimately develop strategies to manage overall supply chain performance.

This chapter describes a model that resulted from an empirical study, which focused on the activity of procurement. Procurement decisions are fundamental to describing the underlying structure and behavior of the industry, as this represents the initial starting conditions of a project for all firms. The industrial organization economics theory was examined for its contribution, and the structure-conduct-performance methodology was modified to develop a project-oriented industrial organization economic model for procurement in the construction supply chain. The model defines entities, such as firms, projects, markets, and firm–firm relationships, and their relative associations. The model was static and nomothetic in approach, and lacked the capacity to represent the duality of structure and behavior of entities and individual procurement and project scenarios. The object-oriented methodology was used to reinterpret the construction supply chain using the Unified Modelling Language [Alhir 1998]. A Class Model of Procurement Relationships was developed. The model is interdisciplinary and defines a framework for multiplicity of associations between entities, interaction between structural and behavioral characteristics of entities, and types of entities that have common characteristics. Structural and behavioral model views of real world procurement in construction supply chains were developed based upon six major building projects in an Australian city. One thousand two hundred and fifty three procurement relationships were mapped using data collected from 47 structured interviews and 44 questionnaires. Many supply chains were mapped in this study, and

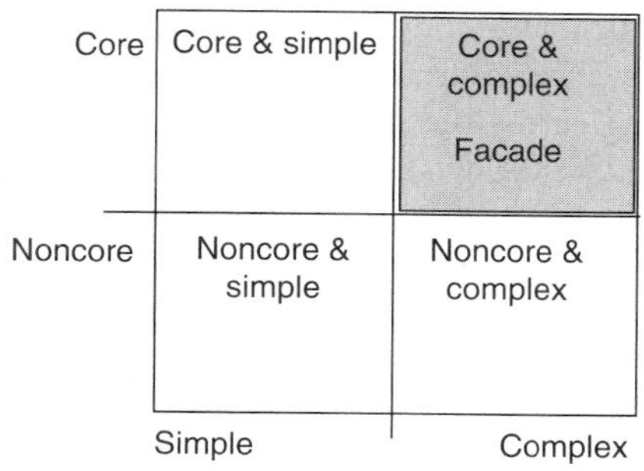

FIGURE 13.1   Categorization of chains.

this chapter presents the results of those supply chains associated with the building of a complex facade which involved integration of various supply chains including: steel, glazing, aluminium, zinc, and stone materials by the facade subcontractor—with a particular focus on glazing and aluminium supply chains. Supply chains can be categorized as core and noncore, and simple and complex [London 2008], and this chain is categorized as core and complex (refer to Figure 13.1).

The data type was qualitative and quantitative. Data display techniques were used to describe common themes and differences to develop an ideographic view of procurement. A statistical categorical data analysis provided a nomothetic view by comparing observed procurement results versus likelihood of expected results.

13.2   Model

Interest in the supply chain management concept by the construction research community arose from the successful implementation by manufacturing sectors to resolve firm and industry performance problems. Construction industry policy makers have also appropriated the concept [London 2004]. Researchers tend to develop normative models to improve industry performance through supply chain integration [Barrett and Aouad 1998; Olsson 2000; Taylor and Bjornsson 1999; Tommelein and Yi Li 1999; Vrihjhoef and Koskela 1999; Saad and Jones 1998; Nicolini, Holti, and Smalley 2001; Cox and Townsend 1998]. Typically, such models are based upon the assumption of a homogenous industry, which is fragmented and composed of numerous small to medium-sized enterprises. However, policy makers are seeking positive economic models [London 2004], yet current policies are not based upon an explicit understanding of the nature of the industry nor an explicit model of firm and industry performance. A positive economic model, as defined by economists [Scollary and John 2000], would accept that the industry is specialized and heterogeneous with varied structural and behavioral characteristics across individual markets. The widespread implementation of supply chain management has proven difficult in construction [Dainty, Briscoe, and Millett 2001; London 2004; Briscoe et al. 2004], and one of the greatest difficulties with supply chain management in terms of construction research theory and practical application is that it relies on both interdependent *management* of firm to firm relationships, and corresponding holistic *information* about the characteristics of these relationships along the chain by large market leaders.

Procurement modeling across the supply chain is fundamental to describing the underlying structure and behavior of the industry. Such a model would merge the elements of the accepted industrial organization economics model of market structure, firm conduct, and market performance; the concepts of supply chain structure [Minato 1991; Hines 1994; Nischiguchi 1994; Bowersox and Closs 1996; Lambert, Pugh, and Cooper 1998; Harland 1994] and behavior, and the characteristics of the project-oriented industry. The rationale for this approach to a procurement model is that government construction industry policy is being developed within a vacuum of appropriate economic models. Uninformed and/or simplistic assumptions are being made in regards to the structural and behavioral characteristics of the construction industry, which are often reduced to such simplistic descriptions of the industry as one that is fragmented with numerous small to medium-sized enterprises.

This chapter proposes a new model to describe the structural and behavioral characteristics of procurement along the construction supply chain using a hybrid project-oriented industrial organization economic approach, and does so by assembling the components of the model from first principles. The model is described in terms of three key elements: projects, firms, and firm–firm procurement relationships, and the various associated entities that link these elements. The model abstracts the construction supply chain to an object composed of firms and industrial relationships brought about by construction projects. Each construction supply chain is composed of a contractual chain connecting firms that respond to a construction project. The contractual chain is composed of firms who are providing services and/or products along the chain. The product and/or service is termed: a commodity. Any construction supply chain:

- Forms in response to a construction project which has particular characteristics
- Has firms with various qualities that provide commodities that may or may not be homogenous that reside within different types of markets
- Has firms that are linked through relationships that have certain attributes

Structural characteristics—supply chain entities:

- Project attributes
- Firms, their commodities, and their market structure
- Attributes of firm–firm relationships

Behavioral characteristics—mapping relationships between the supply chain entities:

- Organization of firms
- Firm procurement events

The following discussion describes each of these dimensions in more detail, and serves to highlight that each of the supply chain entities, namely: project, firm, commodity, market and procurement relationships, structural organization, and firm procurement events have certain characteristics that distinguish them. The model assists in developing a framework to describe industrial organization of project-based industries supply chains, and in particular, the construction supply chain. The following section assembles the structural elements of the model in detail. The underlying idea to the development of the model is the assemblage and definition of the elements and the various associations between the elements, which are then central to the attributes of the contractual procurement relationships associated with a project. Object-oriented modeling is used to provide the language to describe the entities in the supply chains and their relationships. In summary, the important aspects used were that in a system, objects are specific instances of a class of objects (e.g., a subcontractor is an instance of a firm and is an object, and there would be many firms which belong to the class of firms). All firms have attributes, for example, size of firm, but each firm may have a different attribute "value." Objects in a system have relationships to other objects and they interact with each other. There is an underlying structure to each object (and system which is being modeled), which can be termed the structural view, and then there is a behavioral view as well. These are often termed model views. Understanding the structural view; that is, objects and the various relationships, and the behavioral view; that is, the sequence and events of the object interactions, at a project level allows us to build the picture of different supply chains within a particular sector; the aggregation of different supply chains allows us to identify the channels of distribution, and this is put forward as part of the model. The duality of an object and the system to have both structural and behavioral characteristics is an important concept in object-oriented modeling.

13.2.1 Assembling the Structural Elements of the Model

The construction project represents the catalyst for construction supply chains. At the most basic level a project represents a market opportunity for a firm to supply its commodity for a return. There is an association between a project and the firm; a firm mobilizes its resources and *works on* a project. Figure 13.2 illustrates this relationship graphically.

However, construction projects are much more complex than this, and on any individual project there are numerous firms that work on the project. Each firm has a relationship with the project (refer to Figure. 13.3).

This simple abstraction, although a useful start, masks the depth and breadth of the construction industry, as it is a narrow and limited view on the number of projects that are occurring concurrently. In reality, there are multiple projects in various stages taking place simultaneously, and although there is one project that has many firms working on it, there are also many other projects in the industry. Despite these multiple connections, each firm usually only forms one relationship that connects them to the project, which is typically through their upstream client on each project. In some cases, there can be different stages to projects, and firms are engaged for different contracts. Firms also have multiple suppliers on projects, and so an individual firm may have many

FIGURE 13.2 Project–firm: one to one relationship.

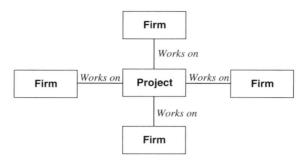

FIGURE 13.3   Project–firm: one to many relationships.

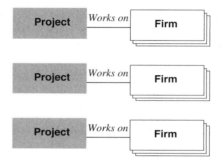

FIGURE 13.4   Associations between isolated multiple projects and multiple firms.

firm–firm relationships on an individual project. These firm–firm relationships are discussed later in Section 13.2.5 Firm-Firm Procurement Relationships.

Figure 13.4 indicates the underlying structure of the construction industry in terms of projects and firms; that is, the industry is composed of many projects and numerous firms working on these projects.

However, this is simplistic as firms typically supply to more than one project simultaneously. To fully appreciate the structural and behavioral characteristics of the industry and the impact that this has on the formation of supply chains and ultimately the performance of the industry as a whole, we need to explicitly account for *multiplicity*. The project is generally not a single entity that occupies all the resources of a firm, in reality, firms work simultaneously on projects and manage their resources and relationships across many projects and with the same or other upstream clients. The construction industry is composed of layers upon layers of individual projects, each populated with numerous firms. Therefore, each firm has many project relationships on multiple projects (refer to Figure 13.5.)

This is similar to Hines' (1994) representation of the supply chain in regional clustering except that the network is made up of many projects and many firms associations rather

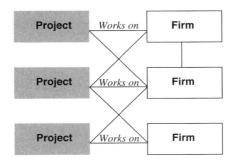

FIGURE 13.5 Network of many firm-project associations.

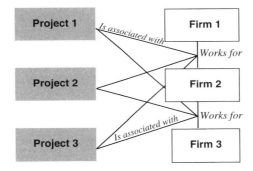

FIGURE 13.6 Network of many firm-firm-project associations.

than simply between many firms, or alternatively Reve's (1990) idea of the firm as a "nexus of contracts." The construction industrial organizational structure has to contend with both firm–firm networks and firm–project networks, and even Figure 13.5 belies the real complexity of the network of multiple firm–firm associations related to multiple projects. Figure 13.5 illustrates the firm–project networks, and Figure 13.6 illustrates a simple firm–firm and firm–project network.

13.2.2 Project Attributes

The types of firms that are drawn to a project are both typical and unique to that project, and the attributes of a project can alter the type of suppliers required. For example, a project requires a firm that supplies glazing and this is typical to many projects; however, the project may require a particular type of glazing or a particular production process that makes the supplier firm more specialized and perhaps unique. There are many product and service-related firms that will be repetitive across projects, for example, few buildings are constructed without some form of steel, concrete, glass, aluminium, and timber-based products, and without some form of design and construction service

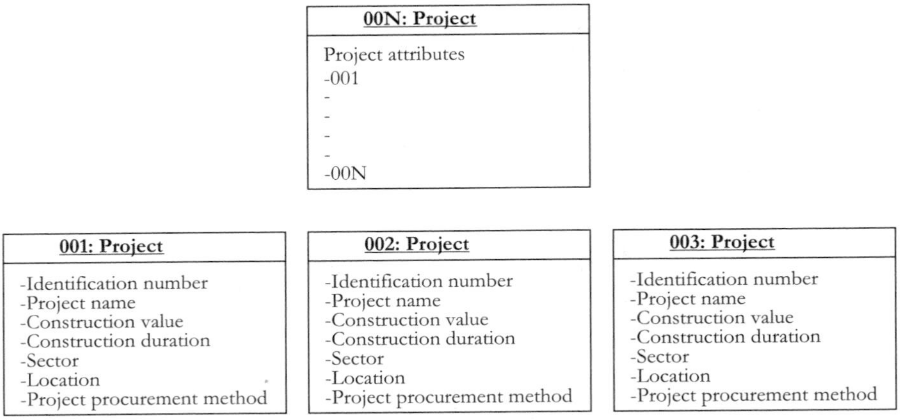

00N: Project
Project attributes
-001
-
-
-
-
-00N

001: Project	002: Project	003: Project
-Identification number	-Identification number	-Identification number
-Project name	-Project name	-Project name
-Construction value	-Construction value	-Construction value
-Construction duration	-Construction duration	-Construction duration
-Sector	-Sector	-Sector
-Location	-Location	-Location
-Project procurement method	-Project procurement method	-Project procurement method

FIGURE 13.7   Project attributes.

specialists. The attributes of the project will help to define which commodities, and correspondingly, which firms are within its project market environment. We of course would refer to this as customized and standardized commodities.

After we move beyond the obvious project attribute that defines the commodity required, and therefore defines the supplier market, there are other project attributes that may define more accurately the boundaries of the project market environments. For example, the construction value, complexity, and duration of a project can differentiate the supplier's markets.

This discussion is beginning to suggest the variety of scenarios possible when representing the firm–firm procurement relationships in the construction supply chain. Figure 13.7 graphically summarizes the project attributes; each of the objects can be represented in this manner. The projects, actual firms supplying to projects, their commodities, and the various markets, nor the actual associations that link firms together has been considered in detail.

13.2.3   Firms, Commodities, and Markets

This section defines the attributes of the following entities: firms, their commodities they provide, and their commodity markets. It also describes the associations between these entities. It then provides a description of the attributes of the object which is at the core of these associations; the firm–firm procurement relationship, and concludes by defining the associations between all the entities.

Firms may be categorized by various means. The key descriptors include: allocated firm numbers, location, scope, size, project types, specialism, market segmentation, firm differentiation, and workload.

Each firm supplies at least one commodity and may supply more than one. There are potentially numerous attributes that can describe a commodity; for example, a product can be described by many physical descriptors, such as weight, size, and color. However, for the supply chain procurement model the attributes of a commodity have been narrowed to those that are primarily concerned with describing the type of commodity that is being transacted at a higher level. There are various building information models and product classification models that are largely interoperable which, ultimately, the model described in this text would connect with. Typically such attributes of a firm could include: location, scope, specialism, market segmentation, and firm differentiation.

In the past, descriptions of firms in the construction industry have been limited to generic groupings, such as subcontractor, consultant, etc. The generic groupings describe little except that consultants supply a service, subcontractors supply a product and a service, and manufacturers supply products. These generic groupings may also bring with them various stereotypes that are ill-defined. At times, these terms may also be too broad and become meaningless. The commodity attributes allow more specific descriptions of what is supplied by the firm for each transaction.

The firm supplies a product, a service, or a product and a service within a transaction. Firms may supply more than one commodity type (refer to Figure 13.8), therefore firms may be competing in a number of markets. Each firm may have a set of commodity objects; each commodity object with its own attributes.

Each firm also has an association to a project by virtue of the commodity it supplies. Each project requires a generic commodity, which a group of firms may supply; however, the project and the upstream customer may then demand unique attributes for a particular transaction. It is the characteristics of the commodity supplied that is important to the project and the firm–firm procurement relationship. It is a combination of the firms' commodities and market attributes and the project attributes that are critical to the customer's demands and the ultimate composition of the supply chain that forms for each project. Each commodity is located in one sector market, although may be associated within more than one project market, and this is discussed in the next section on associations between various elements. The uniqueness of each commodity supplied for each project is related to the firm–firm procurement relationship object, which is also described later in this chapter.

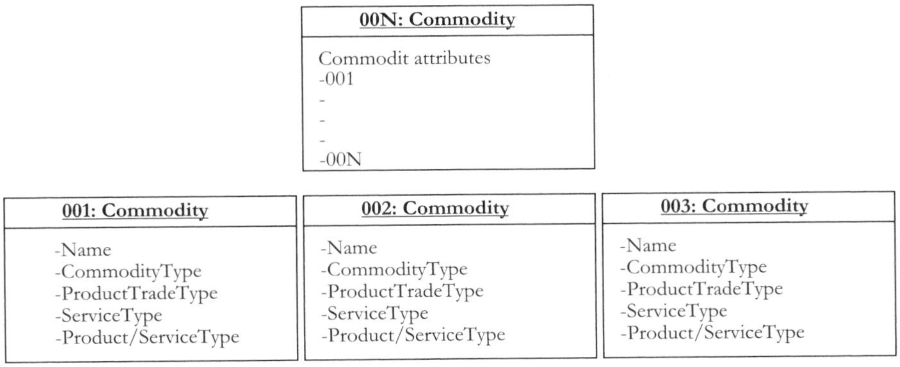

FIGURE 13.8 Commodity attributes.

Commodities are located in markets, and four key attributes have been identified to describe the markets. All markets can be classified according to the International Standard Industrial Classification (ISIC) system or corresponding system for that country (for example, in Australia and New Zealand, ANZSIC). The ISIC system [ABS 2003] is based upon commodities. Coupled with this is the market structural attribute which is denoted by the number of competitor firms and/or the concentration ratio if applicable. Markets may also be segmented.

The industry classification numbers and concentration ratios are useful to a certain extent; however, they do not describe completely the competitive nature of the industrial market. In simple terms, concentration ratios are a measure of the market share of the leaders of the market (note there are numerous permutations and fine-tuning of definitions of ratios). Similar to how a firm locates itself in a generic sector that may be segmented, correspondingly, each commodity may be located in a market that is segmented. Although not considered at all in the construction supply chain literature, it is suspected that there are levels within these markets. The levels are related to how a firm differentiates it. This is difficult to capture through the national industry classification systems, and this is what is really interesting in relation to the nature of competition in the market.

According to ANZSIC, a commodity (and a firm) could be classified as E4222; which would mean that the firm supplies a commodity related to brick laying services. This immediately locates the commodity within that industrial market; however, industrial organizational theory suggests that firms perform in certain ways as they compete in the market, which produces various levels within a market. There are many degrees of service levels of brick laying services possible in the construction industry. A further attribute: segmentation is proposed for the industrial market entity of a commodity which accounts for this quite dynamic attribute.

The nature of competition within the markets can be more specific. In the market attribute for segmentation, this specificity of competition is accounted for. Describing industrial market competition characteristics in a more responsive manner than the national statistical system is important to describing the context of the relationship between upstream and downstream suppliers in the chain. It is a fundamental property of the structural and behavioral characteristics of the construction supply chain. Segmentation identifies the following: facade fabrication for special projects, facade fabrication for standard projects, architectural design service major projects, architectural design services small-scale residential projects, etc. The segmentation criteria can be based upon, for example, product complexity, service scope, firm scope, price, etc. The same sort of categorization occurs in the subcontracting group; for example, there are many steel fabricators, yet a more select group may be focused upon less complex projects such as small-scale residential projects, or more complex projects such as a national stadium (refer to Figure 13.9 for a summary of the industrial market attributes). We can think of segmentation in terms of tender invitations and expressions of interest. A key attribute of the market that allows us to understand the nature of competition is profit margins.

The firm–firm relationships that form for individual projects have not yet been considered. The firm–firm relationship brings with it an even greater clarity of

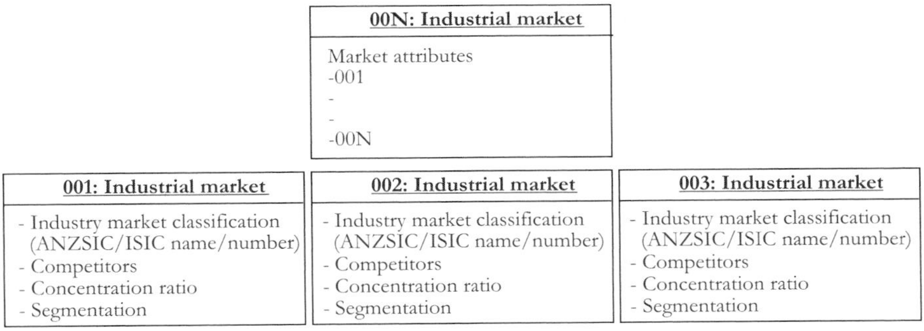

00N: Industrial market
Market attributes -001 - - -00N

001: Industrial market	002: Industrial market	003: Industrial market
- Industry market classification (ANZSIC/ISIC name/number) - Competitors - Concentration ratio - Segmentation	- Industry market classification (ANZSIC/ISIC name/number) - Competitors - Concentration ratio - Segmentation	- Industry market classification (ANZSIC/ISIC name/number) - Competitors - Concentration ratio - Segmentation

FIGURE 13.9 Market attributes.

commodity market and our understanding of supply chain industrial organization, as there are often project-specific markets. Project markets may provide a further level of segmentation according to the individual characteristics of the specific project. In the construction industry, firms may supply commodities that vary according to the project requirements, and thus the accepted concept of vertical integration is challenged because within one firm the level of integration up and down the production chain changes with each project. This high level of variability and uniqueness that arises with each project may be a unique characteristic of the industrial organization of the construction industry. For example, a firm may typically supply a commodity that is a product and a service involving the design, fabrication, and installation of a facade. This particular firm may also manufacture the aluminium extrusions and conduct 2nd order glass processing. This represents a degree of vertical integration. If the firm also supplies these extrusions to its competitors, then this represents another commodity and can be clearly identified. The project, the firm, the commodity, and the industrial market objects are all interlinked, and these associations are now explicitly considered. The association manifests itself in the firm–firm procurement relationship.

13.2.4 Associations: Projects, Firms, Commodities, and Markets

Each firm *works on* a project. Each firm *supplies a* commodity on a project. Each commodity *competes in* an industrial market. Figure 13.10 represents these relationships as simple, linear, and static associations that occur between these entities. Each entity has a number of attributes and each attribute has descriptors; for example, the commodity type can either be a service, a product, or a product and a service, and the firm scope can either be local, regional, national, international, or multinational. Each attribute requires a set of criteria that assists in establishing distinctions between the objects; for example, commodity type can be service, product, or product and service; firm scope is local, regional, national, international, or multinational; and industrial market class can be sourced from the national statistic classes and subclasses.

The next stage of the model development is to explicitly focus on the objects that tie the supply chain together, that is the firm–firm procurement relationships; this is discussed in

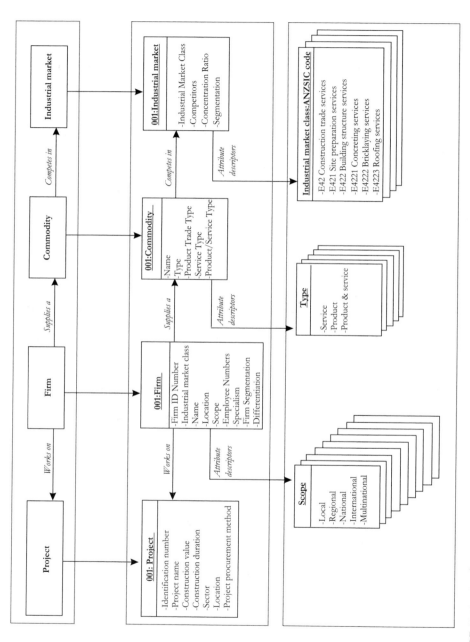

FIGURE 13.10   Attributes and associations of projects-firms-commodities-markets.

Section 2.5. Before this is done, however, it is possible to consider to some extent the dynamic nature of these relationships and the multiplicity of relationships across many projects.

There are numerous projects that make up the construction industry and firms typically work on one or more projects at any one time. On each of these projects firms supply a commodity. The commodity may differ from project to project. Figure 13.11 illustrates the multiplicity that underpins the industrial organization of the construction industry. The complexity and high volumes of transactions that take place in reality in the industry are now beginning to be realized through this model.

The supply chain organization is reliant upon the firm–firm procurement relationships that arise on projects. This object draws together the following connexions between the:

- Upstream and downstream firm objects as they are linked in the supply chain
- Supplier firm object and the commodity object
- Firm–firm procurement relationship object and the project object

13.2.5 Firm–Firm Procurement Relationships

The development of the model thus far has considered that the supply chain is composed of the basic entities or objects: projects, supplier firms, their commodities, and the industrial markets. The commodities and markets were considered in relationship to the broad industrial market sector. Now the upstream firm–downstream firm procurement relationship is examined in detail to bring this association directly to the project. Firms don't typically transact with a project, instead they transact with a "client" or a "customer." However, firms do have an association with a project. Supplier firms are associated with a project through their upstream relationship. The next section will deal with discussing firstly the procurement relationship and its attributes from the perspective of the *supplier market environment*, and then will identify the attributes from the *customer demand environment*.

The supplier market environment includes the following attributes: project market, supplier location, transaction complexity, and transaction significance. The firm–firm procurement relationship also has attributes dependent upon the customer behavior, including: sourcing strategies, supplier choice, transaction type, transaction frequency, supplier management, number of parties, payment method, and financial value.

At this point it might be worthwhile remembering the industrial organization descriptors for distribution of market power, namely, seller and buyer concentration. The distribution of power in the construction supply chain should consider that much of the power is derived from actual volume and purchasing power. The more work that is won by the firm, which translates into more projects and/or higher contract values, then they are, in turn, able to exert more influence when purchasing commodities downstream. The greater the need of the downstream firm to win the work then the weaker their position in the market. Let us call this market vulnerability and it is evident during tendering, negotiation, and during the life of the contract. It is a particularly dynamic factor—market vulnerability and need to win work can change within a month, so that when a firm submits a tender to when it begins a contract can be a very different level of market vulnerability.

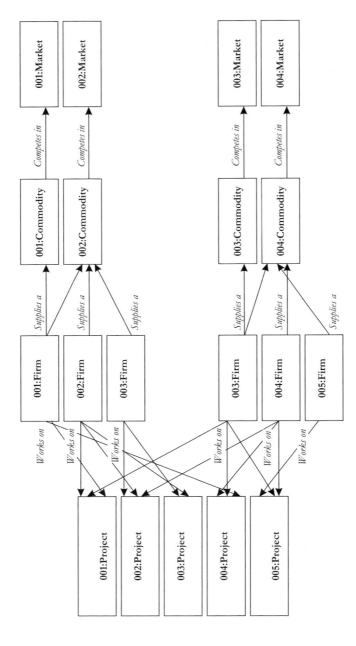

FIGURE 13.11   Multiplicity of projects-firms-commodities-markets.

Alternatively, in some cases there are only a few materials, component suppliers, consultants, and it is then that the downstream firm is able to exert a higher degree of negotiating power in the relationship. These are important considerations in understanding the organization of the supply chain and the different types of relationships that develop. It is speculated that much of what preoccupies the chain participants is their behavior in relation to this strategic positioning for market power. Figure 13.12 is a graphical representation of the procurement relationship attributes and the associations with the previously discussed entities.

13.2.6 Supply Chain Organization

Firms generally require a number of suppliers to assist in their ability to fulfill contracts. For example, a concrete subcontractor may require contracts with a concrete manufacturing firm, a steel reinforcement supplier, a waterproofing specialist supplier, a labor contracting firm, and a concrete formworker. Each of these suppliers then, in turn, may have a number of suppliers that they require to fulfill their contract to the concrete subcontractor. Each firm appears to be a node for a cluster of firms supplying various

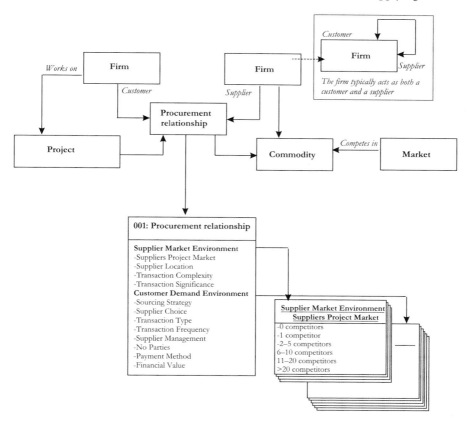

FIGURE 13.12   Procurement relationship attributes.

commodities. The chain has often been represented primarily as a linear entity when, in reality, it may be a series of clusters of firms. Each firm acts as a node in a network of its own suppliers or as the nexus of contracts [Hart 1989; O'Brien 1998; Reve 1990].

Figure 13.13 indicates one scenario of the arrangement of firms supplying on three projects from the focal firm, the client. Each connecting line represents an individual

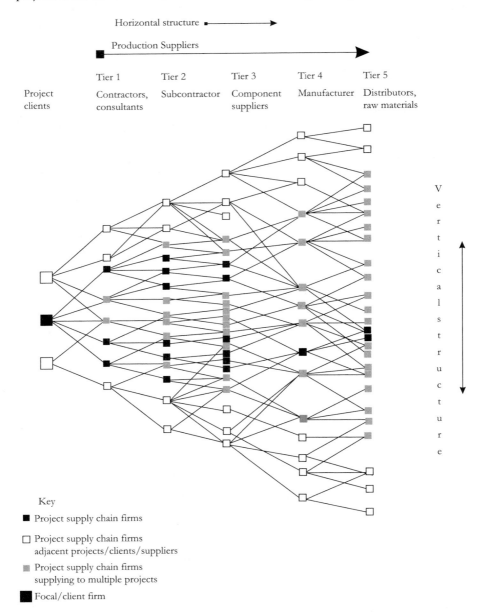

FIGURE 13.13 Complex construction supply chain organizational structure—multiple projects.

firm–firm procurement relationship between firms. This graphical representation maps firms and firm–firm procurement relationships, and gives an indication of the transfer of commodities from one firm to another. Past representations of the supply chain in construction literature have not attempted to indicate this volume of transactions that takes place, nor the number and variety of firms involved in the supply chains. For example, the impression of the supply chain that is often given is that there is a homogenous group of firms supplying at tier 1, and they are subcontractors who behave in the same manner, and who are located in the same type of markets.

The firms represented in Figure 13.13 are those that have been selected. At any one position in any of the multitude of chains, a different firm may be chosen for the project, which would alter the composition of the supply chain. A permutation of one firm can dramatically alter the supply configuration.

13.2.7 Aggregated Project Supply Chain Organization: Supply Channels

Comparisons between supply chains of the one commodity type and supply chains of different commodity types; that is, differences or similarities with respect to such attributes as distributions of firm size, degree of horizontal and vertical integration, and number of relationships, could well be indicators for such characteristics as competitiveness, power distribution, innovation, effectiveness, and efficiency—concepts related to market performance.

There are a range of comparative analyses that would be available to the construction research community and policy makers alike once a deeper understanding of the underlying structural and behavioral characteristics are mapped.

Within the context of this model it would be possible to locate such types of relationships and the likelihood of such relationships occurring between particular types of firms on projects with certain characteristics. The value of this model is to understand change in the structuring of relationships and, in time, model the residual impact of change on the industrial market and the industry as a whole. The model also creates a mechanism to understand and locate any differentiation that occurs between firms and create new relationships accordingly. Figure 13.14 graphically summarizes the aggregation of project market supply chains.

13.3 Empirical Results

The remainder of the chapter is organized in three main sections:

- Firm attributes
- Industrial markets and commodities
- Supply channels

The study also identified the way in which firms in the supply chain categorized their suppliers, and then the criteria upon which they based their procurement decisions related to this categorization. This chapter does not report on this and only maps selected supply channels.

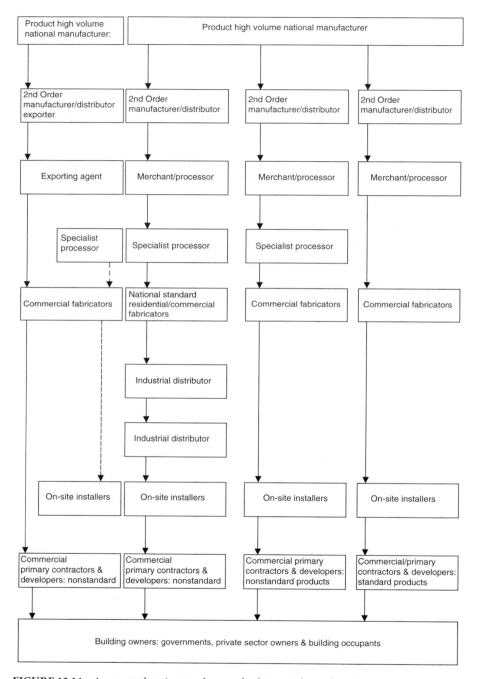

FIGURE 13.14   Aggregated project market supply chains: industrial market chains.

After the clients and/or contractors were interviewed, specific subcontractors were interviewed, and various chains were followed in detail related to a commodity product. For the facade subcontractor, this also involved interviewing the structural steel fabricator, merchants/distributors and manufacturer, the glazing supplier, the glazing distributor and manufacturer, the aluminium fabricator, and the steel painter. The suppliers related to the steel supply chain are discussed in Chapter 8. During the interviews, the range of projects that firms were involved in was also uncovered and these are listed:

- Facade for project 1
- Structural steelwork fabrication for project 1, 2, and 4
- Glazing supplier for project 1
- Glazing manufacturer for project 1–6
- Aluminium for project 1–5
- Steel distributor, merchants, manufacturers for project 1–6

As it was discovered, many of the commodities are found in all six projects, feeding into numerous other chains. The firms were all allocated a number and will largely be referred to by that number. A large database was developed for all the firms, which was used for the statistical component to the study.

One of the most significant factors of project 1 was the complex building form: angular in three dimensions. The outer skin of the building was a curtain wall facade with triangular infill panels that were either stone, zinc, or glazing. The facade was not only complex in its three-dimensional geometry, but also in detailing. As a result of this complexity the facade was a significant element in relation to design, construction, and project management, and many firms were involved in some manner. As well as this, the facade subcontractors were required to source and integrate suppliers that they typically did not coordinate and manage on projects.

The structural organizational map in Figure 13.15 indicates four clusters of firms that the client procured: design, project, construction, and marketing. In many ways, all clusters relate to the facade, even marketing in this case, since the facade was controversial. However, the design, project management, and construction clusters are more directly related to the construction of the facade than the marketing cluster. Within these clusters, the architectural consultant (Arc), the quantity surveyor (QS), the building surveyor (BS), and the contracts lawyer (Law) are significant to the facade and, as the facade was particularly complex, a specialist facade consultant (Fac) firm supplied a service of specialist advice to the Arc as well. The detailed design, construction, and installation were provided by the facade subcontractor who was procured by the primary contractor.

Figure 13.15 indicates the chain of contracts, from the facade subcontractor at tier 2 through to tier 4 suppliers. The facade subcontractor is Firm 191. To deliver the facade to the client, the facade subcontractor, Firm 191, contracted 12 suppliers.

These suppliers to the facade subcontractor are organized in three groups: common products, commons services, and unique products. These categorizations were offered by the facade subcontractor in the following manner:

- Common products: aluminium extrusions, specialized glazing (seraphic), gaskets, silicon, pop rivets.

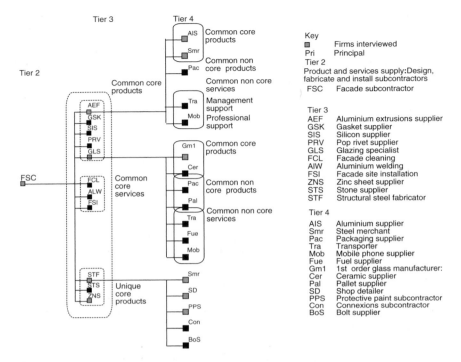

FIGURE 13.15   Facade structural organization map.

- Common services: facade cleaning, aluminium welding, facade site installation.
- Unique products: steel, stone, zinc.

This section describes selected firm attributes, their firm suppliers, and the market environment for the following key firms in the chain: the facade subcontractor, aluminium extrusions fabricator, specialist glazing manufacturer, and the glazing manufacturer. The steel fabricator was also part of this project supply chain; however, they are not discussed in this chapter.

13.3.1 Firm Details

The matrix in Table 13.1 summarizes attributes for the following firms: the facade subcontractor, the aluminium extrusions fabricator, the specialist glazing supplier, and the glazing manufacturers. The discussion focuses on describing the characteristics of the facade subcontractor, aluminium fabricator, and the glazing suppliers firms, and then the industrial market and commodity types, and although not presented in this chapter, this provides the context for the firm–firm procurement relationships developed.

TABLE 13.1 Matrice of Supplier Details for Project 1 Facade

Firm Type and Project	Structure and Scope, Size—Employees and Turnover	Commodities and Competitors	Competitive Advantage
191 Specialist façade subcontractor Project 1	12 manufacturing plants located in Asia, Australia, Europe and US. Within Australia there are two divisions both in capital cities. Multinational, $6–25M t/o. >1000 employees and approximately 60 employees in this division	Designs, manufactures and tests prototypes of curtain walls and then manufactures offsite and installs onsite. Two types of markets: • Standard curtain wall: Five competitors & highly competitive • Complex customized facades: One competitor and is a niche market Competitive advantage: International scope. Integrated firm. Innovative.	
532 Aluminium extrusion supplier Project 1	Two divisions in capital cities with joint ventures in Pacific Asia; international, 4500 employees and >$250M t/o.	Designs and fabricates metal and plastic extrusions. Three main divisions: copper, brass and aluminium extrusions. Plastics was an acquisition and diversification. Two types of markets for the construction industry: 1. project specials and 2. standard extrusions: high volume work supplied to large national manufacturers (for eg window mfrs) Four main competitors including a general import competitor—Refer to table on aluminium extrusions fabricator for specific discussion on the commodities and competitors	The competitive threat is high. *It is a hugely competitive market that we operate in. To give you an idea from the dollars and cents just quoting some figures. An average prices in Melbourne for an extrusion at the moment form the mill might be say $4.80/kilo. Chinese can get it in for say $3.80/kilo. No one—firm 146, Capral, Boral or us can do this at the moment.* The importers have to have an agent in Australia. These companies tend to be Chinese based. However the majority of these extrusions go towards the standard high volume and the automobile industry markets and not in the project work. For Australian firms in this market the competitive advantage is lead time and service. The project is the same generally across the board. The ability to respond to the unique project, customer requirements and/or customer changes is particularly important for this market.

(continued)

TABLE 13.1   Matrice of Supplier Details for Project 1 Facade (Continued)

Firm Type and Project	Structure and Scope, Size—Employees and Turnover	Commodities and Competitors	Competitive Advantage
532 Aluminium extrusion supplier Project 1			In the project type work Firm 191 are competing against Firm 146. When 191 win the project we are competing against Capral or imports. Imports for project work is very risky. I don't think Rocco would be too comfortable having half his load stuck on a ship. It would be a huge risk. In project work there is the ongoing "oh and we need this". And they have forgotten about it. It is the responsiveness that is important. I think physical closeness is important, for eg, the project is in Melbourne and we are in Melbourne, also if the project was in Sydney or vice versa that sort of closeness is ok. Because you always have an architect in inverted commas. You know no architect ever wants the same shape even though it is exactly the same in our opinion, they will want a little tweak of it. Something different.... to the human eye there is no difference but and to us as manufacturers there is a huge difference. And you need to be able to respond to that quite quickly or they will change their mind half way through and they do that as well.
472 (170/640) Specialist glazing firm	Division operating independently within a multinational. 3000 employees nationally; 125 in division. $26–75M t/o.	Processing of raw glass for production of seraphic glass. Four markets: 1. Appliance glass; that is, kitchen appliances, cooking appliances which is the Australian market. 2. Export market appliance glass. 3. Architectural products (export and local). 4. Other products primarily for Japanese market: automobile industry, solar panels, street lighting—various products. In the architectural market, there are three primary products: cladding, internal partitions, and internal/external signage. Highly specialized product; aesthetic quality.	

473 (170/64)
Raw glass
processor and
merchant

Firm 472 have a large share of the market although it is noticeable that they do not produce high-performance glazing products and that Firm 146 is the only supplier of these types of products in Australia.

We are in a commercial market to a certain extent but not the level that we used to be in. We don't have a glass which is suitable for the high rise building, for example, the Rialto or any of those in Collins St. We used to produce a glass for that type of building; that is, it would be a coated glass to get that semireflective surface. Not very high but semi which of course need now to keep your air conditioning loads down. The glasses that we would produce these days and this is as a result of the building bust that came about in the late 80s where we sold our coater. The coater puts the reflective surface on. We sold it to Firm 146 up in Qld our customer up there. We used to sell it under the Suncor name that is the brand name for it. It went on numerous buildings around Australian. We sold that and concentrated more upon the low-rise buildings be it in suburban area or otherwise apartment blocks with what I would call a medium performance glass using tinted glasses and those sorts of things. As distinct from what I would call high performance glass. In summary we are in the commercial market for low-rise buildings, factories to a certain extent, apartments blocks to quite a big extent but not in the 30/40 high rise building. That tends to now be serviced by either Firm 146 but mainly by the US suppliers. By that I mean people like Virocon is a typical example where they have large production runs of the appropriate product and therefore can get economies of scale.

When Firm 767 (external to Firm 170) and Firm 473 (division of Firm 170 at the state level) act in competition at the state level they are treated the same as customers.

From a national viewpoint we overcome that by ensuring that our Sales Centres buy at no better but they buy at the best available price within that particular state. In other words the prices that I sell to my major independent, for example, in Victoria that is Firm 767, our Firm 473 Sales Centre would buy at that same price. I am not treating one more favourably than the other. This seems to work out well.

(continued)

TABLE 13.1   Matrice of Supplier Details for Project 1 Facade (Continued)

Firm Type and Project	Structure and Scope, Size—Employees and Turnover	Commodities and Competitors	Competitive Advantage
170 (640) Raw glass manufacturer	Multinational, 3000 employees nationally and >30,000 overall. >$250M t/o. To manufacture glass this firm has three building product plants: two interstate (Sydney), and one located in this state (Melbourne). Another plant does laminating in Geelong (southwest of Melbourne) for the automotive industry, and one of the plants does laminating, one does toughening, and one does patterned glass. There is an encapsulation plant (which reprocesses), and a small plant interstate (South Australia) for automotive. The patterned glass plant in Sydney does a number of distribution centers in each state.	Manufacturer of raw glass. Firm 170 has a large share of the market but they do have competitors depending upon the product. If you take raw float glass which is the glass that generally makes up most of the residential windows they have about 75–80% market share. However that product could go down to be cut to size for window or go through another process; that is, laminated, etc., or it could go to another path. At the level of laminated glass Firm 170 tend to have sales of laminated glass they tend to hover around a bit less than 50% of market share. In terms of mirror glass they tend to hover around a bit less than 45% of market share. In terms of toughened product it is probably about 55 or 60%. *There is quite a difference in market share for Firm 170 between the products. Apart from the firms that are their competitors in Australia the other major threat is from overseas supply and this tends to concern the firm more than the other local processors as they tend to supply the majority of the raw glass at the beginning of the chain. The main concern with imported glass is the low price. However Firm 170 claim that their competitive advantage is a principle that they have described as "mother warehousing" which is simply that they have large warehouses which means that the customer can buy from them without having to wait for long lead times. Lead times are at least 4–6 weeks shipping times. In reality the lead time is actually a lot longer than that because there is the lead time for the manufacturer as well. Firm 170s major competitor is from Asian imports and to a less extent European glass. The glazing market in Australia is characterized by firms competing and also supplying to each other referred to as the customer/competitor dilemma.* *Firm 146 and the others are purchasing Pilks glass and then processing it to form another product that Pilks actually produce as well. We have this customer competitor dilemma that people like myself would have to deal with. However somebody at Mike Gleeson's level don't have to. Mike in Victoria sees 767 as a competitor to me it is a customer. We have to walk this very fine line of selling to and having a very close partnership with them. But at the same time they are competing with the people that buy from them. It is tricky diplomatic game to say the least.*

13.3.2 Commodities and Industrial Market Details

Firm 191's market for project 1 was the niche market for complex curtain walls. The subcontractor competes in two markets: standard and complex curtain walls. The division is a medium-sized enterprise embedded within a large multinational. There is one other major competitor in Australia for the complex curtain wall. The facade subcontractor sources from 12 suppliers, and the strategies and manner in which the firm groups the suppliers is discussed later in this section. They are indicated graphically in Figure 13.15.

The aluminium extrusions fabricator, Firm 532, has four other major competitors including "imports" as a general category. Table 13.2 indicates the firms and the product types in the Australian market, and was developed by the general manager of the aluminium extrusions fabricator during the interview. Similar to how the facade subcontractor competes in two types of markets based upon level of complexity, there are five aluminium extrusions markets based upon degree of complexity. The facade subcontractor competes in two markets: standard and specialized. The aluminium extrusion supplier competes in the construction industry in two markets: project specials and standard extrusions. The five aluminium extrusion markets include the following: standard high-volume work (typically residential window sections and standard commercial sections for shopfronts/light industrial/offices/showers, etc.), project work

---

BOX 13.1   DEGREE OF PREDICTABILITY

I t appears that there are strong historical ties between various firms in supply chains, which relate the procurement relationships that are formed to the nature of the industrial market, to the project market, to the commodity type. Once a decision is made at the subcontractor level of a particular supplier, then there is a small number of typical supplier procurement paths that will follow based upon the interdependency between these four objects: namely commodity type, industrial market, project market, and procurement relationship. This does not mean that there is only one path, more that the choice of supplier is not entirely random and that there will be a likelihood of certain suppliers being chosen; that is a degree of predictability.

The strategic direction of firms, firm governance and decisions regarding firm boundaries impact upon the nature of the market and eventually the types of procurement relationships that form. Such long-term characteristics of markets and movements may evolve over many years and be structurally embedded within the sector and the supply chains. The historical context assists in explaining the eventual structural organization of the supply chain.

TABLE 13.2　Key Suppliers and Customers Market Leaders by Commodity Type Allied to the Aluminium Extrusions

Firm	Type by Production Phase		Type by Customer/Sector		
	Commodity Type 1 Distribution	Commodity Type 2 Extrusions	Commodity Type 1 General Market	Commodity Type 2 Standard high-Volume Work	Commodity Type 3 Project Work
Firm 146 (customer and supplier)	•	•	•	•	•
Firm 191 (customer)			•	•	•
Firm 765 (customer)			•	•	•
Firm 766 (customer and supplier)	•	•	•	•	•
Firm 141 (customer)			•	•	•
Firm 194 (supplier and customer)	•	•	•	•	•
Firm 532 (supplier)		•		•	•

(customized sections), general market (supply directly to contractors), distribution (standard sections mass production to distribution centers), and extrusions (fabricate special extrusions to various customer types). This table does not include all other firms that purchase extrusions; however, it includes the major players in the market or what is commonly referred to as market leaders.

The table includes two of Firm 532's major customers, including a customer in project work and the other in standard high-volume work—the previous upstream facade/curtain wall subcontractor supplying to project 1 (Firm 191), and another national window manufacturer (Firm 765). Firm 765 also competes against Firm 194, and they are both national manufacturers and suppliers of standard windows. They are market leaders in the supply of residential windows. Firm 194 is also a market leader in the supply of commercial standardized aluminium window sections, and also a market leader in the supply of numerous other construction materials and components as well, such as concrete, quarry, and timber products. Firm 146 and Firm 141 are the two firms that compete in the specialist curtain wall glazing market (Firm 191 is the facade subcontractor for project 1). Firm 141 is closely allied to Firm 146. Firm 532 does not supply to Firm 146 and, in fact, Firm 146 and Firm 766 (another commercial window section manufacturer and supplier) are their major competitors. The table was developed based upon descriptions of the market by Firm 532, Firm 141, Firm 191, and Firm 003 (the primary contractor for projects 1, 2, and 3). The descriptions were provided by company documents and interviews. Each interviewee confirmed the data in the table.

As previously discussed, the commodity type and market characteristics impact upon the procurement relationships that are formed in the supply chain. The previous discussion gave examples for the aluminium fabrication market in Australia.

The customer and supplier firms were all easily identifiable as separate firms, and yet there is an interdependent nature between firms based upon procurement relationships; that is, some firms always align themselves with another firm. Therefore, if a choice of one firm is taken then there is another firm that will most likely be sourced. As noted, Firm 146 and 141 were closely allied. Firm 141 was historically a division of Firm 146 until approximately a decade earlier. Such alliances existed in the glazing supply chain and the steel supply chain as well.

The competitive advantage for the aluminium fabricator is price and lead times, followed by the service they offer—particularly on complex project work. Price for all metal products is based upon the London Metals Exchange and fluctuates daily, therefore the ability to hold prices for a period of time on behalf of their customers is a significant advantage. This is essentially risk management.

The fabricator, Firm 532, was historically a part of the national aluminium manufacturer, and so they too have close ties with the only Australian aluminium manufacturer. The aluminium market discussion would be enhanced by an understanding of the aluminium manufacturer and its supply channels, however, this firm was undergoing major upheaval in terms of mergers and acquisitions at the time of the study, and no one was available for interviews. The following discussion describes the nature of the glazing market in Australia and the various commodities supplied in a similar manner to the previous aluminium extrusions market discussion.

13.3.2.1 Glazing Commodities and Industrial Markets

Firm 472, specialist glazing supplier, sources from the glazing distributor Firm 473 (state sales division) of the glazing manufacturer, Firm 170. Each of these are considered as separate firms as the interviews clearly indicated that, with regards to procurement and pricing, they are treated no differently to any other firm purchasing in the marketplace.

Firm 170, the glazing manufacturer produces four glass products: annealed, laminated, toughened, and mirror. They are the only suppliers of the raw glass material (annealed glass) in Australia. However, three other firms laminate glass including: Firm 146 (previously discussed in the section on aluminium extrusion suppliers and a competitor to the facade subcontractor), and two others—let us call them Firm 767 and Firm 777. There are also a large number of importing agents who supply laminated glass. Table 13.3 summarizes the types of products that are supplied by each player in the glazing market and their role, that is, whether or not they are a window fabricator, a merchant, or involved in 2nd order processing as well.

Firm 473 has a large share of the market although it is noticeable that they do not produce high performance glazing products. Firm 146 is the only supplier of these types of products in Australia.

Standard high-volume work typically includes both the residential market and the commercial market. In the commercial market it may include products such as glazing for retail shopfronts, standardized curtain walls glazed balustrades, internal/external screens, and shower screens. In the residential market it includes primarily standard aluminium window sections and shower screens. The project work typically includes atriums and curtain walls.

TABLE 13.3   Key Glazier Suppliers Market by Commodity/Production

Firm	Anneal. Glass	Lam. Glass	Tough. Glass	Mirror Glass	High perf. Glass	2nd Order Processor	Merchant	Stand. High-Volume work	Project Work
	Commodity					Production Phase		Sector (Direct Supply)	
170	•	•	•	•					
473						•	•		
146		•	•		•	•	•	•	•
767		•	•			•	•		
777		•							
779			•						
780			•			•			
Imports	•	•	•	•	•	•	•		
781								•	
782							•	•	
778								•	•
191								•	•
765								•	

Glazing product supply on projects in Australia generally takes one of three major procurement chain routes. The first is the most direct route, which takes its roots from Firm 473 or the National Sales Division of the only glass manufacturer (Firm 170), to the major nationally operating window residential fabricators/glaziers (of which there are three). The second is more circuitous and involves smaller firms that generally operate at the state level; that is, the State Sales Division (473) purchases unprocessed glass from the National Sales Division of the glazing manufacturer (170), and either simply repackages the product in smaller volumes, or phase/sector alternatively provides secondary processing and then supplies to glaziers/fabricators. Finally, the third route is where glaziers purchase glass from an importing agent who is acting for distributors in another country, and currently this is most typically China. Having described the major paths it was then also acknowledged that there were many permutations in this chain, as noted by the Australasian Supply Manager for the glass manufacturer, Firm 170.

For example, the glazing for project 1 is somewhat different again—the specialist glazing supplier, supplying seraphic glass (472), purchases from the National Sales Division (170). However, there are only three customers that the seraphic glazing supplier (472) will allow to purchase their product direct from them, including Firm 191, the facade subcontractor. All other curtain wall or window manufacturers must purchase through the States Sales Centers (e.g., 473), which adds another link in the chain and, in reality, adds to the price.

Similarly, the National Sales Division, Firm 170, restricts supplies of glass to the major residential window fabricators of which there are three and some state-based fabricators.

Box 13.2   Intricate Supply Chains

It is very competitive in the marketplace. It is also very intricate as well.

(Australasian Supply Manager, Glazing Manufacturer)

Box 13.3   Countervailing Market Power

They would do it two ways. Usually Firm 191 comes here direct. There are about two or three companies that we deal with direct and one is Firm 191, Firm 141 glazing and a couple of others. The reason we do that is because they can not be bothered going through Firm 473 and they would rather deal direct. If they go through Firm 473 they will get priced out anyway. It is just adding another commission on it. We have picked two or three or four of those types of people that we will deal with direct.

(Australasian Supply Manager, Glazing Manufacturer)

Coupled with this is the residential "hack and glaze" market, which is window replacement. They also deal with many small window fabricators or processors as well. The other major category is the customer that is both a residential and commercial window fabricator. The common element amongst these customers is not the size of the firm, but whether or not they have the capability to process the glass as it comes to them in a raw state. Although there are processed products that this firm supplies as well, which in turn means that they supply to their competitors at times.

Contrary to expectations the highly vertically integrated firm in this situation does not necessarily equate to a lack of external competition. As noted previously in the matrix that described the attributes of the firms and their commodities and competitors, at the next tier the division Firm 473, within Firm 170, competes on equal terms with the firms outside, therefore each tier operates purely on a transactional basis from the supplier side. This appears to work out well for the supplier, however the "firm" within the firm, 473, who is the customer upstream does not seem to have the same market independence.

Box 13.4   Thou Shalt Not Stray

A major independent customer is one that is external to Firm 170 and they can purchase from us or one of four competitors, whereas our Sales Centers being part of Firm 473 must purchase from us they do not have that choice.

'Thou shalt not stray!'

(State Manager, Glazing Distributor)

Box 13.5   Market Categories

Although seemingly straightforward the market is further categorised. In terms of building glass we either have the window fabricators on the residential side, the hack and glaze people, and the commercial side is the Firm 767, Firm 146 and Chevron. This is in terms of the building area but you need to look a bit further to split the market up in terms of what is the main function of that customer:

Is he a merchant? Is he a processor of glass? (i.e., making toughened, laminated, etc.)

Is he simply there supplying cut to size glass for smaller people to glaze with?

(Australasian Supply Manager, National Sales Division)

This customer–supplier relationship between the 1st order (glazing manufacturer, Firm 170) and 2nd order manufacturers (glazing distributor Firm 473 and Firm 472—processors) is discussed in more detail now, followed by a description of the downstream suppliers to the 1st order manufacturer.

Firm 170 has three major types of construction industry customers: the residential window fabricators, the "hack and glaze people," and the commercial sector. The "hack and glaze" are window replacement.

Within these three categories the traditional merchant would buy glass from a manufacturer (either this firm or from an importing agency). They would typically purchase 20-tonne loads and then would onsell that in block form. The next extension to that role would be to cut that product to size to whatever someone would require and sell it as a cut piece of glass for a product. The merchant has largely evolved into value adding glass; that is, he became a processor—either furnacing the glass, printing the glass, putting in a laminating line to supply laminated glass. That is adding value to the glass he buys from this firm so that he can, in turn, get a better margin when the firm onsells to their customer.

For the commercial sector projects there tends to be two main supply chain routes, one through a processor and then the glazier, or direct to the glazier. If it went to the glazier then the chain would include the State sales division. If it went through the processor then it would not include the state sales division. He continued with...

Another product market that has emerged for the merchant is the fabricating of double glazed units. A merchant customer of Firm 170 has recently entered this market. This firm merchandises and also processes. Then there are other firms that have developed various stages of production.

The following diagram Figure 13.16 is for Firm 170, the glass manufacturer, and it summarizes the primary types of project, firm–firm procurement relationships, and industrial markets based upon customer types.

It is worthwhile also describing Firm 146. This firm is a major player in the glazing market in terms of processing, fabricating, and installation. It is interesting that when the State Manager explained Firm 146, the description melded descriptions of competitor firms and markets.

Firm 146 competes with Firm 191 for the highly complex facades for projects, and although for project 1, Firm 191 purchased from Firm 170, this is not the usual case as the following quote from the Chief Executive Officer of the specialist glass supplier, Firm 472, indicates.

Box 13.6   Market Competition— Changing Structure of Supply Chain

That merchant/ processor is tending to meld one into the other now. In other words they are not making much money on merchandising. They find that they have to get into value added to stay alive principally. There are still merchants hanging around as distinct from processors. They buy from us in 20-tonne loads. A typical example is Davis Glass who is a merchant.

(Australasian Supply Manager, National Sales Division)

Box 13.7   Three Main Supply Channels— Diverse Supply Chains

N ow with our commercial projects. The processor, bearing in mind that for most projects in the commercial project they need to have some sort of value added glass. So be it either laminated or toughened, we would tend to be supplying either the middle man; that is, the processor or we would supply direct to the glazier. Just depends upon the contract. We would quote for a contract to supply glass for a project XYZ Hotel and it would be quoting the glazier direct via the Sales Centre if the sale went via that route, if it was a Firm 146 or a Firm 767 supplying they would be buying the glass from us and they would then be processing it. So there could be two paths.

If it went to a subcontractor glazier it would go through our Sales Centre and in turn the customer of the Sales Centre may well be a merchant who in turn may well supply into that same building or compete with our own Sales Centre because they would be supplying another glazier or another glazier would be competing for that job.

You have all these subsets that are occurring in the marketplace. It is very competitive in the marketplace. It is also very intricate as well.

(Australasian Supply Manager, National Sales Division)

Box 13.8   Alternate Supply Chain Paths—Value Adding

T hen you might have of course a pure processor like Mowen Glass here in Victoria who set themselves up in the window game… making windows and then they value added further by putting in a double glazing lite. So they now produce double glazed units and in turn put them into windows and out into the marketplace. Of course they sell double glazed units to Boral, Stegbar or whatever. Ever so recently they have put in a toughening furnace and again because toughened products are required in their windows and it is this continual value adding process.

(Australasian Supply Manager, Glazing Manufacturer)

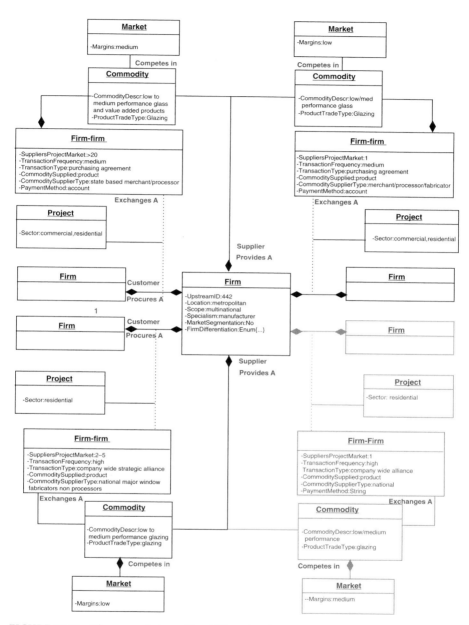

FIGURE 13.16   Glass manufacturer Firm 170 markets by customer.

There are four major national customer accounts for Firm 170, three of whom are residential window fabricators. They require processing and are serviced by the state-based processing/distribution centers; however, the account is managed by the National Division. Likewise, the national glass replacement firm is dealt with by the National Sales Division; however, since they have their own processing facilities they are supplied with the product as well, that is, the raw glass product.

Box 13.9   Integrated Aluminium and Glazing Supply Chains

Firm 146 are a many, many faceted company. They are huge company. They are a merchant, they are a processor, they make toughened, printed, laminated, double glazed commercial and residential units glass, they also are into glazing both low rise and high rise, they are also into aluminium extrusions. They have grown and diversified tremendously.

But they are a bit unique in the glass industry. Firm 767 is going down that track too in as much they do most of that, they merchandise, they process, laminate, print, toughen, glaze, the only thing that they haven't gotten into which Firm 146 has is the aluminium side of things.

In other words you could get Firm 146 to do the complete outside of your building, the complete facade.

(State Manager, Glazing Distributor)

Box 13.10   Hidden Connected Supply Chain Paths

Now we used to deal with Firm 191 who do the same sort of things, we used to deal with them when we produced the Suncor type product, the high performance type product. But we don't anymore. They would tend to import and perhaps buy a bit from Firm 146 and Firm 767 in the way of laminate products. The great majority of stuff they import. If they wanted glass in Australia they would not tend to come to us they would go to merchants or processors.

(Chief Executive Officer, Specialist Glazing Manufacturer)

Box 13.11

We deal with the major window fabricators, people like Boral, Stegbar, James Hardie, plus various state-based fabricators. So we would sell them the glass and they in turn would cut it up or buy it from us in cut size and put a bit of aluminium around it and there is your window.

(Chief Executive Officer, Specialist Glazing Manufacturer)

Box 13.12

There is no point in having a tied in contract with the likes of Stegbar etc. because a contract is only as good as what either party wants. If either party is dissatisfied whether they will let the other one know and request a change and come to another agreement. They will only buy from you if they are happy and so frankly I don't see any point in national agreements.

When they do purchase they purchase from all our Sales Centres all around Australia and they put in daily orders and they would be delivered on a daily basis to their appropriate sites.

I personally deal on the national level and deal with customers who principally buy truck loads of glass from the glass plants, in other words they buy 20-tonne loads of glass. In other words they are the big guys. I then deal with the four major accounts that I mentioned before, that is Boral, Hardies, Stegbar and O'Briens. We deal with them on a national basis. But they are actually serviced on a day to day basis by the State Sales Centres.

(Chief Executive Officer, Specialist Glazing Manufacturer)

It would be expected that national agreements would be in place for these accounts based upon volume and pricing arrangements; however, there are no national agreements between this manufacturer and the four major national residential window fabricators.

Pricing and margins are an interesting issue in this market and these tend to rely upon world supply trends, even though it appears that Firm 170 has a monopoly. The monopoly is only as good as the prices that the firm provides. The existing upstream customer relationships were somewhat surprising given the multinational scope of the firm, and for this reason the entire quote (albeit somewhat long) has been extracted from the transcripts and provided in the following text.

Box 13.13   Gentleman (or Gentle Woman) Agreements

W e don't tend to work on 12 month agreements. We tend to work more on gentleman agreements because we have dealt with them for so long. At times our relationships are rather rocky like they are at present but that is because we are trying to put price increases through.

No we don't have official contracts with them in terms of thou shalt purchase X tonnes at Y dollars. They obviously have price schedules from us and they to date well… we have been very very lucky they have purchased around about 95 to 100% of their volume from us. We are very lucky in that we are in that unique situation that we have some very loyal customers.

Prior to November last year we tended to work on decreases we could not even spell the word increase mainly because of the world glass glut particularly in Asia. This came about because of the Asian crisis of some years back. Glass has been freely available it tends to slush down to Australia and reasonably good prices down here and therefore we were always fighting uphill to get increases.

Since about the beginning of this year glass has become short on the world scene. Asian economies have been improving. America and Europe have been very, very buoyant and have tended to use glass in their own confines. As a consequence glass has become shorter we then saw an opportunity to follow world trends by putting prices up. From November last year we started to have increases on our customers. So we had a very small increase in November we then had the next increase on 1st April and we are about to put through another increase on August 1st. That is very very unusual to have those sorts of increases and of course that is the reason why some of our customers are upset with us. We are basically following world trends.

(Chief Executive Officer, Specialist Glazing Manufacturer)

Firm 170 supplies through to all six projects. Firm 170 either supplies a product or a service in the relationship. The product is a common core product, and the service is a common core service. The service is a management service that is provided to their four major national accounts (Firms 781, 778, 765, and 194). As stated earlier, these major window fabricators do not process and order daily from each state division.

Figure 13.17 and Figure 13.18 describe the upstream and downstream relationships of the glazing manufacturer, and indicates the various relationships to the database which describes in detail the attributes of the procurement relationships, firms, projects, and markets.

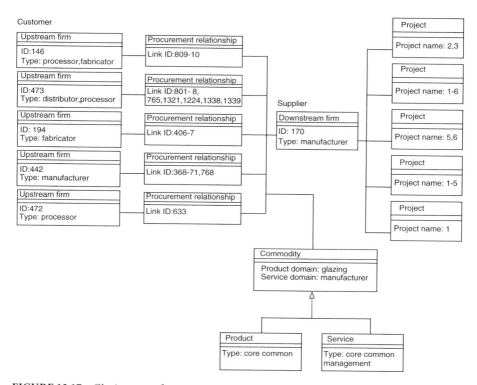

FIGURE 13.17 Glazing manufacturer procurement relationships.

13.3.3 Aggregated Project Supply Chain Organization: Supply Channels

Chain structures map the transfer of ownership of commodities on individual projects, and this section draws together the project supply chains into supply channel structural maps. This section summarizes eight channel structure maps for the construction industry including: aluminium, steel, concrete, glass, fire products, mechanical services, tiles, and masonry. The maps are of the primary material, for example glass, steel, aluminium, and do not include all the subsidiary product and raw material chains required at the manufacturing tier, nor the gathering together that occurs at the subcontractor level for site installation. For example, aluminium windows require numerous suppliers including rubber gaskets, silicon, fastenings, framing, fixings as well as the aluminium window component. The commonly used chain structure that is used is contractor to subcontractor to materials supplier/manufacturer. The following maps clearly develop a much greater picture of the diversity of interactions in channel structure than previously known. This has allowed for a much richer description of industrial market structure. Each channel structure map is now discussed in more detail.

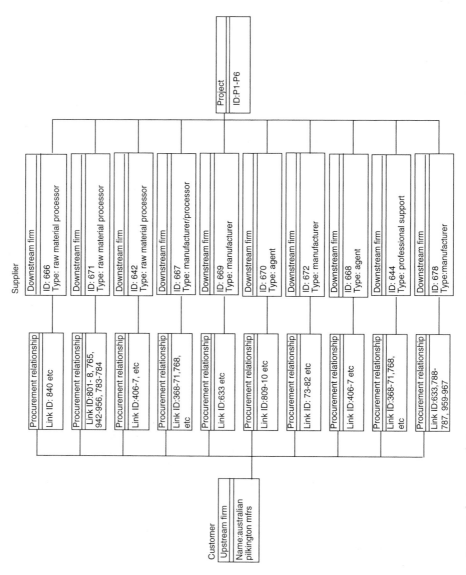

FIGURE 13.18   Firm 170 glazing manufacturer suppliers.

13.3.3.1   Aluminium Chains

Figure 13.19 maps the transfer of ownership of aluminium products in the construction industry, from the manufacturer who extracts raw materials from the ground, to the building owners. The channel maps indicate the structural organization of primary commodity suppliers and the types of firms that are involved. There are currently eight tiers available for the transfer of ownership of the aluminium commodity to flow through disregarding the importation of aluminium billets, however there was no channel identified in the study that included all eight tiers. If the aluminium extrusions fabricator imports aluminium then the channel extends dramatically to 12 tiers. Currently this is the location where imports are introduced into the channel. Thirteen chain options were identified, including one standard import channel.

The organization of the channels extends from three main branches; the extrusions fabricators who manufacture standard as well as special extrusions; the extrusions fabricators who manufacture standard extrusions only; and the large-scale production of residential window extrusions by national manufacturers. Each of these branches has a variety of paths depending primarily upon purchasing power. For example, if the purchasing power of the curtain wall/window fabricators is high then an industrial distributor will not be involved, however if the purchasing power is low then the fabricator procures extrusions from a distributor. Similarly, if the purchasing power of a primary contractor or developer is high then they can procure directly from the national window fabricator or if is lower then the purchase route will be through the industrial distributors.

13.3.3.2   Glass Chains

Figure 13.20 and Figure 13.21 maps the transfer of ownership of glazing products in the construction industry, from the manufacturer who extracts raw materials from the ground, to the building owners. The channel map indicates the structural organization of primary commodity suppliers and the types of firms that are involved. There are currently 10 tiers available for the transfer of ownership of the glazing commodity to flow through including an import channel. Importation currently occurs at the commercial window/curtain wall fabricator tier in the channel. The length of the tier does not extend unduly (six levels) as the commercial fabricators are multinationals and large enough that tier purchasing power by virtue of purchasing volume ensures that they procure glass more directly. Eleven channel options were identified. We can begin to develop interesting questions for further research in relation to cash and information flow and also delivery times. For example, does the commodity flow at the same rate in different chains? How and when can we simulate and optimize time performance?

The organization of the channels extends from three main branches; the extrusions fabricators who manufacture standards as well as special extrusions; the extrusions fabricators who manufacture standard extrusions only; and the large-scale production of residential window extrusions by national manufacturers. If the purchasing power of the curtain wall/window fabricators is high then an industrial distributor will not be involved, however if the purchasing power is low then the fabricator procures extrusions

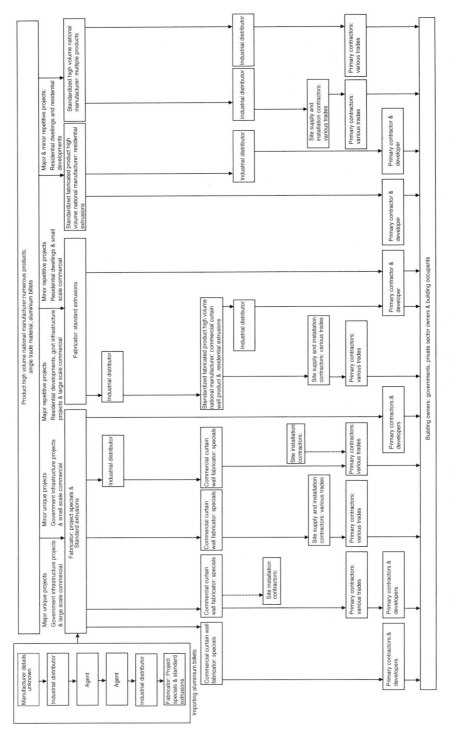

FIGURE 13.19   Structural organization channel map for primary commodity aluminium.

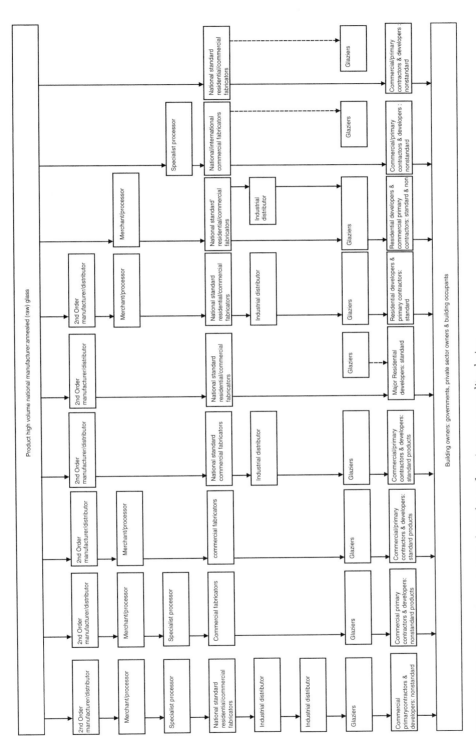

FIGURE 13.20   Structural organization channel map for primary commodity glazing.

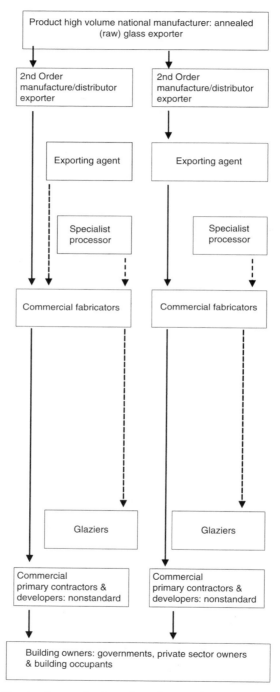

FIGURE 13.21 Structural organization channel map for primary commodity glazing (continued) inclusive of export chain.

from a distributor. If the purchasing power of a primary contractor or developer is high then they can procure directly from the national window fabricator, or if it is lower then the purchase route will be through the industrial distributors.

13.4   Implications

The implications of the results of this case study are discussed in relation to two broad groups: industry practitioners and academic researchers.

The knowledge of the overall structural organization of a sector documented in an explicit manner provides a framework for industry practitioners to make decisions in relation to their own organization's downstream supply chain and upstream clients.

Firstly downstream supply chains, it appears that there are strong couplings between various firms in supply chains; and as such there is a certain degree of predictability. This inherent predictability allows for firms to manage in the first instance their relationships with their suppliers. Of course it is assumed that firms are doing this in the best possible manner already.

The greater transparency of the pressures and impacts from a firms' suppliers' suppliers allows a more strategic approach to managing relationships—monitoring and measuring key relationships that are critical to their own performance rather than letting suppliers manage their own relationships. A more proactive approach can be taken. Once a decision is made at the subcontractor level of a particular supplier, then there is a small number of typical supplier procurement paths that will follow based upon the interdependency between four entities, namely, commodity type, industrial market, project market, and procurement relationship. This does not mean that there is only one path, more that the choice of supplier is not entirely random and that there will be a likelihood of certain suppliers being chosen; that is a degree of predictability. The strategic direction of firms, firm governance and decisions regarding firm boundaries impact upon the nature of the market and eventually the types of procurement relationships that form. In some cases the couplings are based on historical corporate ties between the two firms that have arisen because one of the firms was originally part of the other firm and typically as a division. Such long-term characteristics of markets and movements may evolve over many years and be structurally embedded within the sector and the supply chains. The historical context assists in explaining the eventual structural organization of the supply chain. Other coupling ties that were identified were related to purchasing volume; that is, countervailing power which was typically related to size of firm, and then commodity complexity; that is, standardized vs. unique commodities.

If firms are better equipped in understanding the chains that they are typically involved in and the various players and their relationships to each other, then it may not only allow improvements for supply to them but also allows supply chain management to be achieved.

The results of this case study has also provided an explicit map of one entire sector, albeit sectors are often changing and have implications for government agencies involved in property and construction, either as a client or as a policy maker. There are a wide variety of functional categories of suppliers within the glazing market; that is, merchant, processor, and raw material supplier, and then various combinations of these functional categories; that is, merchant/processor, merchant/raw material supplier, etc.

Thus giving rise to a high level of complexity in the various supply chains. The structural organization is dynamic. Market competition changes the structure of supply chain in glazing supply as the various different functional types of firms take on different functions along the chain to improve their competitive advantage and value add.

Eight tiers for the transfer of ownership of the aluminium commodity were identified and 12 importation chains were included with 13 chain options identified. There were three main branches identified including: the extrusions fabricators who manufacture standard as well as special extrusions, the extrusions fabricators who manufacture standard extrusions only, and the large-scale production of residential window extrusions by national manufacturers. Each of these branches has a variety of paths depending primarily upon purchasing power.

There are currently 10 tiers available for the transfer of ownership of the glazing commodity to flow through including an import channel. Importation currently occurs at the commercial window/curtain wall fabricator tier in the channel. The length of the tier does not extend unduly (six levels) as the commercial fabricators are multinationals and large enough that tier purchasing power by virtue of purchasing volume ensures that they procure glass more directly.

This level of transparency of the inner workings of the industry is perhaps something that government agencies have never enjoyed before. There are opportunities to explore how various policies, processes, and practices impact upon the underlying structural and behavioral characteristics of whole sectors. Targeted procurement strategies to change behaviors and underlying structural characteristics are achievable if a benchmark is known. Rewards and incentive structures can be put in place to support a change in industry culture towards long-term strategic behavior that is concerned with supply chain management to achieve social, environmental, and economic objectives.

The work described in this chapter has implications for the research community who are involved in supply chain research and beyond—those involved in a wide range of topics in relation to the property and construction industry. For example, the rise of building information modeling is clearly aligned with the way in which the model has been described in this chapter. There is much scope to develop further ideas regarding supply chain collaborative practice between firms. Future data sharing between firms could see such sophisticated techniques as data mining within internal organizational databases used when working on projects.

Perhaps one of the significant outcomes of this research is the move forward for the research community in tracing the link between construction supply chain economics and construction supply chain management, and this chapter has only just begun to formalize an area of study which can be referred to as construction supply chain economics.

References

ABS. 2003. Australian Bureau of Statistics Statistical Concepts Library ANZSIC. Chapter 1: About the classification, classification principles. Canberra: Commonwealth of Australia. Available at http://www.abs.gov.au/AUSSTATS/abs@nsf/ Cat. 1292.0- Australian and New Zealand Standard Industrial Classification (AUZSIC), 1993. (accessed 8th June 2008).

Alhir, S. 1998. UML in nutshell: A desktop quick reference, 1st ed. Sebastopol, CA: O'Reilly & Associates Inc.

Barrett, P., and Aouad, G. 1998. Supply chain analysis for effective hybrid concrete construction. *Proceedings of Joint Triennial Symposium CIB Commissions W65 and W55*, eds. P. Bowen and R. Hindle. Cape Town, South Africa. ISBN 0620 239441. Document Transformation Technologies, SA.

Bowersox, D., and Closs, D. 1996. Logistics management: *The integrated supply chain process*. New York: McGraw-Hill.

Briscoe, G. H., Dainty, A. R. J., Millett, S. J., and Neale, R. H. 2004. Client-led strategies for construction supply chain improvement. *Journal of Construction Management & Economics*, 22, 193–201.

Cox, A., and Townsend, M. 1998. Strategic procurement in construction: *Towards better practice in the management of construction supply chains*, Vol. 1, 1st ed. London: Thomas Telford.

Dainty, A., Briscoe, G., and Millett, S. 2001. Subcontractor perspectives on supply chain alliances. *Journal of Construction Management and Economics*, 19, 841–48.

Harland, C. M. 1994. Supply chain management: *Perceptions of requirements and performance in European automotive aftermarket supply chains*. PhD thesis, Warwick Business School, UK.

Hart, O. 1989. An economists perspective on the theory of the firm. *Columbia Law Review*, 89, 1757–1774.

Hines, P. 1994. *Creating world class suppliers: Unlocking mutual competitive advantage*. London: Pitman Publishing.

Lambert, D. M., Pugh, M., and Cooper, J. 1998. Supply chain management. *The International Journal of Logistics Management*, 9 (2), 1–19.

London, K. 2004. Construction supply chain procurement modelling. PhD diss., thesis, University of Melbourne.

London, K. 2008. *Construction supply chain economic*. London: Taylor and Francis.

Minato, T. 1991. The development of Japanese parts supply relationships: Past, present and future. Customer-Supplier Relationship Study Tour, EC-Japan Centre for Industrial Cooperation.

Nicolini, D., Holti, R., and Smalley, M. 2001. Integrating project activities: The theory and practice of managing the supply chain through clusters. *Journal of Construction Management and Economics*, 19, 37–47.

Nischiguchi, T. 1994. *Strategic industrial sourcing: The Japanese advantage*, Vol. 1, 1st ed. New York: Oxford University Press.

O'Brien, W. 1998. Capacity costing approaches for construction supply chain management. PhD thesis, Department of Civil and Environmental Engineering, University of Stanford.

Olsson, F. 2000. Supply chain management in the construction industry: Opportunity or utopia? Unpublished licentiate thesis, Department of Engineering Logistics, Lund, Sweden.

Reve, T. 1990. The firm as a nexus of internal and external contracts. Chapter 7 in *The firm as a nexus of treaties*, eds. M. Aoki, B. Gustafsson and O. Williamson. London: Sage Publications.

Saad, M., and Jones, M. 1998. Improving the performance of specialist contractors in construction through a more effective management of their supply chains. Proceedings of the Seventh International IPSERA Conference, London, UK.

Scollary, R., and John, S. 2000. Macroeconomics and the contemporary New Zealand economy, 2nd ed. New Zealand: Pearson Education NZ Ltd.

Taylor, J., and Bjornsson, H. 1999. Construction supply chain improvements through internet pooled procurement. IGLC Seventh Annual Conference, Berkley, CA.

Tommelein, I., and Yi Li, E. 1999. Just-in-time concrete delivery: *Mapping alternatives for vertical supply chain integration.* IGLC Seventh Annual Conference, Berkeley, CA.

Vrihjhoef, R., and Koskela, L. 1999. Roles of supply chain management in construction. Proceedings of IGLC Seventh Annual Conference, Berkeley, CA.

14

Innovation Management in the Construction Supply Chain

Bart A.G. Bossink
VU University Amsterdam

Ruben Vrijhoef
*Delft University
of Technology*

14.1 Introduction

Previously, drivers of the development, design, and realization of innovations in construction have been explored under headings like "innovation leaders" and "innovation champions" [e.g., Nam and Tatum 1997; Bossink 2004a], "interfirm innovation networks" [e.g., Bröchner and Grandison 1992; Korczynski 1996; Lampel, Miller, and Floricel 1996; Robertson, Pearson, and Ball 1996; Bossink 2004b], and "innovation stimulating regulations" [e.g., Bernstein 1996; Larsson 1996; Gann, Wang, and Hawkins 1998; Guy and

Kibert 1998; Bon and Hutchinson 2000; Ngowi 2001; Seaden and Manseau 2001]. Several drivers applying to different levels in the network of cooperating organizations in the construction supply chain have been identified [e.g., Tatum 1989; Nam and Tatum 1992a; Winch 1998; Gann and Salter 2000; Bossink 2004b]. These levels include three levels: the industry level [Pries and Janszen 1995], the institutional and firm level [Winch 1998], and the construction project level [Lampel, Miller, and Floricel 1996]. The key question here is how innovation can be managed effectively in the construction supply chain at these three levels [Nam and Tatum 1989; Veshosky 1998; Winch 1998; Mitropoulos and Tatum 1999; Barlow 2000; Koskela and Vrijhoef 2001; Miozzo and Dewick 2002]. This paper reports a research project that was aimed at finding answers to this question. The main question of the research was: How can innovation be managed at the various levels of cooperation in the construction supply chain? The two related subquestions were: What are, according to the literature, the main innovation management tools in the construction supply chain? At which levels of cooperation do these innovation management tools stimulate innovation in construction practice?

In the second and third section of this chapter, the research approach and the characteristics of the construction supply chain are investigated in order to understand the context of applying innovation management tools, and how to observe the effectiveness of applying those tools in construction practice. In addition, a literature study has been carried out to provide an overview of the innovation management tools applicable to the construction supply chain. The results of the literature study are presented in the fourth section. The empirical part of the research has been carried out in the context of the Dutch construction industry, resulting in an analysis of the feasibility of the innovation management tools on different levels of cooperation in the construction supply chain. The interview results of the empirical research are presented in the fifth section. Based on the research findings, conclusions are presented in the sixth and last section of this chapter.

14.2   Research Method

The research method consisted of a literature study on innovation management tools, an initial analysis of the construction supply chain and its characteristics, and an interview survey among managers in the Dutch construction industry. The literature study and the interview survey were carried out to provide an overview of the innovation management tools in the construction supply chain. Another purpose was to determine the levels to which these tools could be used to stimulate innovation in construction practice. Based on the literature study, an overview of innovation management tools was made. The overview has been verified in the interview survey [Emory and Cooper 1991; Kumar, Stern, and Anderson 1993]. The results of the interview survey have also been used to analyze to which level the innovation management tools had been applied in practice [Borch and Arthur 1995]. In the interview survey, 66 experts in construction innovation were interviewed. The experts represented various institutions and firms in the Dutch construction industry, varying from universities and research institutes (3), governmental bodies (20), and non-commercial knowledge centers (1), to commercial consultants (6), architects (12), real-estate agents (12), and construction companies (12).

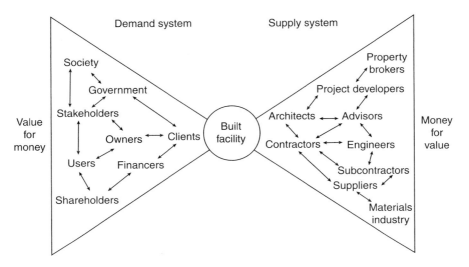

FIGURE 14.1 Schematic representation of the demand and supply systems in construction. (From Vrijhoef, R., and De Ridder, H., *Proceedings QUT Research Week*, 4–6 July, Queensland University of Technology, Brisbane, 2005. With permission.)

The interviews had an open structure. The innovation management tools that were found in literature were the main subjects of the interviews. The people interviewed were asked to reflect on the background and characteristics of these tools [Creswell 2003].

14.3 Understanding the Construction Supply Chain

This section describes some of the basic characteristics of the construction supply chain emphasizing its complexity in terms of networks of stakeholders, their interests, and cooperative ties. This forms the background of the analysis of innovative management tools on three levels of cooperation in the construction supply chain in the next sections of this chapter. Often the construction industry has been characterized by high levels of complexity, referring to the demography of the industry with many Small and Medium Sized Enterprises (SMEs) and specialist firms, as well as the configuration of the construction supply chain with a vast network of relations between firms and stakeholders (Figure 14.1). Indeed construction as such is a less structured industry compared to other industries, characterized by a loosely coupled network of actors, which is hampering learning and innovation [Dubois and Gadde 2002].

14.3.1 Make-To-Order Delivery and Craftsmanship

Generally, the construction supply chain has been referred to as a make-to-order supply chain, and even as a concept-to-order supply chain (Figure 14.2). In such a supply chain, project management and engineering are important issues, and design is often disconnected from production leading to various problems of production.

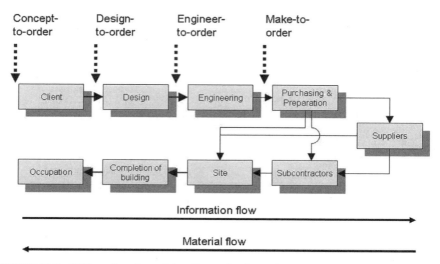

FIGURE 14.2 Different locations of the client order decoupling point in the construction supply chain.

The producer is not the designer, and production is very much influenced by craftsmanship. Moreover, production involves many crafts and many relatively small firms. This causes problems originating upstream the supply chain to persist and often become worse downstream, because of the mechanisms of causality and interdependence within the supply chain.

14.3.2   One-of-Kind and On-Site Production

Most production in construction is one-off and done on site. The "factory" is organized on site, and mostly very few materials and components of the end product are prefabricated or preinstalled off site. The logistics in construction are converging, meaning relatively many suppliers are directly involved in the production of an end product for one or very few customers. On the other hand, buildings are physically, economically, and socially linked with their environment in terms of use of space and surface, availability of local resources, and local design and construction practices.

14.3.3   Role of the Client and User

In most construction projects the end-customer is the start as well as the end of the entire process, and therefore the customer and end-users play a dominant role in construction. This causes the make-to-order mechanism and the need for responsiveness in construction supply chains. This is also the reason why in construction products are rarely "launched" and "marketed" as in other industries, and why construction is different than most other industries, for example, consumer goods. Most contractors are not manufacturers of integrated end products. Most products are not standard, and processes are not repetitive, and often cause high levels of waste [Vrijhoef and Koskela 2000].

14.3.4 Complex and Interdependent Exchange Mechanism

Much more than merely exchanging products, the firms within the supply chain exchange services, knowledge, competences, and other capabilities. This is particularly true for the construction supply chain because of the characteristics of one-kind and make-to-order production in a temporary project setting, and the reliance on the client and the physical and social context of a project. This results in the effect that the exchange mechanisms in construction are highly complex and uncertain, in relation to most other industrial sectors. This causes high levels of instability and problems in the production environment of construction. Due to the interdependency and causality existing in the supply chain, the problems flow downstream through the supply chain, and often cause additional problems with them (Figure 14.3).

14.4 Innovation Management Tools in the Construction Supply Chain

In this section, an overview of innovation management tools is presented on three analytical levels of cooperation in the network of organizations in the construction supply chain. The overview is based on the literature study, which was part of the research reported. The first level is the intrafirm level. This level comprises the issues that are important for an individual organization in the supply chain, like for example a contractor, an architect, or a client organization [Tatum 1989; Slaughter 1993; Lampel, Miller, and Floricel 1996; Toole 1998; Veshosky 1998; Bossink 2004b]. The second level is the interfirm level. This level represents the projects in which organizations cooperatively develop and build objects [Tatum 1989; Bröchner and Grandison 1992; Nam and Tatum 1992b; Winch 1998; Bresnen and Marshall 2000; Bossink 2004b]. The third level is the transfirm level. The transfirm level exceeds the intra and interfirm level, and comprises issues that are important for the construction supply chain as a whole of interrelated firms and institutions [Pries and Janszen 1995; Winch 1998; Goverse et al. 2001; Seaden and Manseau 2001; Bossink 2004b].

Four categories of innovation management tools can be distinguished based on their respective functions and effects. The first is environmental pressure. This category comprises the influences that force and stimulate organizations in the construction supply chain to innovate [Tatum 1989; Pries and Janszen 1995; Toole 1998; Winch 1998; Gann and Salter 2000; Bossink 2004b]. The second category is technological capability. This category consists of technical factors enabling organizations in the construction supply chain to develop innovative products and processes [Bröchner and Grandison 1992; Nam and Tatum 1992a; Nam and Tatum 1997; Gann and Salter 2000; Mitropoulos and Tatum 2000; Bossink 2004b]. The third category is knowledge exchange. This category consists of the arrangements that facilitate the sharing of knowledge and information supporting innovations in and between organizations in the supply chain [Slaughter 1993; Kangari and Miyatake 1997; Toole 1998; Veshosky 1998; Gann and Salter 2000; Goverse et al. 2001; Bossink 2004b]. The fourth category is boundary spanning. This category represents the initiatives to co-innovate across the boundaries of departments,

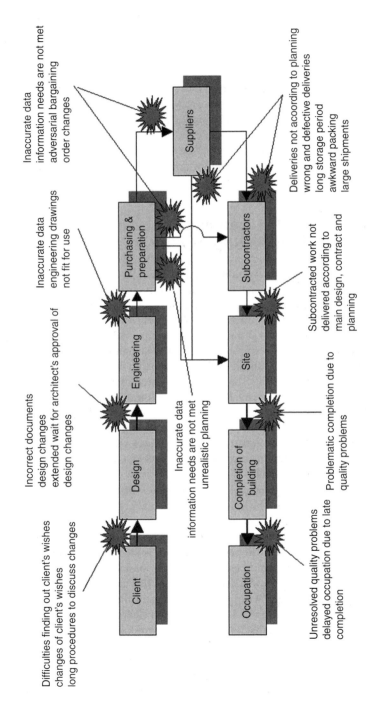

FIGURE 14.3 Problems progression due to interdependency in the construction supply chain. (From Vrijhoef, R., and Koskela, L., 1999. Roles of supply chain management in construction. *Proceedings of the 7th Annual Conference of the International Group for Lean Construction, IGLC 7*, Berkeley, CA, 133–46.)

organizations, and partnerships in the construction supply chain [Tatum 1989; Arditi, Kale, and Tangkar 1997; Barlow 2000; Gann and Salter 2000; Slaughter and Shimizu 2000; Bossink 2004b]. In Table 14.1, the innovation management tools in the construction supply chain are categorized.

14.4.1 Environmental Pressure

14.4.1.1 Market Pull

Nam and Tatum (1992a) studied the innovations that were developed in 10 construction projects in the USA and concluded that innovations were developed because of market pull. Pries and Janszen (1995) analyzed the innovative characteristics of the Dutch construction industry from 1945 to 1995, and argued that although some companies' strategies in the construction industry were traditionally technology oriented, these strategies changed because of the market exerting pressure to innovate. Arditi, Kale, and Tangkar (1997) investigated the innovation rate in construction equipment in the United States over a period of 30 years, and found that the innovation rate increased due to market forces.

14.4.1.2 Governmental Guarantee for Markets for Innovative Firms

Miozzo and Dewick (2002) explored the development of strategic innovations by, and the operational capabilities of the largest contractors in Germany, Sweden, Denmark, France, and the United Kingdom. One of their conclusions was that governments were able to stimulate innovation in the construction supply chain by guaranteeing markets for innovative firms.

TABLE 14.1   Innovation Management Tools in the Construction Supply Chain

Environmental Pressure:	Knowledge Exchange:
Market pull	Stimulation of research
Governmental guarantee for markets for innovative firms	Creation of knowledge networks
	Programs promoting collaboration
Governmental clients with innovative demands	Broad view of risk
	Integrated and informal R&D function
Innovation stimulating regulations	Effective information gathering
Subsidies for innovative applications and materials	Training of workers on site
	Lateral communication structures
Technological capability:	Boundary spanning:
Product evaluating institutions	Integration of design and building
Programs promoting access to technology	Client involvement
Finance for pilot projects	Mechanisms for sharing financial risks and benefits
Technology fusion	Coordination of participating groups
Technology leadership strategy	Empowerment of innovation leaders and champions
Technology push	Innovations from other industries
	Explicit coordination of the innovation process
	Strategic alliances and long-term relationships

14.4.1.3   Governmental Clients with Innovative Demands

Seaden and Manseau (2001) reviewed the national policies of 15 countries in Europe, North and South America, South Africa, and Japan towards innovation in construction. They presented an overview of public policy instruments driving innovation. One of the main innovation-stimulating policy instruments was considered to be the definition of innovative wishes, demands, and specifications by governmental bodies who were in the position of a powerful client in construction projects.

14.4.1.4   Innovation-Stimulating Regulations

Mitropoulos and Tatum (2000) investigated the forces driving construction firms in the United States to adopt new information technologies. One of the forces they distinguished was regulation. In addition, Goverse et al. (2001) studied the opportunities to increase the use of wood in the Dutch construction industry, and concluded that to stimulate the use of wood an important focus point of policy making should be the development of new regulations. Bossink (2007) presented a similar conclusion. Based on a research project into the diffusion of sustainable innovations in the Dutch house building industry, the conclusion was that legislative pressure is a powerful tool to speed up the diffusion of innovations.

14.4.1.5   Subsidies for Innovative Applications and Materials

Subsidies for innovative applications and materials were mentioned by Goverse et al. (2001) as an innovation management tool, and can be part of regulatory measures. Gann, Wang, and Hawkins (1998) studied the effects of regulations on innovativeness in the construction industry in the United Kingdom. They argued that traditional regulations forcing organizations to innovate in conformance with detailed specifications are too prescriptive and hamper creativity. On the contrary, performance-based building regulations stimulating organizations to innovate in a certain direction have a positive effect on innovation.

14.4.2   Technological Capability

14.4.2.1   Product Evaluating Institutions

Seaden and Manseau (2001) mentioned the evaluation of new products, processes, and technologies before market launch, as an instrument to guarantee the quality of innovations. Objective institutes, responsible for the testing of the newly developed products, processes, and technologies, can assure that the industry is confronted with high quality innovations.

14.4.2.2   Programs Promoting Access to Technology

Seaden and Manseau (2001) as well as Goverse et al. (2001) have stressed the importance of programs and bridging institutions facilitating access for organizations to the technology needed to innovate. In addition to this, Miozzo and Dewick (2002) stated that long-term relations between firms and external knowledge centers facilitated access to, and adoption of new technologies in the construction supply chain.

14.4.2.3   Finance for Pilot Projects

Miozzo and Dewick (2002) and Bossink (2002a) stated that innovation could be stimulated by financial support for innovative pilot projects in which technologies are tested and evaluated before market launch.

14.4.2.4   Technology Fusion

Kangari and Miyatake (1997) described factors that contributed to the development of innovative construction technology in Japan. One of the factors they described was construction technology fusion, which means that diverse technologies from various disciplines are integrated to develop a new construction technique, object, or process.

14.4.2.5   Technology Leadership Strategy

Nam and Tatum (1992a) stated that firms that want to be innovative should follow a technology leadership strategy. Kangari and Miyatake (1997) supported this statement and argued that a prominent reputation is built with innovation leadership. Nam and Tatum (1997) also studied the role of key individuals in 10 successful innovation projects in the construction industry in the United States. They concluded that technological competence is a prerequisite for effective leadership in construction, which is essential for technological innovation in construction.

14.4.2.6   Technology Push

Nam and Tatum (1992a), Bröchner and Grandison (1992), and Mitropoulos and Tatum (2000) all concluded that advanced technology of designers and contractors in many cases shapes the demands of the client. Technological capabilities of organizations in the construction supply chain push the implementation of the solutions by these organizations in their construction projects.

14.4.3   Knowledge Exchange

14.4.3.1   Stimulation of Research

Goverse et al. (2001) stressed the importance of stimulation of research in construction innovation. Bossink (2007) found that most of the inventions in sustainability in Dutch house building were initially developed at universities and research institutes funded by government.

14.4.3.2   Creation of Knowledge Networks

Goverse et al. (2001) and Bossink (2008a) underpinned the importance of the creation, stabilization, and upgrading of knowledge networks consisting of universities, research institutes, and knowledge-intensive business services. Exchange of knowledge facilitates the development of new knowledge that can be used to innovate.

14.4.3.3   Programs Promoting Collaboration

Seaden and Manseau (2001) and Bossink (2008b) mentioned programs promoting collaborative arrangements between organizations as an innovation management tool.

14.4.3.4   Broad View of Risk

One of the innovation management tools Tatum (1989) described is a company with a broad and long-term view of risk. Such a view stimulates innovation by reducing the risks associated with innovations, and thus increases the chances of successful implementation of innovations.

14.4.3.5   Integrated and Informal R&D Function

Nam and Tatum (1992a) stated that an integrated and informal R&D function in a firm strengthens its innovative capability. The R&D function initiates new developments, and creates attention for these developments in parts of the organization working under market conditions.

14.4.3.6   Effective Information Gathering

Kangari and Miyatake (1997), Toole (1998), and Veshosky (1998) mentioned effective information gathering as an important innovation management tool. In an empirical investigation in the construction industry in the United States, Toole (1998) found that a firm's capacity to gather and process information about new technology contributed to its innovativeness. Toole (1998) also concluded that building material manufacturers and retailers providing information about new products take away the designers and contractors uncertainty about the new possibilities, and thus contribute to innovation in the industry. Veshosky (1998) conducted a research project in the engineering and construction industry in the United States, and focused on the information seeking behavior of managers. Veshosky (1998) found that firms that recognized the importance of innovation facilitated their project managers in obtaining innovation information.

14.4.3.7   Training of Workers on Site

Slaughter (1993) conducted a study in the construction industry in the United States and reported that builders developed innovations on site by changing the methods of assembling components and integrating them with other elements of the construction. Innovation by workers on site is highly stimulated when the workers are informed about new products and procedures, and get the space to develop their own working methods with these innovations.

14.4.3.8   Lateral Communication Structures

Barlow (2000) studied innovation in offshore construction in the United Kingdom, and concluded that innovation is made possible by nonhierarchical internal and external communication structures. In organizations and construction projects, lateral communication facilitates the exchange of knowledge and stimulates innovative solutions in the process.

14.4.4   Boundary Spanning

14.4.4.1   Integration of Design and Building

Tatum (1989) mentioned the integration of design and construction disciplines as an important innovation management tool. It prevents construction projects from innovative designs that cannot be constructed. It also secures that the knowledge of builders is used in the design stage. This supports and improves the constructability of the design.

14.4.4.2 Client Involvement

Nam and Tatum (1992b) investigated several construction innovations to establish which instruments could be used to stimulate innovation in construction projects. One of the instruments they reported to be useful is the involvement of the client in the construction project. Bröchner and Grandison (1992), and Mitropoulos and Tatum (2000) mentioned client involvement as an innovation management tool.

14.4.4.3 Mechanisms for Sharing Financial Risks and Benefits

Barlow (2000) and Bossink (2002b) concluded that the establishment of financial mechanisms for sharing project risks and benefits is needed to ensure that innovations are defined and it is clear how costs and revenues are shared between project participants.

14.4.4.4 Coordination of Participating Groups

Tatum (1989) reported that the coordination of all participating groups in the innovation process is an important innovation management tool. The effect of the coordination is increased common understanding often leading to a positive decision to actually implement an innovation.

14.4.4.5 Empowerment of Innovation Leaders and Champions

According to Tatum (1989), Nam and Tatum (1992b), Barlow (2000), and Bossink (2004a) staffing for key positions, such as innovation champions and leaders, is important in the innovation process. Innovation champions act as individual drivers of innovation in organizations and projects. Innovation leaders function as initiators and managers of the innovation processes in organizations and projects.

14.4.4.6 Innovations from Other Industries

Arditi, Kale, and Tangkar (1997) and Ben Mahmoud-Jouini (2000) illustrated that construction is heavily dependent on other industries, such as the construction equipment industry. Suppliers of innovative construction materials and construction processes contribute to the innovative capacity of organizations and their output in construction projects.

14.4.4.7 Explicit Coordination of the Innovation Process

Slaughter and Shimizu (2000) analyzed the nature of the innovations in the development of 11 bridges, situated in Denmark, Sweden, Japan, France, China, Canada, and the United States. They concluded that the innovation process should be explicitly coordinated in order to achieve multiple and interrelated innovations.

14.4.4.8 Strategic Alliances and Long-Term Relationships

Nam and Tatum (1992b) stressed the importance of the establishment of long-term relationships between organizations in the construction supply chain to increase the level of innovativeness. Kangari and Miyatake (1997) mentioned the strategic alliance as a means to leverage scientific and technical capabilities, and to share the financial risks.

Most of the innovation management tools apply to specific levels of cooperation in the supply chain. For instance, the phenomenon of market pull concerns the industry as a whole, and thus addresses the transfirm level. In contrast, client involvement will normally take place on a project level, which is the interfirm level. On an intrafirm level, one can imagine that information sharing and R&D are particularly effective. Table 14.2 summarizes whether the distinctive innovation management tools apply to the transfirm, interfirm, and intrafirm level of cooperation.

14.5 Application of Innovation Management Tools in the Construction Supply Chain

In this section, based on the outcomes of the interview survey held among managers in the Dutch construction industry, we analyze how the innovation management tools presented above can be applied to stimulate innovation in the construction supply chain. In addition, we will give an overview of a few examples of innovation management tools as they have been applied in recent years in the Dutch construction industry.

TABLE 14.2 Application Levels of the Innovation Management Tools

Innovation Management Tools	Transfirm	Interfirm	Intrafirm
Environmental pressure:			
Market pull	X		
Governmental guarantee for innovative firms	X		
Governmental clients with innovative demands		X	
Innovation stimulating regulations	X	X	X
Subsidies for innovative applications and materials	X	X	X
Technological capability:			
Product evaluating institutions	X		
Programs promoting access to technology	X		
Finance for pilot projects	X		
Technology fusion		X	
Technology leadership strategy		X	X
Technology push	X	X	X
Knowledge exchange:			
Stimulation of research	X		
Creation of knowledge networks	X		
Programs promoting collaboration	X		
Broad view of risk			X
Integrated and informal R&D function			X
Effective information gathering			X
Training of workers on site		X	X
Lateral communication structures		X	X
Boundary spanning:			
Integration of design and building		X	
Client involvement		X	
Mechanisms for sharing financial risks and benefits		X	
Coordination of participating groups		X	X
Empowerment of innovation leaders and champions		X	X
Innovations from other industries	X	X	X
Explicit coordination of the innovation process		X	X
Strategic alliances and long-term relationships	X	X	X

14.5.1 Environmental Pressure

14.5.1.1 Market Pull

In case the market requires innovations of building products or processes, this can be seen as a pressure on the transfirm level affecting the construction supply chain as a whole, and pressing firms involved to innovate. Managerial action that supports market pressure for innovation is a client or real-estate developer, for instance, who include these new market demands directly in the brief of requirements of new projects.

14.5.1.2 Governmental Guarantee for Markets for Innovative Firms

When the government creates a market for innovative building projects or new building materials, this can be categorized as a pressure to innovate on the transfirm level. Managers who want to use this force can direct their projects towards the markets that are created by the government. They can also apply new materials promoted by governmental guarantees.

14.5.1.3 Governmental Clients with Innovative Demands

In many projects, the government acts as a client of a building, bridge, or road. In these projects, the government and its related departments and institutions have the opportunity to demand innovative designs, materials, and building processes. This pressure on the interfirm level, at which governmental clients direct the innovative activity of designers and contractors, provides key opportunities to building partners with innovative goals.

14.5.1.4 Innovation-Stimulating Regulations

The government has the ability to introduce new laws and regulations to stimulate, or even force innovative processes, products, and procedures in the construction supply chain. It can impose legislation that enforces innovations on the intrafirm, interfirm, as well as transfirm level.

14.5.1.5 Subsidies for Innovative Applications and Materials

In cases regulations do not work, the government can also provide subsidies for innovative activity in the construction supply chain. These subsidies stimulate the application of new products and building processes by individual firms (intrafirm), in projects (interfirm), and the industry as a whole (transfirm).

14.5.2 Technological Capability

14.5.2.1 Product Evaluating Institutions

When new building products are introduced to the market, many architects and construction companies often hesitate to apply them. One of the main reasons is fear for claims in case the products appear to be deficient. The government and the building industry have the opportunity to work with product evaluating institutions who test new products, and provide a guarantee on its functioning, or protect firms applying the products against possible claims.

14.5.2.2 Programs Promoting Access to Technology

The construction supply chain generally consists of a fair amount of small and medium-sized companies that are primarily interested in collecting sufficient orders and projects. During the course of the projects, they concentrate on making a reasonable profit. Innovation and innovative behavior is not always a guiding motive. A transfirm, governmental program, in which employees of these small and medium-sized firms are taught how to be profitable and innovative at the same time, can overcome this impediment to innovation.

14.5.2.3 Finance for Pilot Projects

Most of the activities in the construction supply chain are based on traditional building practice. To encourage innovations, the industry needs pilot projects to experiment with new building products, cooperative processes, and governance structures. On the national, transfirm level, the government and institutions have the opportunity to allocate financial resources to stimulate innovative pilot projects.

14.5.2.4 Technology Fusion

The construction supply chain is a diverse and complex cooperative structure in which clients, developers, architects, contractors, advisors, and suppliers exchange products and services. In most cases, one cannot build alone, but has to cooperate. The cooperative nature of building stimulates that, on the interfirm level, these practices and technologies come together and fuse into innovative processes and products.

14.5.2.5 Technology Leadership Strategy

Many firms concentrate on a cost leadership strategy, and focus on cost reduction, efficient processes, and a position in the market based on lowest price propositions. A strategy that can also be profitable is a technology leadership strategy in which the firm competes with its capability to renew its products and processes continuously. Firms can apply this strategy both on the intrafirm and interfirm level, and innovate internally and in cooperation with other organizations.

14.5.2.6 Technology Push

Technological development seems to be unstoppable. It is an engine for innovation on the transfirm, interfirm, and intrafirm level of operation and cooperation in the construction supply chain. Firms that have a sustainable competitive advantage today can be out of competition tomorrow when they neglect new technological options that are used by competitors, or demanded by clients.

14.5.3 Knowledge Exchange

14.5.3.1 Stimulation of Research

On the transfirm level, the government has the ability to fund and stimulate scientific research at universities and research institutes. These research environments produce the innovations and new technologies that may not be profitable or desirable yet, but can be the options the industry and the market need in the near future.

14.5.3.2 Creation of Knowledge Networks

In competitive settings, firms protect their innovative knowledge, technologies, and processes. Companies that interact under competitive constraints, approach their counterparts in the supply chain as competitors who aim for the same profits. On the transfirm level, the government can overcome this situation by stimulating and creating networks in which representatives of firms that do not compete, for example, from different industries, exchange knowledge and information.

14.5.3.3 Programs Promoting Collaboration

When firms in a competitive setting have found partners with which they have profitable projects, they tend to continue the cooperation in future projects, and protect the cooperative status quo. To stimulate new cooperative structures and renew old ones, the government can introduce transfirm programs that promote new collaboration structures between institutions and firms in the construction supply chain.

14.5.3.4 Broad View of Risk

On the intrafirm level, firms can be risk averse or risk seeking. Firms that have chosen to be innovative often accept the risk of failure of new processes and products. Innovative processes and products that become a success, contribute to the future turnover of the firm, and subsidize the innovative initiatives that failed.

14.5.3.5 Integrated and Informal R&D Function

Some large firms have an R&D department. However, most of the firms in the construction supply chain do not have a department that is responsible for research and development. This does not mean that they neglect R&D. Innovative firms need an R&D function that is loosely integrated in the tasks of most of the employees. In case a company's representatives have a responsibility in developing new products and processes, a firm has an informal R&D function that can contribute to future innovation.

14.5.3.6 Effective Information Gathering

Firms need to develop the capability to gather innovative information. They need project managers that are in contact with various possible partners in the construction supply chain. It enables them to obtain information on new building processes and products that can be used when the building process demands rapid innovativeness.

14.5.3.7 Training of Workers on Site

Innovation demands change, and change is difficult when colleagues do not have the capabilities to implement or operate change. Managers and employees have to develop the skills that are needed on the cooperative, project level, and on the level of their individual operations.

14.5.3.8 Lateral Communication Structures

The division of a firm into functional departments, structuring of firms in hierarchies of managers and operators leads to an efficient and effective organization. A firm that aims to be innovative also needs to stimulate lateral communication in and between departments in a firm, and between firms in the supply chain.

14.5.4   Boundary Spanning

14.5.4.1   Integration of Design and Building

In the construction supply chain, the design function and the build function are often divided. In most cases, the design function is performed by an architect, and the building function by a construction company. When these two functions are separated, the risk that designs cannot be realized increases. An innovative building process integrates the design and building activities.

14.5.4.2   Client Involvement

In the development of buildings, it is often the case that a construction supply chain develops a building for unknown clients and users. The design is based on their experience with the wishes and demands of previous clients. In case a client is known, and introduced as one of the advisors, or principals of the design process, the participants in the supply chain develop on customer demand, and often have to increase their innovativeness.

14.5.4.3   Mechanisms for Sharing Financial Risks and Benefits

In most cases, innovation requires additional resources, and these investments must be earned back. It often happens that the costs have to be made by a partner in the supply chain who does not receive the revenues. To neutralize this imbalance the partnering firms have to negotiate and decide on a fair distribution of the costs and benefits of the innovations.

14.5.4.4   Coordination of Participating Groups

In construction projects, the activities of specialists and components coming from different firms have to be coordinated and integrated in a temporal production setting. This is also true for any innovations brought in by the firms, which have to be comprehensively integrated in the project.

14.5.4.5   Empowerment of Innovation Leaders and Champions

Most of the interfirm and intrafirm innovative projects need one or more team members who have the power to allocate sufficient financial resources, and often take risky decisions. In addition, these projects need team members with a creative attitude who develop the innovations, and convince the team members who are more averse to risk and change, and doubtful about the application of innovative ideas to the project.

14.5.4.6   Innovations from Other Industries

The construction supply chain consists of various organizations that work together. Many of these firms are also part of supply networks with organizations from other industries. An innovative industry integrates these influences on both the interfirm and the intrafirm level.

14.5.4.7   Explicit Coordination of the Innovation Process

Organizations who have embraced innovation also have managers who coordinate the innovation processes of their interfirm building projects and intrafirm innovation projects. Large firms sometimes have one or more innovation managers. Small and medium-sized companies who aim to be innovative often expect their line managers to coordinate the effective and efficient application of innovations within the company.

14.5.4.8   Strategic Alliances and Long-Term Relationships

To be innovative, in the long term, it can be effective to develop one or more alliances with companies that have competencies that are complementary and provide a long-term competitive advantage in the market.

The above results have been gathered based on interviews with managers in the Dutch construction industry. Implicitly the tools that have been discussed have been observed by those managers on different levels of cooperation, dependent where the tools are most effective. The effective use of some tools can be illustrated with examples from the Dutch construction industry. Those are listed in Table 14.3. For each category of management tools; that is, environmental pressure, technological capability, knowledge exchange, and boundary spanning, and on each cooperation level; that is, transform, interfirm, and intrafirm, Table 14.3 presents an example of a tool used in the Dutch situation.

Clarification of Dutch terms and abbreviations in Table 14.3:

- IFD = Industrial fexible demountable government program aimed at the development of more efficient, customer-focused, and sustainable buildings
- TNO = Dutch national technological research center
- KIWA = Independent organization for certification and accreditation of products
- PSIBouw = National innovation program in the Dutch construction sector 2004–2009
- Bouwend Nederland = Umbrella organization of employers organizations in the Dutch construction industry
- BOB = Centre for vocational education and training for professionals of all levels in the Dutch construction industry
- Regieraad Bouw = National taskforce stimulating innovation and reform in the Dutch construction sector

14.6   Summary and Conclusions

Several innovation management tools have shown to be useful at different levels of cooperation in the network of firms in the construction supply chain. This chapter addressed the question how innovation can be managed at three levels of cooperation: the intrafirm, interfirm, and transfirm level. The intrafirm level is the level of the individual organization. The interfirm level represents the projects in which groups of organizations cooperate. The transfirm level is the construction supply chain as a whole. In the research reported, a literature study was carried out first to provide an overview of innovation management tools. The literature study was followed by an interview survey

TABLE 14.3   Examples of Innovation Management Tools on the Three Levels of Cooperation in the Dutch Construction Industry

Categories of Innovation Management Tools	Transfirm	Interfirm	Intrafirm
Environmental pressure	EU regulations prescribing procurement routes for (public) clients and contractors	Ministry of Finance demanding DBFMO bids from contractors bidding for government projects	IFD building government program to stimulate innovations to increase efficiency, sustainability, and flexibility
Technological capability	Technical research centers, for example, TNO and KIWA, testing and approving new construction materials and methods	National innovation program PSIBouw supporting innovation in projects and groups of firms	An increasing number of individual firms reengineering their business and taking on innovation leadership
Knowledge exchange	Symposiums, debates, etc., organized by employers organization Bouwend Nederland	Projects increasingly starting with project start-ups explicitly aiming to find room for innovation in the project	Educational centers, for example, BOB, for practitioners, particularly aimed at staff and workers from SMEs
Boundary spanning	National construction task force "Regieraad Bouw" stimulating firms to investigate possibilities to adopt techniques and concepts from other industries	Particularly construction firms and suppliers teaming up to unify their innovative strengths to develop new technological solutions and methods	Particularly architects and project developers empowering (young) employees to be innovative and change the firm's business and practices

among 66 experts on construction innovation in the Dutch construction industry. The people interviewed were asked to reflect on the characteristics and the implications of the innovation management tools applied in construction practice, on the three levels of cooperation.

The research project showed that 27 innovation management tools could be distinguished. The tools could be divided in to four categories based on their respective functions and effects. Five tools create environmental pressure to innovate. Six tools improve the technological capability of firms that need to innovate. Eight tools improve firms' management of knowledge that is needed to innovate. Eight other tools enable firms to extend the boundaries of their organization and stretch their innovative capabilities.

The literature study and the interview survey showed that the collection of innovation management tools cover all three levels of cooperation in the network of organizations in the construction supply chain. All 27 innovation management tools stimulate innovation on one, two, or three levels. Each level has its own set of useful innovation management tools. Selective and comprehensive use of the tools enables a focused as well as a comprehensive stimulation of innovation on each level. In addition, depending on the extent to which a firm wants to apply innovation, managers can decide which of the innovation management tools are appropriate to use in their situation and to achieve their goals.

To conclude, the collection of 27 innovation management tools provides a powerful set of instruments to stimulate innovation simultaneously on all three levels of cooperation in the construction supply chain, which increases the chances of successful application of innovations. In fact, the interview survey showed that organizations in the Dutch construction industry innovate by using all innovation management tools found in the literature on all three levels of cooperation. On the intrafirm level, the innovation management tools are used by firms to defend, sustain, or enforce their strategic positions in the marketplace. On the interfirm level, the innovation management tools are primarily used to improve and strengthen the cooperative ties and procedures among project partners. On the transfirm level, the tools are used as measures of governmental and institutional incentives putting pressures on firms to innovate.

References

Arditi, D., Kale, S., and Tangkar, M. 1997. Innovation in construction equipment and its flow into the construction industry. *Journal of Construction Engineering and Management*, 123 (4), 371–78.

Barlow, J. 2000. Innovation and learning in complex offshore construction projects. *Research Policy*, 29 (7–8), 973–89.

Ben Mahmoud-Jouini, S. 2000. Innovative supply-based strategies in the construction industry. *Construction Management and Economics*, 18 (6), 643–50.

Bernstein, H. M. 1996. Bridging the globe: Creating an international climate of engineering and construction innovation. *Industry and Environment*, 19 (2), 26–28.

Bon, R., and Hutchinson, K. 2000. Sustainable construction: Some economic challenges. *Building Research and Information*, 28 (5–6), 310–14.

Borch, O., and Arthur, M. B. 1995. Strategic networks among small firms: Implications for strategy research methodology. *Journal of Management Studies*, 32 (4), 419–40.

Bossink, B. A. G. 2002a. A Dutch public–private strategy for innovation in sustainable construction. *Construction Management and Economics*, 20 (7), 633–42.

———. 2002b. The development of co-innovation strategies: Stages and interaction patterns in interfirm innovation. *R&D Management*, 32 (4), 311–20.

———. 2004a. The effectiveness of innovation leadership styles: A manager's influence on ecological innovation in construction projects. *Construction Innovation*, 4 (4), 211–28.

———. 2004b. Managing drivers of innovation in construction networks. *Journal of Construction Engineering and Management*, 130 (3), 337–45.

———. 2007. The inter-organizational innovation processes of sustainable building: A Dutch case of joint building innovation in sustainability. *Building and Environment*, 42 (12), 4086–4092.

———. 2008a. Leadership for sustainable innovation. *International Journal of Technology vManagement and Sustainable Development*, 6 (2), 135–50.

———. 2008b. Assessment of a national innovation system of sustainable innovation in residential construction: A case study from the Netherlands. *International Journal of Environmental Technology and Management*, 8 (x), accepted for publication.

Bresnen, M., and Marshall, N. 2000. Partnering in construction: A critical review of issues, problems and dilemmas. *Construction Management and Economics*, 18 (2), 229–37.

Bröchner, J., and Grandison, B. 1992. R&D cooperation by Swedish contractors. *Journal of Construction Engineering and Management*, 118 (1), 3–16.

Creswell, J. W. 2003. *Research design: Qualitative, quantitative and mixed methods approaches*, 3rd ed. Thousand Oaks, CA: Sage.

Dubois, A., and Gadde, L. E. 2002. The construction industry as a loosely coupled system: Implications for productivity and innovation. *Construction Management and Economics*, 20 (7), 621–31.

Emory, C. W., and Cooper, D. R. 1991. *Business research methods*. Boston, MA: Irwin.

Gann, D. M., and Salter, A. J. 2000. Innovation in project-based, service-enhanced firms: The construction of complex products and systems. *Research Policy*, 29 (7–8), 955–72.

Gann, D. M., Wang, Y., and Hawkins, R. 1998. Do regulations encourage innovation? The case of energy efficiency in housing. *Building Research & Information*, 26 (4), 280–96.

Goverse, T., Hekkert, M. P., Groenewegen, P., Worrell, E., and Smits, R. E. H. M. 2001. Wood innovation in the residential construction sector; opportunities and constraints. *Resources, Conservation and Recycling*, 34 (1), 53–74.

Guy, G. B., and Kibert, C. J. 1998. Developing indicators of sustainability: US experience. *Building Research and Information*, 26 (1), 39–45.

Kangari, R., and Miyatake, Y. 1997. Developing and managing innovative construction technologies in Japan. *Journal of Construction Engineering and Management*, 123 (1), 72–78.

Korczynski, M. 1996. The low-trust route to economic development: Inter-firm relations in the UK engineering construction industry in the 1980s and 1990s. *Journal of Management Studies*, 33 (6), 787–808.

Koskela, L., and Vrijhoef, R. 2001. Is the current theory of construction a hindrance to innovation? *Building Research & Information*, 29 (3), 197–207.

Kumar, N., Stern, L. W., and Anderson, J. C. 1993. Conducting inter-organizational research using key informants. *Academy of Management Journal*, 36 (6), 1633–1651.

Lampel, J., Miller, R., and Floricel, S. 1996. Information asymmetries and technological innovation in large engineering projects. *R&D Management*, 26 (4), 357–69.

Larsson, N. 1996. Public–private strategies for moving towards green building practices. *Industry and Environment*, 19 (2), 23–25.

Miozzo, M., and Dewick, P. 2002. Building competitive advantage: Innovation and corporate governance in European construction. *Research Policy*, 31 (6), 989–1008.

Mitropoulos, P., and Tatum, C. B. 1999. Technology adoption decisions in construction organizations. *Journal of Construction Engineering and Management*, 125 (5), 330–38.

———. 2000. Forces driving adoption of new information technologies. *Journal of Construction Engineering and Management*, 126 (5), 340–48.

Nam, C. H., and Tatum, C. B. 1989. Toward understanding of product innovation process in construction. *Journal of Construction Engineering and Management*, 115 (4), 517–34.

———. 1992a. Strategies for technology push: Lessons from construction innovations. *Journal of Construction Engineering and Management*, 118 (3), 507–24.

———. 1992b. Non-contractual methods of integration on construction projects. *Journal of Construction Engineering and Management*, 118 (2), 385–98.

———. 1997. Leaders and champions for construction innovation. *Construction Management and Economics*, 15 (4), 259–70.

Ngowi, A. B. 2001. Creating competitive advantage by using environment-friendly building processes. *Building and Environment*, 36 (3), 291–98.

Pries, F., and Janszen, F. 1995. Innovation in the construction industry: The dominant role of the environment. *Construction Management and Economics*, 13 (1), 43–51.

Robertson, H., Pearson, A. W., and Ball, D. F. 1996. The development of networks between engineering contractors and their clients: The special case of partnering. *R&D Management*, 26 (4), 371–79.

Seaden, G., and Manseau, A. 2001. Public policy and construction innovation. *Building Research & Information*, 29 (3), 182–96.

Slaughter, E. S. 1993. Builders as sources of construction innovation. *Journal of Construction Engineering and Management*, 119 (3), 532–49.

Slaughter, E. S., and Shimizu, H. 2000. 'Clusters' of innovations in recent long span and multi-segmental bridges. *Construction Management and Economics*, 18 (3), 269–80.

Tatum, C. B. 1989. Organizing to increase innovation in construction firms. *Journal of Construction Engineering and Management*, 115 (4), 602–17.

Toole, T. M. 1998. Uncertainty and home builders' adoption of technological innovations. *Journal of Construction Engineering and Management*, 124 (4), 323–32.

Veshosky, D. 1998. Managing innovation information in engineering and construction firms. *Journal of Management in Engineering*, 14 (1), 58–66.

Vrijhoef, R., and De Ridder, H. 2005. Supply chain integration for achieving best value for construction clients: Client-driven versus supplier-driven integration. *Proceedings QUT Research Week*, 4–6 July, Queensland University of Technology, Brisbane.

Vrijhoef, R., and Koskela, L. 2000. The four roles of supply chain management in construction. *European Journal of Purchasing & Supply Management*, 6 (3–4), 169–78.

Winch, G. M. 1998. Zephyrs of creative destruction: Understanding the management of innovation in construction. *Building Research & Information*, 26 (4), 268–79.

———. 2003. Models of manufacturing and the construction process: The genesis of re-engineering construction. *Building Research & Information*, 31 (2), 107–18.

II

Commentary

Will Hughes
University of Reading

T HE PAPERS IN THIS SECTION adopt different theoretical stances, and each tends to look at the construction supply chain through a different lens. These range across management theory, economic contract theory, game theory, and organization theory, taking in diverse influences from all of them. The way that research is carried out in this sector is not homogeneous, and this section of the book demonstrates this very clearly.

Towill starts from the premise that the construction sector is often perceived as unique, even though it has individual characteristics that on their own are not, resonating strongly with early work on construction economics by Patricia Hillebrandt who was writing in 1974. Because of the individual commonalities with other sectors, it would seem practical to apply techniques from other sectors where the characteristics match. He also emphasizes the widespread negative image of the construction sector, picking up themes of opportunism, delay, waste, exploitation, and price escalation, edging this cloud with the silver lining of market leaders who are punctual, productive, and responsible. Towill explains the origin of a concerted effort to compress total cycle time in business processes, and argues that there is an emerging formalism and regularity to such improvement programs that would justify calling this approach an emergent part of management theory. It is interesting to see how the ideas of business process reengineering are continuing to produce tangible benefits for industry. He picks up ideas from various parts of management literature and shows how they can be synthesized into a practical approach to improving business performance in the construction sector, such that rewards can be shared by all those taking part, with the greatest benefits going to the businesses for whom the construction projects are being carried out. A concerted focus on improving the overall business process needs to be carefully managed and, if it is, the evidence will be clear from all metrics, not just from one aspect such as improved cycle time, improved health and safety record, or reduction in defects on handover. This is an interesting example of innovating by introducing ideas from other industrial sectors, and draws upon ideas from a wide range of sources.

Cox takes a strikingly different approach from Towill, and begins with the notion that there cannot be one best way of organizing the process of construction, an idea that resonates strongly with the contingency theorists, tracing their roots back to Woodward (1965). He makes clear that any advice about how to procure construction should be based on a clear understanding of the market structure within which the exchange relationship takes place. Moreover, he places the capital and operational expenditure decisions in the context of how the relationships between buyers and sellers change at the point that the contract is formed. Precontract negotiations characterize capital expenditure (CAPEX) decisions, whereas operational expenditure (OPEX) is more dependent on postcontractual negotiations. As Cox points out, the one-off nature of CAPEX decisions leads to a very different approach to contracting than would be found in the repeating nature of relationships in OPEX decisions. In other words, not only do procurers vary in their relationship to the construction market, but each procurer would have a different relationship for different types of construction procurement. One very important aspect of Cox's chapter is that construction procurement decisions can be viewed in the context of the well-developed and widely tested theoretical perspectives that form the basis of how procurement decisions are understood. Against this backdrop, the recommendations of many government reports from diverse countries look absurd; specifically the recommendation that all procurers should develop proactive sourcing approaches to their supply chains. While it may seem appealing to insist that a win–win approach is always the best, this widely held belief is not supported by the kind of theoretical analysis to which it is put by Cox. By taking account of the relative power of the parties and the inherent contention in their fundamental business aims, he provides a much richer picture than we have been used to.

London begins with similar theoretical frames to those used by Cox and approaches the task of object-oriented modeling of a specific supply chain. Like Cox, London echoes the theme that governments are developing their policies for the construction sector in a vacuum, with no real understanding of the behavioral characteristics of the sector. One of the interesting things about applying organizational theory to projects rather than firms is that some of the basic tenets of organizational theory are transformed. For example, the widely accepted assertion that individuals are not merely members of one organization is transformed into a very interesting assertion that firms who work on construction projects are not merely working on one project. A moment's reflection confirms that this must indeed be the case, but, as London points out, it is so rarely pointed out that there is a danger that policy makers and analysts are working under the assumption that any one project holds the full and undivided attention of the participants in the project. Another idea usually missing from the supply chain literature is the structure of markets and how the market for each participant in a supply chain is specific to that particular category of supplier, or even to that particular supplier. This is one of the reasons why economists rarely look at the construction sector; it is far too heterogeneous to suit the kind of models they seek to develop. London's assertion that "firms do not typically transact with a project" is a very strong way of encapsulating the fundamentally fresh view that she brings to this area. The essential dynamism of her model is very compelling. For example, the market position of a firm varies rapidly, as the significance of a single transaction can be so great as to change completely their need to win further work for the time being. There are some fascinating insights into what happens in supply chains when the detail of her research is exemplified. The diverse academic disciplines that underpin this work, coupled with the robust approach to modeling what happens in real supply chains highlights the importance of this kind of research to our collective view of how construction work should be designed and procured.

Finally in this section, Bossink and Vrijhoef consider the extent to which innovation can be managed in the context of these vastly complex processes. Given the multiple layers and interconnections in construction supply chains, the notion of managing innovations is a severe challenge. They begin by identifying three levels at which this might occur: the industrial sector, the institution or firm, and the project. Their thorough review of the literature on innovation tools produces a useful list of different techniques that may be used to encourage innovation to happen, and these are categorized into the levels at which they may be applied. They interviewed a range of industry participants to find out how each of the techniques might be used in practice, and conclude that there are many ways that innovation can be managed in the construction sector. This is useful for those who need to ensure that innovation takes place.

One of the interesting things that this brief overview of papers has revealed to me is the way in which management perspectives have been used to produce prescriptive accounts of how managers ought to behave in doing their work, whereas the economic perspectives tend to focus on explaining why things are happening the way that they are. Each approach is equally important and indispensable, and the task for all those interested in the procurement of construction work is to appreciate the huge complexity of the processes with which they are confronted in order that they can make choices about what is important. The proving ground for all of these ideas is at the intersection between management, economics, and law, as empirical research in this area has shown (Hughes et al. 2006). From a theoretical perspective, the great value of this work is in showing theorists the impact in practice of their ideas. This is important because it is the only way of trying out new ideas and theories. Without application, it is hard to see the value of developing such theories. The most interesting thing about working in this area is the need to develop and strengthen theoretical frameworks that tend to be developed in much more straightforward situations. The construction sector tends to push such theoretical models to their limits. This is why research in this area can be so rewarding.

References

Hillebrandt, P. M. 1974. *Economic theory and the construction industry.* London: Macmillan.

Hughes, W. P., Hillebrandt, P., Greenwood, D. G., and Kwawu, W. E. K. 2006. *Procurement in the construction industry: The impact and cost of alternative market and supply processes.* London: Taylor & Francis.

Greenwood, D. J. 2001. Power and proximity: A study of sub-contract formation in the UK building industry. Unpublished PhD thesis, University of Reading, UK.

Woodward, J. 1965. *Industrial organization: Theory and practice.* London: Oxford University Press.

III

Information Technology

15

Overview of IT Applications in the Construction Supply Chain

Kalyanaraman
Vaidyanathan
i2 Technologies

15.1 Introduction

Supply chain management (SCM) as a means to improve operational efficiency has been recognized since the early '90s within the various manufacturing industries. After several successful implementations and studies, today SCM is a proven, well-understood theory for managing several aspects of a manufacturing business, from finance to operations management. Companies are now openly adapting their business processes, and adopting technology-based SCM tools to reduce inefficiencies and get better visibility into their operations, better manage their manufacturing planning and scheduling, and better interact and integrate with their suppliers. Concurrent with improved SCM practice, a new breed of enterprise software vendors emerged to sell "Manufacturing Advanced Planning and Scheduling" or manufacturing SCM solutions (e.g., i2 Technologies, Inc., http://www.i2.com; SAP AG, http://www.sap.com; Oracle Corporation, http://www.oracle.com; among several other vendors). Benefits from successful adoption of SCM techniques and information technology (IT) solutions include reduced inefficiencies, reduced work-in-progress (WIP), increased throughput, and increased productivity, all of which lead to reduced bottom-line operating costs, and increased topline revenue and associated profits. Section 15.4 has examples of value realized from specific projects.

The cost savings and performance improvements in manufacturing have not gone unnoticed in construction. There have been increasing calls by academics [e.g., Arbulu and Tommelein 2002; Vrijhoef and Koskela 1999] and practitioners [e.g., Eagan 1998; Vaidyanathan and O'Brien 2003] to apply SCM principles to construction practice. FIATECH (http://www.fiatech.org) and European Construction Technology Platform (ECTP) (http://www.ectp.org) are two such instances of industry/academic initiatives. The FIATECH initiative in particular is developing an extensive technology roadmap for construction that calls for an "integrated supply and procurement network" supported by extensive IT applications. The current version of the FIATECH roadmap calls for integrated procurement throughout the supply chain, links from design to procurement, autonomous supply and delivery to site, and integration of site production activities with off-site production, delivery, and materials staging [FIATECH 2003]. The FIATECH roadmap also calls for a vision for implementation. The ECTP is a similar initiative to transform the construction industry to be increasingly client-driven, sustainable, and knowledge-based [ECTP 2005]. Based on the experience of designing and implementing manufacturing supply chain solutions, while technically it is possible to realize the FIATECH and ECTP vision, the special challenges of construction projects and the organization within the construction industry make implementations difficult. This chapter will explore the issues within the construction supply chain, and propose development of coherent sets of information technologies to support construction SCM based on the knowledge from manufacturing and extending the same to construction. This provides a basis for an implementation guideline for firms to adopt improvement supply chain practices supported by IT.

The chapter has six major sections. Each major section will be split into two subsections—one on manufacturing to discuss the current state of the world in manufacturing, and a second one on construction to compare, extend, draw analogies, and propose ideas for construction SCM practice and/or IT tools. Where possible, the construction subsection highlights and discusses differences in the supply chain issues and solutions for general contractors (GCs) and subcontractors, architects, and engineers. Section 15.2 will start with the description of the construction supply chain as seen from the perspective of the various stakeholders in the industry, and identify the areas of operational inefficiency. Section 15.3 will then review the existing landscape of applications, and also identify areas that are currently not being addressed by IT. Within the scope and objective of the chapter, where applicable, Section 15.3 will discuss the business processes that need to be modified/adapted to take advantage of the solutions to realize true supply chain operating efficiencies. Section 15.4 will describe case studies from manufacturing and construction that highlight the values companies have realized from modifying business processes and adopting associated IT solutions to realize supply chain operating efficiencies. Using the case studies as the basis and the experience from other similar manufacturing industries, a phased approach to implementing technology-enabled applications for better SCM in construction is proposed as a guideline in Section 15.5. Section 15.6 will draw some conclusions on the review of IT applications and guidelines for implementation of construction SCM solutions.

15.2 Description of the Manufacturing and Construction Supply Chains

15.2.1 Manufacturing Supply Chains

The "supply chain" refers to the flow of goods, services, money, and information in the process of converting raw materials to end-products that are consumed by the end-user [Simchi-Levi, Kaminsky, and Simchi-Levi 2000]. Figure 15.1 shows a typical manufacturing supply chain. As shown in Figure 15.1, goods flow from the raw material producer to the end-consumer. Money flows from the end-consumer to the raw material producer in the opposite direction. At each step along the flow, money is distributed as a function of the value provided by the stakeholder in the network. Information also flows from the customer to the raw material producer. SCM broadly refers to the management of materials, information, and financial flows to effect performance improvements across the supply chain [Simchi-Levi, Kaminsky, and Simchi-Levi 2000].

It is important to note that Figure 15.1 depicts both a manufacturing supply chain and an extended supply chain. Much of the implementation of improved supply chain practice in manufacturing falls within a single large firm, such as Hewlett Packard [Beyer and Ward 2002; Davis 1993] that owns several factories and distribution centers. This single firm is noted as the manufacturing supply chain in Figure 15.1. Generally, improvements in practice are easier to make inside a firm as any cost savings or growth in market share go to a single bottom line.

For many products, the manufacturing process is the same from order to order, and hence the processes and stakeholders (for instance, suppliers for the manufacturers in Figure 15.1) in the supply chain remain the same. This means that the various stakeholders often have long-term relationships with each other. However, firms are still learning to leverage these relationships. Due to lack of shared information and trust, estimates of demand are often padded (or lowered) by the manufacturer in as much as the delivery date is padded (or increased) by the supplier. For instance, since the manufacturer does not know the supplier's lead time to supply raw materials or, more importantly, supplier's ability to react to changes in demand, the manufacturer orders more material than needed and stores it as inventory. This padding, which amplifies as information flows up each tier of the supply chain, is called the bullwhip effect [Lee, Padmanabhan, and Whang 1997]. The bullwhip effect leads to inefficiencies in the system in the form of excess inventory or stockouts at various stages of the supply chain, and the corresponding locked up capital can lead to delays in order fulfillment and loss of business. The overall cause of the bullwhip effect and related problems is a lack of end-to-end visibility

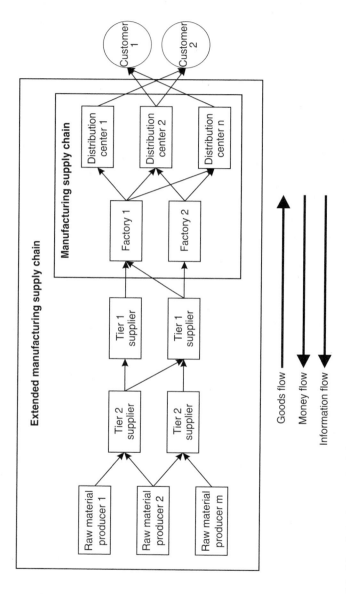

FIGURE 15.1   Manufacturing supply chain.

in the supply chain, along with appropriate incentives for firms to improve system level performance. Much current research and implementation efforts in manufacturing SCM involve sharing information and system level incentives [e.g., Corbett and Tang 1999; Gallego and Özer 2002; Lee and Whang 2001]. In the past 10 years or so, there has been a realization that value can be derived from interfirm communication, providing suppliers visibility into the customer demand from the manufacturer, and automating procurement between manufacturers and suppliers [Narayanan and Raman 2000].

15.2.2 Construction Supply Chains

The construction supply chain, by direct analogy with manufacturing practice, refers to the goods, information, and money flow for a particular construction project (see Figure 15.2). Goods flow from the raw material producer (or building products manufacturer) to the end-customer, and money flows in the opposite direction from the end-customer to the raw material producer. Money is distributed as a function of the value-added services provided and the amount of risk taken by each trade. Information flows in all directions as indicated between all the stakeholders in the project.

Unlike the manufacturing supply chain in Figure 15.1, Figure 15.2 shows value-added services provided by several trades like the architect, engineer, trade contractors,

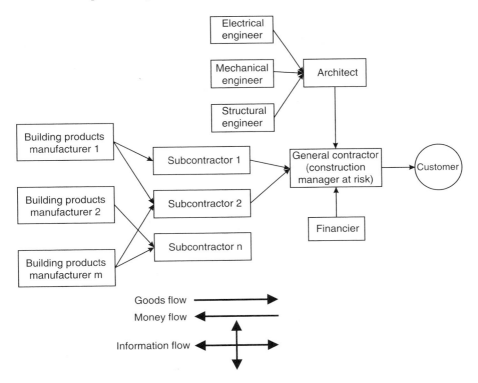

FIGURE 15.2 Construction supply chain.

construction managers, financiers, insurers, etc. This is not to say that equivalent parties are not involved at some level in a manufacturing supply chain, but rather that such parties in construction play a much more active role in the construction process than they do in the manufacturing. The short-term and prototype nature of construction projects implies that the designers and engineers will play a large role in the working of a construction supply chain.

As the construction project is typically a one-off prototype, information flows in all directions between the trades, the GC, and the customer to complete the project. Information exchange is often through phone, fax, and/or printed drawings exchanged through mail. Information is typically recreated several times between these various trades for trade-specific analysis. For instance, the architect sends drawings to the engineer, who recreates the CAD drawings with engineering information added. After completion of design, the construction manager recreates the drawings to add construction ready details and associated information. The reason for this duplication of effort is twofold—first is the lack of economic incentives to encourage sharing, and the second is the lack of tools and technologies. Each of the architect, engineer, and the construction manager in the example above is an enterprise of its own and there is no economic incentive for them to share the effort for creating and sharing information (CAD drawings). In fact, there is a legal disincentive in that if information is shared and that leads to construction faults in the project, it makes it easy for the various parties to pass the blame around [Macomber 2003]. Add to that the fact that until recently there were no tools or technologies to share information (tools and technologies are further described in Section 15.3).

It is also important to realize that information is often recreated even within an organization across functional boundaries. For instance, within a mechanical subcontracting firm, the estimation department has to hand count or manually input quantities by looking at the construction ready CAD drawings generated by the drafting department. This is primarily because the tools for generating CAD drawings and estimation do not yet seamlessly integrate with each other. Structural, mechanical, and other engineers feel the same inefficiency when it comes to translating between engineering analysis tools and CAD tools. There is a similar manual coordination between scheduling and estimation as well. The recent effort around development of interoperability standards (discussed in Sections 15.3 and 15.5) is a step in the right direction to address this, but the industry is long way away from widespread adoption of the standards.

Sharing (or lack of sharing) information is a major source of delays, errors, and duplication on projects. Every time information is passed from one party to another, there is a loss of time and productivity, all of which leads to increased costs and increased inefficiency. Thus, problems of sharing information on projects can lead to bullwhip effect type problems [Lee, Padmanabhan, and Whang 1997], and also to much broader issues of errors, and omissions, and incompatibilities. The various ways and means to reduce these inefficiencies is at the core of construction SCM.

A further difference between manufacturing and construction is the degree to which participants work simultaneously in multiple supply chains. As shown in Figure 15.3a, the GC is involved in multiple projects with multiple clients. In each project, the GC is likely to interact with different architects, engineers, financiers, end-customers, and subcontractors. But given the GC's business model, they typically employ very few direct laborers, and are present primarily to coordinate across all the trades and act as an agent for the customer to orchestrate the project. Employees of GCs are typically dedicated to the project for the duration of the project and, hence, their view of the supply chain is primarily single project oriented (similar to the customer's view of the project), and their focus is around coordination and information sharing across the various stakeholders in the project, as described above. Herein, even though the GCs might have sophisticated tools and technologies, they may not be able to use them effectively since

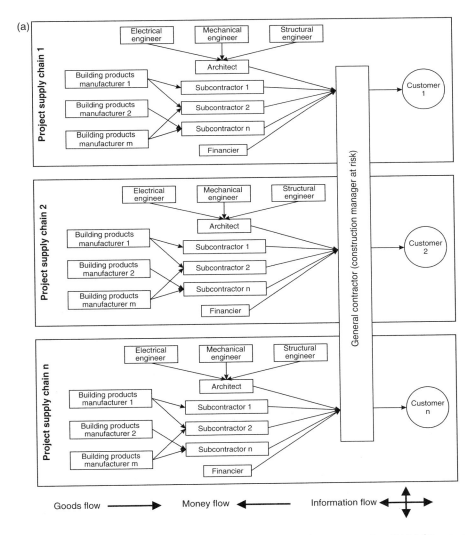

FIGURE 15.3 (a) Multiproject network supply chain—general contractor view. (b) Multiproject network supply chain—subcontractor view.

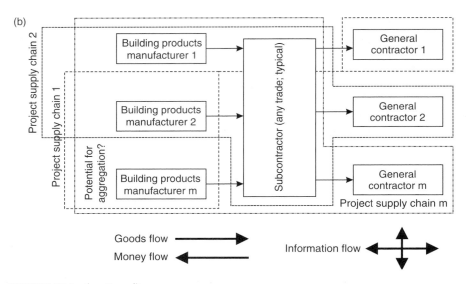

FIGURE 15.3   (continued)

the subcontractors usually do not have the requisite compatible tools to enable a seamless two-way electronic communication. There is a similar lack of information sharing across construction projects (representing different construction supply chains) within the GC firm.

Each of the other stakeholders, particularly subcontractors, on a given construction project are also participating in several concurrent projects (see Figure 15.3b). In their situation, since few projects are so large as to dominate a firm's activities (or a supply chain's) and given that their involvement is limited at best to the trade(s) that they work on, there is a need in construction to take a multiproject viewpoint [O'Brien and Fischer 2000]. The subcontractors typically employ direct labor and are looking to maximize utilization of them. To that extent, coordination of work *across* projects within the firm is a key pain that needs to be addressed. This coordination is primarily felt in the schedule interactions between projects and the constant schedule changes that happen between them by virtue of the fact that all the projects share the same labor resources. This multiproject stance, together with the short-term nature of projects, generally implies that a large number of the relationships on any given project are short term with little to no information sharing and also no long-term strategies to address the schedule interactions. The schedule interaction is further acerbated by the fact that current scheduling tools and techniques are inadequate at best (see details in Sections 15.3 and 15.5). Schedule coordination is typically done on an ad hoc basis, likely through cell phones and walkie-talkies.

Building materials suppliers within the construction industry have their own supply chains, which are similar to the manufacturing supply chains described in Section 15.2.1. Depending on the type of building materials, the manufacturing process could be make-to-stock, build-to-order, or engineer-to-order. For instance, the manufacturers of commonly used building materials (like carpet, rebar, sheet metal, etc.) have a supply chain that mirrors the manufacturing supply chains discussed above, which is a make-to-stock manufacturing environment. From a product manufacturer's perspective, they only need to ensure that they have adequate material in stock. They typically do not manufacture for a specific construction project. They manufacture products based on historical sales forecasts and not based on orders placed on a project. On the other hand, manufacturers of capital intensive equipment, like generators and air-conditioning units, will tend towards being an engineer-to-order manufacturing environment. They do not undertake to manufacture until they have a firm order. And this largely shields them from the vagaries of the construction project and its changes.

Finally, in direct contrast to manufacturing SCM, in construction there is not a single stakeholder who stands to benefit from an efficient SCM solution. Each firm is interested in increasing its efficiency, but no large firm exists that currently has incentives interested in improving the whole system—or even a significant subset of the system as large manufacturers have. The nature of the construction supply chain is decentralized and multienterprise-oriented. It is not clear who will reap the benefits of the increased efficiency. And given that each of the participants is concurrently dealing with several projects (see Figure 15.3a and Figure 15.3b), they are much more likely than manufacturing firms to focus on improving the efficiency of their work to realize immediate economic benefit rather than to improve system performance.

Table 15.1 summarizes the inefficiencies in the construction supply chains as a result of the current organization and way of doing business. IT and IT-enabled applications coupled with business processes that look at supply chain efficiencies have the potential to address some or a lot of these inefficiencies. The next section will review the current IT tools landscape that the construction industry uses to manage its supply chains.

15.3 Review of IT Applications

The comparison of manufacturing and construction supply chains suggests that the problems of implementing IT solutions in construction are larger than in manufacturing. At the same time, there are also enough parallels to suggest that the construction industry can learn from the implementation efforts in the manufacturing industry. In this section, IT tools that support SCM in manufacturing and construction are reviewed.

15.3.1 Manufacturing Supply Chains IT Applications

IT tools for manufacturing SCM has grown and matured since the early '90s. These tools were a logical extension to the then existing enterprise resource planning (ERP) and manufacturing resource planning (MRP) systems that had become entrenched in manufacturing. ERP and MRP systems were designed to primarily help automate single business unit functions in accounting, finance, billing, and the handling of payroll for employees. And for those purposes, the systems served well as the repository of

TABLE 15.1   Inefficiencies in the Construction Supply Chain

General Contractor	Subcontractor	Supplier
1. Typically focused on coordinating all stakeholders in a single construction project	1. Typically focused on managing labor (business) across multiple construction projects	1. Typically has adopted some form of manufacturing SCM solutions
2. Work is coordinated usually through phone/fax leading to translation errors and omissions	2. Nonintegrated business processes within the firm leading to manual recreation of data (CAD, estimation, scheduling, design, engineering)	2. Direct incentives to improve operational efficiency within organization
3. Largely there is a lack of incentive to share information across the supply chain that in turn leads to duplication of data creation	3. Typically lack tools and technologies to aid business process management	3. Largely unable to use electronic tools to communicate with GCs and subcontractors both due to lack of the same at the other end and lack of integration data standards
4. Largely unable to use tools and technologies to communicate with subcontractors due to unavailability of compatible tools at the subcontractor end	4. Inadequate scheduling tools to address multiproject interactions	4. Usually unable to gain visibility into demand for capital equipment to reduce lead time
5. Typically there is a lack of visibility and need (incentives) to aggregate procurement across projects	5. Trade subcontractors lack mobile collaboration tools that simplifies communication between field workers and office	

historical warehousing of data and to log planning and scheduling issues after the fact. The systems did not have any requirement to support planning and forecasting; rather, actuals of parts that were procured, orders that were manufactured, and hours people worked were all that mattered.

With the need to improve operational efficiency, manufacturers realized that ERP/MRP systems were lacking in forward-looking decision support capabilities. They needed tools to help a production planner at a factory decide how to run a shop floor, a factory, or an entire manufacturing organization looking into the future. New tools were developed to address this requirement. These tools were designed to be a layer of decision support on top of the ERP and MRP data. These commercially available tools are broadly classified into SCM tools and supplier relationship management (SRM) tools (see Figure 15.4).

15.3.1.1 SCM Tools

These include a suite of tools to improve productivity, efficiency, and provide visibility across the manufacturing supply chain that typically are business units of a single man-ufacturing organization (supply chain). The SCM software vendors (e.g., i2 Technologies Inc., SAP AG, Oracle Corp., among others) have modules to cover one or more of the

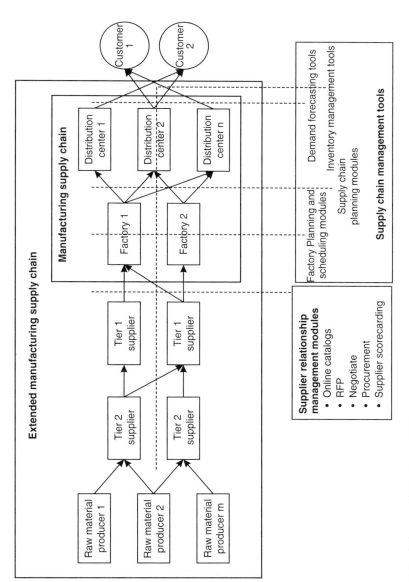

FIGURE 15.4   IT tools for manufacturing supply chain.

functionality described below as part of their SCM solution. Known as advanced planning and scheduling (APS) solutions, they include several modules, including:

- *Demand forecasting:* this module looks into past historical sales data, and forecasts demand for products into the future. Sophisticated algorithms are capable of accounting for various situations, including seasonality, and isolating the effects of promotions while predicting future sales.
- *Factory planning and scheduling:* these tools generate capacity and material constrained plans that are feasible to execute and provide forward visibility into the manufacturing plan looking ahead a few days, weeks, or months, depending on the nature of the particular business. Their advanced optimization techniques can, in addition, take feedback from the shop floor and readjust the plan periodically. The tools provide decision support capabilities that allow a production planner (scheduler) to perform what-if analysis and manage unplanned scenarios (e.g., a rush order, a machine break down, a part shortage, labor availability disruptions).
- *Supply chain planning:* these include solutions and optimizers to load balance production capacity of factories (producers of end-products) with the demand variations across sales distribution centers (consumers of end-products) to ensure that all factories are producing at optimal capacity and that the distribution centers are stocking inventory per their ability to sell. Their decision support capabilities permit a corporation to better manage excess inventory and stockouts over a longer time horizon, in turn helping them release locked-in capital for growth and investments.
- *Inventory management:* these include tools and solutions to track and manage inventory at various parts of the supply chain. The purpose of these solutions is to provide visibility and track inventory along the supply chain. In addition to software, inventory management solutions can include hardware solutions, like RFID tag readers, automatic bar code scanners, etc., to track inventory and provide alerts when there is inventory shortage (or excess). Historical data on inventory along the supply chain can be used to set inventory policy (maximum and minimum inventory holding) for optimal operation of the supply chain.

As tool providers gained experience in manufacturing SCM, they realized that each of the above capabilities, although generic across manufacturing, needed different solutions/approaches depending on the type of manufacturing—discrete vs. process, or make-to-stock vs. engineer-to-order, etc. Experience and knowledge in SCM has also led to some unique business models from the manufacturers' side. One such business model is a vendor managed inventory (VMI) business model, wherein manufacturers of consumer goods and retailers share the risk of inventory holding in excess or shortage of consumer [Kaipia, Holmstrom, and Tanskanen 2002]. For instance, one of the manufacturers of high-end consumer electronics has developed a VMI model with one of their retailers for sales of their newer high-end TV. By linking point of sale data at the retailer to forecasting and ordering systems at the manufacturer, the manufacturer has been able to quickly adjust the manufacturing of Plasma and other high-end flat screen televisions to variations in demand (vs. forecasted demand) to reduce the possibilities of stockouts [Gilmore 2006].

15.3.1.2 SRM Tools

Once the in-house supply chain of a single manufacturing organization had been optimized, manufacturers have started to look for better ways to collaborate with their tier 1, tier 2 suppliers outside the boundaries of their firm. The goal is to improve the system level operational efficiency of the extended manufacturing supply chain (see Figure 15.1), and this led to the development of SRM tools. (See the case studies in Sections 15.4.1.1 and 15.4.1.3 to realize the value of SRM tools.) SRM software vendors (e.g., i2 Technologies, SAP AG, Oracle Corporation, among others) have modules and solutions to facilitate information sharing and collaboration with suppliers, and also modules to facilitate procurement. Broadly, SRM tools include:

- *Supplier collaboration:* these include tools to electronically communicate information with supplier, be it manufacturer's demand, request for expedited shipping, or supplier delivery schedules, confirmations, shipment tracking, etc. Collaboration aided reduction in communication via phone/fax and permitted digital communication via electronic data interchange (EDI).
- *e-Procurement:* these include modules to facilitate procurement, including electronic submission of proposals, bid processing, and payment processing. All the paperless transactions simplified processing and reduced the cash collection times between manufacturers and suppliers. Suppliers were able to release locked-in capital and manufacturers, in turn, had a better pricing from the suppliers leading to a win–win situation for both.
- *Supplier scorecarding:* these include tools for the manufacturers to rate suppliers based on their capabilities, customer service, due-date compliance, quality and reliability of service, etc., and use it to negotiate better procurements terms.
- *e-Catalogs:* electronic catalogs have emerged as part of the SRM suite to help suppliers publish their product specifications to manufacturers. Manufacturers, on the other hand, can compare products from various suppliers before making an informed choice about procurement.

The SRM suite of products is still an emerging and maturing industry within manufacturing SCM. Once the majority of the manufacturers are over the adoption curve in the basic SRM suite of products, there is emerging talk about "design-to-availability" in the academic circles, wherein a manufacturer can specify parts for its products during their design cycles by looking into the cost, availability, and technical specifications from their suppliers. For instance, CATIA™, a leading CAD vendor for manufacturing CAD, has a partnership with product sourcing and design modules of the SRM suite to include parts design specifications and, potentially, availability, pricing, etc., during the product design cycle—particularly for aerospace, defense, and automotive manufacturers.

Interoperability of the various tools and modules discussed above is another topic of discussion in this context. Manufacturing SCM tools evolved without the explicit need for interoperability standards. Manufacturers and tool vendors could afford for it to be so since the data integration was initially and also largely within the manufacturing organization and, hence, could either be custom and proprietary, or since a manufacturer typically bought all of its SCM tools from one vendor, the vendors figured out ways to integrate their SCM modules. But when the tools, data, and workflows started going beyond the manufacturing organization (systems) into suppliers' organization (systems), particularly for SRM tools, the need for standards and ways to simplify interoperability became increasingly higher. Standards organizations like RosettaNet™ (http://www.rosettanet.org) [RosettaNet 2006], Organization for the Advancement of Structured Information Standards (Oasis™) (http://www.oasis-open.org) [OASIS 2006], and Supply Chain Operating Reference (SCOR™) (http://www.supply-chain.org) [SCOR 2006] evolved to develop these standards. These are largely industry academic collaborations and they have designed data standards to facilitate interoperability. Technologically, there has been a class of integration tools, known as enterprise application integration (EAI), which have evolved to simplify the ability to integrate data from one system to another. But recently, the advent of service-oriented architectures (SOA) has given manufacturers the added benefit of being able to integrate data and workflows from various tool vendors with relative simplicity [Sprott and Wilkes 2004; Bierberstein et al. 2005]. All of the major vendors that offer manufacturing SCM and/or SRM tools today offer their products with SOA architectures, although this is an evolving technology in manufacturing SCM/SRM.

To summarize, manufacturing IT tools evolved in three distinct phases:

- Phase 1 involved the development of ERP/MRP systems to *automate a single business unit function*
- Phase 2 involved the development of SCM tools to provide supply chain visibility *across business units but usually within a manufacturing supply chain*
- Phase 3 involved the development of SRM tools to provide visibility *across organizational boundaries into the extended manufacturing supply chain*

15.3.2 Construction Supply Chains IT Applications

The construction industry has its share of IT tools available to manage the construction business. However, in direct analogy to the evolution of tools in the manufacturing SCM, most of the tools that dominate the industry are the tools that can be classified within the Phase 1 category above. They serve to automate a single business unit function, but provide little value to aid information flow and visibility across business units and/or across organizations. In short, as construction SCM is a relatively new theory in construction, there is a corresponding lack of construction IT tools that explicitly support supply chain improvement. The tools that have gained widespread adoption are discussed below. See also Table 15.2 for mapping of tools available by business process.

- *CAD:* 2D CAD tools that help digitally render drawings are commonly used by various stakeholders in the industry. Digital rendering makes it easy to accommodate changes to the drawings for any reason (architectural, engineering, or constructability). Use of 2D CAD packages is common among all stakeholders in the construction supply chain including subcontractors. AutoCAD™

TABLE 15.2   Mapping of Tools and Associated Business Processes

Business Process	Tools Available[a]	Advantages	Limitations
Design	AutoCAD, Microstation	Electronic rendering Easy to accommodate changes 3D and 4D CAD improves constructability and coordination	Document sharing not widespread across stakeholders in the supply chain No integration to other tools like scheduling, project management, analysis, and estimation (with limited exception of 4D CAD)
Analysis	Autodesk, Bentley, ANSYS	Easy to analyze and design various options/scenarios Simplifies code compliance verification	Until recently, lack of tools that integrated CAD into engineering analysis Lack of robust tools to integrate cost considerations, decision support, constructability into analysis
Project planning/ scheduling	Microsoft, Primavera, SAP	Electronically create and manage schedules, track progress, generate reports including earned value management, etc.	Schedules are static and rarely replanned during execution Schedules are not capacity and material constrained Schedules do not account for multiproject interactions
Estimation	Timberline, WinEstimator	Easy to compute costs, reduces scope for human errors in calculations	Largely lack of robust tools to automate quantity take-offs from CAD drawings Lack of integration of material costs from supplier into estimation tools for updated costing information Lack of robust tools to integrate procurement across projects for cost savings
Accounting, finance, HR (ERP)	Intuit, Sage, Microsoft, Oracle, SAP	Manage accounting, taxation, finance, billing Manage HR, payroll, etc.	Lack of integration of the various ERP functionality Lack of integration between planning and estimation to accounting, procurement, and billing
Collaboration	Project extranets	Single location to share data Coupled with 3D and 4D CAD tools, improves supply chain productivity	Data is still potentially recreated multiple times
Field tech	RFID, Mobile phones, etc.		Lack tools to share information across various stakeholders Not widespread due to cost and other technological reasons

[a] These are meant to be representative rather than exhaustive. There are several other vendors not mentioned here that offer solutions to each of these business processes.

(Autodesk Inc.) and Microstation™ (Bentley Systems Inc.) are the most commonly used CAD applications, but there are several other vendors offering CAD packages depending on the type of project (commercial vs. residential) or size of the firm. Even though the drawings are created electronically, until recently they were printed and shipped from one trade to another. Recently, it is more common for them to be shared electronically as an e-mail attachment or through a project Website. In fact, the tool vendors are even providing capabilities to track the changes in the drawings to easily identify the person (or firm) and purpose of the changes for the sake of accountability.

- *Project management and scheduling:* numerous construction planning and scheduling tools based on critical path method (CPM) of calculations [O'Brien 1965] methods have long been available in the industry. In a supply chain context, scheduling tools attempt to provide a centralized place to coordinate production activities of a supply chain. But, in order for them to be effective, they need to be shared across the supply chain and/or projects. Today, schedule interactions are at best manually coordinated. Schedules are also typically static and not maintained dynamically as changes happen during the course of the project. Smaller subcontractors tend to use paper-and-pencil scheduling rather than an electronic tool, since the project management tools available in the market are too complicated for their needs.

- *Estimating:* analogous to demand forecasting tools in manufacturing, estimating tools provide quantities necessary for cost and resource management purposes. Smaller subcontractors tend to use paper-based estimating techniques rather than electronic tools for the same reason they avoid project management tools.

- *ERP systems:* these tools are used to handle accounting, human resources (HR), finance, billing, etc., and are typically available with all the stakeholders in the industry.

It is worth noting that diversity of the industry means that a one-size-fits-all technology solution is not possible within the industry. A GC within the Engineering News Record (ENR) top 400 (http://www.enr.com) might have a need and the money to buy a large enterprise ERP system (e.g., SAP AG or Oracle Corp.), and also have the need to buy the other systems listed above. But smaller trade subcontractors, who are the most dominant members of the construction market, belong to the small and medium business (SMB) market segment and need simpler tools that are inexpensive and easy to use. There are ERP systems that cater to the SMB market (e.g., Intuit, Microsoft, Sage), but other tools are underserved in this market.

The technology boom of the '90s saw another wave of technology tools that were introduced to the construction industry. These tools belong to the Phase 3 category (see Section 15.3.1) that provides e-commerce and collaboration capabilities for the stakeholders in the industry. But with the technology bust, there has been a wave of rationalization in both the tools and the vendors that provide the tools. In some cases, the companies were ahead of their time with few takers for their solutions. Several of them collapsed, others were either acquired or merged. A few tools that have survived after the technology bust are discussed in brief below. The chapters that follow this chapter in the book discuss some of these tools and technologies in further detail, specifically, visualization, e-commerce, collaboration, and field technologies like RFID.

- *Collaboration, project Websites, and file sharing:* numerous tools, from intranets, file transfer protocol (FTP) sites, to project Websites, are used by organizations to share files both within and across organizations. Files shared can include CAD drawings, specification documents, project reports, etc. In addition, it is not uncommon these days to expect submissions of tenders and RFPs as electronic documents.
- *3D and 4D CAD:* the rise of 3D tools, in particular, provides improved design coordination and facilitates construction planning. In the bigger GCs and more expensive projects, the use of 3D CAD is rising, but still not with all the stakeholders in the industry. In fact, in a recent survey done by CADALYST magazine, only 6% of the survey participants are completely using 3D CAD tools compared to 24% that are still using exclusively 2D CAD tools. The others are either using 2D but evaluating 3D (26%), or using a hybrid of 2D and 3D tools (44%). And there is not a significant upward trend in the adoption of 3D tools [Green 2007]. A newer class of 4D applications, which combine 3D CAD models with scheduling and construction coordination, are also starting to be used. 4D CAD applications have the added advantage beyond 3D in that they improve the quality of construction schedules with ensuing benefits to off-site production scheduling.
- *In-house tools:* with the use of spreadsheet tools (like Microsoft Excel) increasing, a lot of firms end up developing in-house tools to solve the problem of automation across business units. For instance, it is not uncommon to find in-house spreadsheets to automate estimation from CAD. Such tools serve the purpose of proving the demand that is yet to be capitalized by the tool vendors and industry.
- *Field technologies:* there is also a class of hardware coupled with complementary software tools that improve the productivity and help track construction equipment and material at the construction site and/or through the construction supply chain. RFID tags, bar code scanners, handheld devices, etc., are being increasingly used in the past few years in major construction projects. Such technologies are yet to gain widespread adoption due to cost and other technological reasons like ensuring reliable performance of the tools at the construction site, sometimes under harsh conditions.

With the limited exception of certain 4D CAD and field technology tools, construction IT tools are point solutions with no real data and workflow integration between them. Data is still being recreated multiple times and transferred manually within and across enterprises. While collaboration tools remove data entry duplication to a limited extent, their primary value-add is the fact that they facilitate electronic information transfer rather than paper-based information transfer. IT tools are largely directed at improving the productivity and efficiency of people within a single business function and within a single firm. Hence, potential efficiency improvements are difficult to identify and/or leverage given current tools. Such limitations have long been understood in construction, and efforts to develop data standards have been made to promote interoperability [IAI 1996]. Tools based on these standards do show promise [Fischer and Kam 2002], but are still largely directed at execution of traditional construction tasks and are not explicitly directed towards supply chain practice.

The construction industry also lacks the tools to manage the extended project supply chain. By that, two things are meant. First is that within the four walls of the firm, there is a lack of integration across processes and tools. For instance, take the case of scheduling tools. CPM-based [O'Brien 1965] schedules' produce plans based on assumptions of infinite capacity [O'Brien and Fischer 2000]. They ignore the interactions between various projects that share resources. While this is probably less of an issue for a GC, it is an extremely important issue for subcontractors that employ direct labor, particularly once they go beyond a certain size. Take also the case of estimation tools. While there are good estimation tools available, there are no tools to automatically pull up the prices of items from standard catalogs given the item specification. It is only recently that under the aegis of FIATECH, projects have been undertaken to develop data exchange for bulk material and engineer-to-order projects. The automating equipment information exchange (AEX) project [FIATECH 2006] is an XML data specification to facilitate such data exchanges. Tool vendors can integrate their tools to exchange data using the specifications to facilitate interoperability. Similarly, tools that provide visibility and/or aggregate procurement needs across the various projects for bulk procurement are not yet widely available, although some work is being done in this area [Taylor and Björnsson 2002].

Second is that across firms as well, there is a lack of widespread availability and use of tools to provide the visibility and collaboration across the stakeholders in a construction supply chain. While there is a commercial and contractual barrier that needs to be crossed for the concept to be adopted, the tools to enable such needs have not yet fully evolved either to eliminate duplication of data creation, or streamline information flow. In some cases, infrastructure availability is another issue. For instance, it is only in the past 10 years or so, that computers and Internet connectivity is starting to be widely used. In most parts of the world, until recently, particularly with smaller subcontractors, phone and fax has been a more common mode of communication than those facilitated with computers and the Internet (e.g., e-mail).

FIATECH recently concluded a research project on emerging technologies that provide value to the construction industry [FIATECH 2005]. The report has identified several areas to improve productivity in the construction industry. One emerging area is the use of mobile applications. The mobile applications are based on the realization that cell phones, PDAs, tablet PCs, and other handheld devices are ubiquitous among subcontractors and other construction field workers, and are an ideal platform to deliver supply chain and other productivity improvement tools. Another emerging area is automated supply management, and is based on the need to do collaboration, e-commerce, and procurement through the use of Internet enabled applications. Certain sectors within the construction industry, like the plant and industrial construction sectors, are further along in the adoption of tools and technologies than other sectors, like residential and commercial office construction. Data models and standards are also evolving for interoperability. FIATECH has been developing standards for data storage and exchange. Parametric modeling and BIM-based [GSA 2006] modeling has recently led to the development of integration of various construction IT tools. IFC is another attempt at integrating CAD with estimation and scheduling [IAI 1996]. CAD is evolving from 2D to 3D and vendors are integrating analysis and design. Academic circles have discussed 4D and 5D CAD models that integrate estimation, scheduling, and construction sequence to the CAD models [Fischer, Haymaker, and Liston 2003].

The FIATECH initiative is by no means unique, but reflective of a growing trend around the world for the need to improve the productivity of the construction industry. For instance, the ECTP (http://www.ectp.org) is a program initiated in the European Union that outlays a roadmap for the use of technology and newer integrated business processes as a means to improve the productivity of the construction industry [ECTP 2005]. Similarly, a committee headed by Sir John Eagan developed a vision for rethinking the construction industry in UK through the use of collaborative and integrated business processes, and the use of the latest technology-enabled tools [Eagan 1998].

The next section will continue this discussion through the use of case studies to illustrate the value of tools and associated processes in SCM.

15.4   Implemented Case Studies

This section highlights some of the case studies of actual implementations of SCM in manufacturing and construction. The next section will extract common characteristics from these case studies and develop strategies for successful IT implementations in construction.

15.4.1   Manufacturing Implementations

15.4.1.1   Case Study: PC Manufacturer

This case study is about one of the world's largest PC manufacturer that believes supply chain excellence is a competitive weapon. They have evolved their tools and processes in phased implementations as discussed in Section 15.3.1. They have had explosive growth since it came into the PC manufacturing business. Their famous build-to-order model has transformed the PC manufacturing industry. Through sharp focus on improving supply chain operation efficiency, they have very lean and efficient production operations [BusinessWeek 2001]. The company has a driven goal for customer service and manufacturing excellence. Everyone from CEO on downwards focuses relentlessly on driving low-cost materials from suppliers through the supply chain to their customers. Their goal is to replace inventory with information. They began their journey to supply chain excellence by first implementing a material and capacity constrained factory planning solution. This helped ensure that their manufacturing plans were reliable. Then they added a demand planning solution that helped plan for expected demand. Then they extended the solution to the entire supply chain, ensuring that all manufacturing units and distribution centers shared information for efficient balancing of manufacturing workload and sales demand load. This helped improve their customer demand fulfillment and reduced the total order delivery time (time when customer places order to when finished goods are delivered to them).

In the next phase, they began sharing their order and demand information with their suppliers. With SRM tools, they effectively streamlined and opened visibility of their customer orders through their supply chain into their suppliers' supply chains. The company works with their suppliers to establish long-term relationships and is loyal to all of them. Suppliers and the company understand the value of shared information for mutual growth, and are willing to work towards it. Today, the company shares demand data with its suppliers near real time, and replans its supply chain once every two hours. Its sales reflect shifting inventory availability from its suppliers, and in the process, customers get the lowest cost with maximum customer satisfaction. At each step of the way, the company and its suppliers measured the value that they got from the solution, either in terms of bottom-line savings, or in terms of reduction in order delivery time. The value derived and the return on investment (ROI) earned paved the way for each additional solution to be added on.

15.4.1.2 Case Study: Engineer-to-Order Beverage Equipment Maker

This case study is described as an example of a company that has a project-based supply chain and its continuing journey to achieve supply chain excellence. Founded more than 55 years ago, this company is a highly successful engineer-to-order company that is in the business of producing automated equipment for use in the beverages industry. Its products range from machines for rinsing, filling, and labeling, to appliances for pasteurizing liquids and molding plastics. Each machine that it produces is unique to its customer's requirements (and hence has design and engineering phases in addition to manufacturing and delivery), and the complexity of its products varies enormously. If a customer is setting up a new bottling plant, it may require multiple machines for inspection, rinsing, filling, labeling, and packing, all linked by conveyor belt. Alternatively, a customer might simply require an upgrade to a single machine. Given their nature, each customer order is managed as a discrete project, with a unique bill of materials to meet specific customer requirements. In short, they have a project-based supply chain business, and they have multiple projects running in parallel that share resources (including labor) and equipment. In this environment, precise supply chain planning is essential, but often difficult to achieve. Like many companies in its industry, they suffered from poor due-date performance and resource overloads that made order fulfillment a monumental exercise that, in turn, had a negative impact on customer satisfaction and revenues.

Their first step to supply chain excellence was to implement an ERP system. They implemented four modules to automate each of HR, materials management (MM), production planning (PP), and project scheduling (PS) processes. These modules helped automate each of those individual business units. At the end of that phase, they realized that they needed to share information across the four modules to make better decisions. They realized that data was being duplicated across the modules, leading to human errors and loss of operational efficiency. For instance, if a material came in late, it impacted the project schedule, worker assignment, and shop floor production operations. So, during the next phase, they wanted to combine information from each of these modules to plan and make better decisions.

As of 2004, the company still computed its due dates primarily based on critical path method (CPM) dates without fully taking into account the interaction between various orders (projects) due to their shared finite resource capacity or material availability. To try to compensate for this problem, the company frequently experienced busy periods, when they either hired extra temporary labor or paid staff overtime. This helped to keep projects on schedule, but lowered margins. There was also an impact on staff moral, as employees got tired and burnt out. Delays, if any, caused ripple effects downstream making interdepartmental coordination a strenuous exercise. In addition, they tended to "pad out" the duration of jobs (a standard technique used in all project-based enterprises) when providing estimates to customers. As a result, however, the company's predicted lead times for projects were unnecessarily long, which contributed to some lost business opportunities. The company's material procurement plan, developed using CPM dates at the beginning of the project and seldom updated, typically was out-of-sync with the actual plan, leading to either an early or late procurement and causing problems. The company realized that all of these business issues stemmed from the lack of an end-to-end visibility across divisions of a project and across projects within a division.

To overcome this problem, they implemented an enterprise-wide multiproject planning and scheduling tool designed for the engineer-to-order industries in 2003 [Vaidyanathan 2003]. The tool is ideally suited to helping companies plan, communicate, and manage the execution of multiple projects from quote-to-delivery in a dynamic and complex environment. The solution gave them the end-to-end supply chain visibility from sales through engineering and delivery. It also gave its employees one central system to track their work and get visibility (schedule and scope) into changes. With the new solution, the company is now able to plan capacity two years ahead and view materials requirements six to nine months in advance. The decision support component of the new solution gives its planners in various divisions a powerful tool that allows them to evaluate the impact of changes (to scope, delivery date, capacity changes, etc.) to plans, as well as perturbations to the plan due to execution dates being different from the plans. In addition, when events occur that cause projects to be delayed (or advanced), planners are now able to use the solution's problem-solving capabilities to automatically assess the impact not only on the individual project, but also on the entire division and company level.

Along the way, the process management group at the company has been modifying their business processes and the solution to work in sync with each other. They had to align people, processes, and IT to effect real change. They undertook a rigorous change management and education program to explain the impact of integrated planning with its employees, and even offered incentives to encourage people to change and provide correct information to the system. As a result of this relentless effort, in the four years that they have had the solutions in production, they have realized an over 20% increase in throughput without practically any increase in headcount. Improved enterprise-wide visibility of projects has also led to more timely procurement and reduced resource overload, overtime, and weekend shifts. The company is fully expecting all of this to lead to cost savings and have a direct impact on the profitability of the company. Currently, they are evaluating reducing the lead time to deliver a project. They expect the tool and associated change in business process (i.e., their supply chain) to become a competitive weapon as they further solidify their leadership in the beverage equipment industry. In the near future, they are also looking to integrate their supplier portal into their planning solution to gain better communication with their suppliers. This, they expect, will give them better visibility and a better capability to manage the extended supply chain outside their organization boundary.

15.4.1.3   Case Study: Aircraft Manufacturer

The next case study presented here describes the value a company gained from implementing an SRM solution. The company is a market leader in manufacturing commercial aircrafts, with a complex engineer-to-order supply chain. They concurrently manage multiple programs for aircraft manufacturing in a resource-constrained environment. Being primarily responsible for the final assembly of the aircraft, 75% of their production costs is in procurement from their suppliers. Their ability to deliver on-time is also extremely sensitive to the abilities of their suppliers. In 2006, they put a supplier collaboration portal solution into production. The purpose was not just to create a Web-based procurement portal, but a new working relationship between the company and its suppliers, including process change, technology introduction, leading to shared cost reduction. The solution involved harmonizing communication modes, rationalizing part numbering across suppliers, providing a common portal, and other data and workflow harmonizing. In addition, they also moved to a VMI business model with their suppliers, with contracts for minimum/maximum stock violations. The solution went into production with over 500 suppliers across the world trading on over 100,000 items valued at over €3 billion managed through the solution. The company realized value from streamlining their supplier relationships that aided their objective of cost reduction, leading to a lean manufacturing model.

15.4.2   Construction Implementations

15.4.2.1   Case Study: Major US Home Remodeling Retailer

There have been few attempts at implementing SCM practices in construction as aggressively and thorough a manner as the manufacturing case studies discussed above. A known exception was a major US home remodeling retailer who wanted to have visibility and control over the home remodeling supply chain. The retailer was in the business of supplying materials, and would subcontract the delivery and installation labor to construction subcontractors [Vaidyanathan 2002]. In its attempt to orchestrate the supply chain, the lack of efficient processes—shifting priorities from client executives, and several other reasons—led to a stalled program. The value in providing end-to-end visibility was proven. In fact, it was estimated that there would be a 67% increase in scheduling capacity, a 60% reduction in project duration, and a 50% reduction in material holding costs. The technology was available, but the company could not get the information or the processes for the system to work. Several reasons led to the situation:

- *Business processes:* the solution hinged on the suppliers and subcontractors sharing data with the retailer. All parties could not work out the contractual details and the economic incentives for adoption successfully.

- *IT processes:* not all subcontractors had adequate IT systems in place to work with the system. And vendors and subcontractors were not comfortable sharing proprietary data without adequate security. Feedback into the system was not through wireless systems and that would have hurt the mobility of the users of the system.
- *Lack of data:* the integrity of the solution was dependent on data. There was no legacy data to feed the planning and decision support system. Without good data from material vendors on product lead times and subcontractors on labor capacity, the whole solution could not work efficiently.

Client executives decided that unless the data and infrastructure backbone needed for the solution was in place, proceeding with the solution rollout did not make justifiable business sense. Currently, they are in the process of creating that infrastructure backbone.

15.4.2.2 Case Study: 4D CAD

Researchers at the Center for Integrated Facilities (CIFE) at Stanford University (http://cife.stanford.edu) and the GC, Dillingham Construction Company, joined hands to produce 3D and 4D models to coordinate the overall master plan for the reconstruction of the San Mateo County Health Center. The 4D model coordinated owner relocation and operations schedules with construction schedules, facilitated client input, eased relationships with community, and also became the best fund-raising tool [Collier and Fischer 1996]. A detailed 4D model was used to verify the constructability of the project, and also to make sure that the design and schedule information were complete and well coordinated. A second and harder requirement of this reconstruction project was to maintain the hospital working conditions with minimum to no disruptions during the course of the five-year construction project. The 4D model and the simulation of the construction sequence (schedule) served this purpose. For instance, in one case it helped provide a basis to redesign the central hub of the new facility in order to ensure that hospital operations were not disrupted. Overall, the 4D modeling, which was completed with a four man-month effort, tremendously improved collaboration between various parties, supported master planning and design coordination, and improved constructability. Fischer, Haymaker, and Liston (2003) is another example of value gained from 4D CAD.

15.4.2.3 Case Study: Lean Construction and Planning

This case study discusses the use of project and material management software and the synchronous use of lean construction techniques to reengineer the construction supply chain and reduce project delays and cost [Arbulu, Koerckel, and Espana 2005]. Several such implementations have happened in the past few years [e.g., Arbulu and Tommelein 2002; Arbulu, Ballard, and Harper 2003; Arbulu and Soto 2006]. One of the largest GCs in the UK that is involved in large infrastructure projects (like the expansion of airport terminals or refurbishment of portions of the UK underground metro line and station) has been strategically using lean construction techniques [Ballard and Howell 1998; Arbulu, Ballard, and Harper 2003] to gain better control of the construction process. They realized that in the traditional method of working in functional silos, the planning department and procurement department did not work hand-in-hand.

Material management and project management were decoupled, and delays and changes were not being communicated from one to the other. All this was affecting the overall project delivery process. Lack of materials when required, inability of workmen to work in areas expected per schedule, and accumulation of material inventories and idle labor hours were some of the types of waste generated by these practices.

They realized that by using the lean IT tools for production control and material management (e.g., Strategic Project Solutions, http://www.strategicprojectsolutions.com), which worked in tandem with lean construction techniques, they could reduce these wastes. Their main objectives were to: (1) improve the accuracy of site demand (increase workflow reliability) by enabling a better planning and production control process where constraints are identified and removed, (2) increase transparency across value streams by working with Web-based production management software tools, and (3) manage physical and information flows in real time by linking production-level workflow with material supply.

They initially used the system to improve the concrete delivery to site on the underground metro project. The results were immediate. Within three months of using the new system, concrete deliveries became more reliable due to improved planning and integrated planning with material procurement. Between improvements in concrete productivity and reduction in overtime labor due to more reliable workflow schedule, this led to a savings of about £350,000.

In the airport expansion project, they tried similar techniques for engineer-to-order polystyrene panels. Prior to the use of lean and associated tools, the supplier delivering the panels had no visibility into schedule changes from the GC and kept making panels by the outdated original delivery schedule. The panels being unique meant that unless the assembly schedule and material delivery was in sync, inventory was either held up in logistics centers or work was being held up due to unavailability of panels. The long lead time on the panel manufacture, immediate deliveries were not possible which further affected workflow reliability and work progress. These inefficiencies implied that an estimated £200,000 worth of panels were sitting for more than three weeks waiting for assembly. By using lean processes and tools, synchronizing material delivery with work schedule, the material inventory was reduced to one day from almost three weeks. Information transparency between the supplier and GC created the savings. Given that the supplier and the GC worked together for over a year, the savings were enormous, and the value to the customer was tremendous. In both cases, traditional business processes needed to be changed, rigorously enforced, and information sharing had to be done between the various stakeholders to effect change and realize the value.

15.5 Phased Implementation in the Construction Industry

As discussed in the case studies above, successful SCM implementations have transformed companies and projects and provided tremendous value. Successful SCM projects have a few common characteristics:

- *Phased implementations:* the projects are phased in small measurable units. First phase involves improving operational efficiency within the firm through better planning processes to streamline their throughput. On that foundation, they improve the input to the planning process through better forecasting processes. Both of these enable better customer demand fulfillment, and increased customer service levels. After all of these, they are in a situation to improve the extended supply chain through better SRM.

- *Synchronized processes and tools:* processes and tools have to be modified in tandem to derive value. Current business processes are sometimes obsolete and need to be reengineered to remove inefficiencies. The adoption of processes and associated change management has to be planned and accounted for.
- *Strong management commitment:* change management is hard, and unless there is strong commitment from client executives to push through process changes, the SCM tools do not tend to get fully adopted. Many SCM projects have failed during implementation or at launch because they tend to be considered as IT projects. IT is just an enabler of a business process. Fundamentally, the business has to be ready to reengineer its business process to remove operational inefficiencies.

Compared to the construction industry, an obvious advantage that the manufacturing industry has is its vertical integration of the core manufacturing process within the four walls of a single organization. Within the construction industry, the core value-added process—the construction of the project itself—is distributed across several organizations between the GC and the trade subcontractors. Unless all of them agree to come and work together in a cooperative fashion, supply chain efficiencies are difficult to realize. While there are examples of cooperative contractual agreements in the construction industry [Matthews, Howell, and Mitropoulos 2003; Wandahl and Bejder 2003], it has still not caught on and gained wider acceptance.

Similarly, as experience from the manufacturing industry shows, the value proposition for SCM initiatives is more easily justified within an organization than across organizations. Also, based on the differences and priorities in the construction supply chains between the various stakeholders in the construction industry, as discussed in Section 15.2, it is likely that the sequence of tools and technologies that they adopt in their journey to supply chain excellence need not be the same. For instance, the building products manufacturers can and might adopt tools like the other manufacturing companies, as illustrated in the case studies in Section 15.4.1. As far as GCs and subcontractors go, complete supply chain excellence can be realized with better management of their supply chains along the three dimensions discussed in Section 15.3.1—various business functions within the firm, various firms representing a construction supply chain, and various projects representing multiple supply chains. And as discussed in Section 15.2, given the differences in the supply chain outlook between the GCs and subcontractors, their paths to supply chain excellence need not be the same. Figure 15.5 shows a conceptual adoption path for GCs and subcontractors, and is discussed in detail below.

15.5.1 GC's Path to CSCM Adoption

For GCs, the primary value proposition to realize operational efficiency is within the context of a single project (Figure 15.3a). Hence, tools that reduce (or eliminate) data

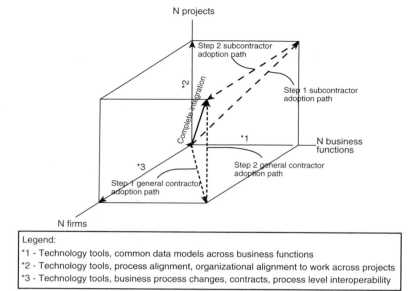

FIGURE 15.5   Supply chain excellence adoption paths for GCs and subcontractors.

re-creation across various business functions, like estimation, procurement, scheduling, and billing will provide the greatest value. And what is needed to realize this is primarily interoperability of existing IT tools. Similarly, collaboration tools provide the greatest amount of value since the bulk of the work for GCs is coordinating work between the customer, the various subcontractors, architect, and engineer. This has to be followed by better project management tools that can incorporate collaborative workflows with the subcontractors, and can respect project and resource constraints between the GC and the subcontractors. Both of these require a new set of IT tools, reengineering and adoption of integrated business processes and process level interoperability (step 1 of General Contractor Adoption Path in Figure 15.5). In short, tools and technologies that simplify information flow, reduce data creation duplication, and automate workflows across multiple business processes within the firm and across multiple firms within the construction supply chain offer the highest value to GCs. This has to be followed with tools that enable sharing information across construction supply chains and by providing value-added services like aggregation of procurement (step 2 of General Contractor Adoption Path in Figure 15.5). All these put together will give the GCs a synchronized set of tools and processes to realize supply chain excellence.

15.5.2   Subcontractor's Path to CSCM Adoption

Subcontractors typically employ direct labor and manage procurement for their projects. Also, typically, they manage multiple projects across GCs (Figure 15.3b), as discussed in

Section 15.2.1. Hence, they will also need interoperability between tools serving various business functions within the firm, like CAD, estimation, and scheduling. But, another equally important need is a superior resource constrained multiproject planning tool that can work across projects representing different construction supply chains. That tool should then provide extensions to handle procurement and collaboration with the GCs. In short, tools and technologies that reduce data creation duplication and increase operational efficiency across business functions within the firm, and across the various projects (construction supply chains) within the firm, provide the highest value (step 1 of Subcontractor Adoption Path in Figure 15.5). That should be followed with the ability to simplify collaboration and information flow across multiple firms within a single construction supply chain (step 2 of Subcontractor Adoption Path in Figure 15.5). All these put together will give the subcontractors a synchronized set of tools and processes to realize supply chain excellence.

15.5.3 Overall

Once the foundation of CSCM is established through these tools and associated processes between GCs and subcontractors, further value creation can be realized through improving system level performance. Together with improved processes, a suggested guideline for implementation will be:

1. A first step, suggested by manufacturing experience, is that firms must improve internal operations. Given the complexity and associated need for human input in the decision process, this step should be taken in two stages: first, with limited use of technology and, second, with more aggressive use of technologies:
   a. Improvements in construction planning can be made using a more rigorous planning and constraint satisfaction process with little, if any, required implementation of new technologies. Lean construction processes—in particular, production shielding [Ballard and Howell 1998]—have been successfully demonstrated for several years, showing broad improvements in schedule reliability, faster and more predictable completion times, and lesser cost overruns.
   b. Following more manual methods that refocus on the process, a second step is to make more aggressive use of 4D CAD technologies to integrate schedule with design. Such tools allow for more experimentation and limited "optimization" that enable production planners to make improved schedules [O'Brien, Fischer, and Jucker 1995; O'Brien 2003].
2. A second step is to move from the increased control of firm processes to better share information and coordinate action with firms across the project supply chain. Technologies such as project Web-sites and Web-enabled scheduling tools are the most appropriate initial tools for this application before firms consider more complex links via Web Services [Alonso et al. 2003] or other sharing architectures. At this point, field technologies like RFID could also be deployed to simplify the tracking and sharing of material flow at the construction site and along the supply chain.

3. With steps one and two, firms are likely to have achieved a highly improved project supply chain practice and operational efficiency. Further integration will likely be achieved with rapid, automatic sharing of information and associated trust. This primarily involves development of data models and standardization of data transfer. It is likely that academic research, like the process connectors work [Shin, Hammer, and O'Brien 2006] and other similar work done through FIATECH, will mature and be commercialized in this time frame. In particular, there is a need to link accounting, estimating, and project controls software. Such linking has many benefits beyond improved supply chain practice. Accomplishment of supply chain performance improvements will hopefully lead other firms to make similar investments, and several more-sophisticated contractors and subcontractors are undertaking such enhancements to their IT infrastructure.

4. Following maturation of technical sophistication within a firm, electronic procurement and associated processes (such as rescheduling and autonomic materials delivery) will be the next step. A first step here is likely to be the tools associated with purchasing and e-procurement—bidding, equipment specifications, product model exchange, on-line catalogs, etc. There is already some movement in the construction industry towards development of these tools, and the attraction of potentially increased sales together with reduced paperwork is likely to provide incentives for development and at least partial adoption of these solutions by the industry.

5. A final step in the evolution of IT solutions for construction supply chains is the integration of design tools with the supply chain operations tools. Elements of 4D CAD tools with associated intelligence for planning, estimating, and trade coordination will be used early in the process to better coordinate supply chain processes.

These five steps are intended to be progressive. Significant realization of each step will take a combination of process change and IT investment that builds on the last step. The logical sequence of adoption of tools and related processes discussed above is further discussed as a Construction Supply Chain Maturity Model [Vaidyanathan and Howell 2007]. To some extent, each step and associated technologies can be achieved semi-independently of the other steps. And there are research efforts, if not early implementation efforts, related to each step. However, building from experience in manufacturing and construction, there is a natural progression to maximize investments in supply chain performance by leveraging performance improvements internal to a firm before making changes to external processes.

15.6 Conclusions

The goal of this chapter is to review IT applications in the manufacturing and construction supply chain. The review and the lessons from the experience in implementing SCM IT solutions in the manufacturing industry are used as a basis to develop a guideline for implementation for the construction industry. The case studies in the manufacturing and construction industries provide insight into the operational efficiency and value that can potentially be achieved through end-to-end supply chain visibility and management. Learning from the experience and drawing analogy into the construction industry, a list of needed capabilities and an implementation guide for the various stakeholders in the construction industry has been proposed. As with all predictions and visions for the future, it is necessarily imperfect. The supply chain characteristics of the various stakeholders in the industry have been highlighted as an indicator to propose different implementation paths for the various stakeholders in the industry. But it is highly likely that the actual implementation path will be different in these segments and within each firm of these segments. It is also useful to briefly note some omissions: other than briefly mentioning in case studies, the need for integrated business processes and their interrelationship between processes and tools has been given the short shift. This is not to downplay their role, but it is beyond the scope of the chapter. Nevertheless, they are discussed in other chapters in the book. As firms in construction begin to practice SCM, the need for newer, reengineered, and integrated business processes will become exigent, and appropriate processes will quickly evolve from existing efforts. Similarly, the review has focused primarily on operational aspects of SCM. However, as firms are better able to control and hence predict their own operational performance (step 1, Section 15.5.1), it may be possible to make predictions at a systems level.

It should be noted that many existing supply chain applications under development for construction fall at steps 2–5, as identified in Section 15.5. See, for example, the numerous (and in some cases now defunct) efforts in online bidding, procurement, etc. While there is value in such efforts, it seems like the extent of their success will be predicated on effective control of internal firm processes (step 1 in implementation guideline list). Certainly, the manufacturing experience has focused first on internal development of controls. Of course, construction is not manufacturing, and the need for extensive coordination and outreach on construction projects may determine that the first "killer application" for SCM may be externally focused. But the success of lean construction thinking applied to internal process coordination (see, for example, the case study of the subcontractor Pacific Contracting in [O'Brien 2003]) suggests that a renewed focus on the firm is a necessary precursor to significant changes in construction SCM practice.

All said, with the maturation of the Internet and increasing levels of business process automation in the manufacturing supply chain, the construction industry is bound to adopt and adapt some of the tools, processes, and systems thinking found in manufacturing SCM. It is not a question of "if", but a question of "when." As such, opportunities for IT are abound in the space, all the way from planning, to procurement, to process automation. The proposed phased solutions for the construction industry will help individual stakeholders to realize the value of SCM within their enterprise, and hence make the case for adoption in the extended supply chain.

Finally, the need for development of integrated (new) business processes should be highlighted again to illustrate its importance. IT and IT-enabled applications are only enablers. Improvements in the productivity and operational efficiency in the construction supply chain can only be realized with business processes that enable end-to-end supply chain visibility, collaborative business practices, and long-term contractual relationships that will lead to no (or at least reduced) data creation duplication and seamless information sharing across the supply chain.

References

Alonso, G., Casati, F., Kuno, H., and Machiraju, V. 2003. *Web services: Concepts, architectures, and applications.* New York: Springer.

Arbulu, R., Ballard, G., and Harper, N. 2003. Kanban in construction. *Proceedings of the 11th Annual Conference of the International Group for Lean Construction,* Blacksburg, VA.

Arbulu, R., Koerckel, A., and Espana, F. 2005. Linking production-level workflow with materials supply. *Proceedings of the 13th Annual Conference for the International Group for Lean Construction,* Sydney, Australia.

Arbulu, R., and Soto, J. 2006. A design case study: Integrated product and process management. *Proceedings of the 14th Annual Conference of the International Group for Lean Construction,* Santiago, Chile.

Arbulu, R. J., and Tommelein, I. D. 2002. Value stream analysis of construction supply chains: Case study on pipe supports used in power plants. *Proceedings of the 10th Annual Conference of the International Group for Lean Construction,* Gramado, Brazil, 183–95.

Ballard, G., and Howell, G. 1998. Shielding production: Essential step in production control. *ASCE Journal of Construction Engineering and Management,* 124 (1), 11–17.

Beyer, D., and Ward, J. 2002. Network server supply chain at HP: A case study. In *Supply chain structures: Coordination, information, and optimization,* eds. J.-S. Song and D. Yao, 257–82. Boston/Dordrecht/London: Kluwer Academic Publishers.

Bieberstein, N., Bose, S., Walker, L., and Lynch, A. 2005. Impact of service-oriented architecture on enterprise systems, organizational structures, and individuals. *IBM Systems Journal,* 44 (4), 691–708.

BusinessWeek. 2001. Q&A: How Dell keeps from stumbling. *BusinessWeek,* May 14.

Collier, E., and Fischer, M. 1996. Visual based scheduling: 4D modeling on the San Mateo country health centre. *Proceedings of the Third ASCE Congress on Computing in Civil Engineering,* Anaheim, CA.

Corbett, C. J., and Tang, C. S. 1999. Designing supply contracts: Contract type and information asymmetry. In *Quantitative models for supply chain management,* eds. S. Tayur, R. Ganeshan and M. Magazine, 269–98. Boston/Dordrecht/London: Kluwer Academic Publishers.

Davis, T. 1993. Effective supply chain management. *Sloan Management Review,* 34 (4), 35–46.

ECTP. 2005. Challenging and changing Europe's built environment – vision 2030. *European Construction Technology Platform (ECTP),* February.

Eagan, J. 1998. Rethinking construction. *UK Department of the Environment, Transport and the Regions,* July 16.

FIATECH. 2003. *Strategic overview: Capital projects technology roadmap initiative* (version 1). March.

FIATECH. 2005. *Emerging construction technologies – a FIATECH catalogue.* October.

FIATECH. 2006. *Automating equipment information exchange (AEX).* December.

Fischer, M., Haymaker, J., and Liston, K. 2003. Benefits of 3D and 4D models for facility managers and AEC service providers. In *Developments and applications: The visual construction enterprise*, eds. R. Issa, I. Flood and W. O'Brien, 101–24. Lisse, The Netherlands: A. A. Balkema.

Fischer, M., and Kam, C. 2002. *PM4D final report. Technical Report 143*. Center for Integrated Facility Engineering, Stanford University, October.

Gallego, G., and Özer, Ö. 2002. Optimal use of demand information in supply chain management. In *Supply chain structures: Coordination, information, and optimization*, eds. J.-S. Song and D. Yao, 119–60. Boston/Dordrecht/London: Kluwer Academic Publishers.

Gilmore, D. 2006. One heck of a supply chain story. *Supply Chain Digest*, May 18.

Green, R. 2007. Analyzing the CAD manager's survey 2007. *Cadalyst Magazine*, September.

GSA. 2006. Building information modeling (BIM) overview guide. *US General Services Administration*, November.

IAI. 1996. End user guide to industry foundation classes, enabling interoperability in the AEC/FM industry. *International Alliance for Interoperability (IAI)*, August.

Kaipia, R., Holmstrom, J., and Tanskanen, K. 2002. VMI: What are you losing if you let your customers place orders? *Planning Production & Control*, 13 (1), 17–25.

Lee, H. L., Padmanabhan, V., and Whang, S. 1997. Information distortion in a supply-chain: The bullwhip effect. *Management Science*, 43 (4), 546–58.

Lee, H. L., and Whang, S. 2001. E-business and supply chain integration. *SGSCMF-W2-2001, Stanford Global Supply Chain Management Forum*, November.

Macomber, J. 2003. Follow the money: What really drives technology innovation in construction. *Proceedings of ASCE Construction Congress 7*, Honolulu, Hawaii.

Matthews, O., Howell, G. A., and Mitropoulos, P. 2003. Aligning the lean organization: A contractual approach. *Proceedings of the 11th International Conference on Lean Construction*, Blacksburg, VA.

Narayanan, V. G., and Raman, A. 2000. *Aligning incentives for supply chain efficiency*. Harvard Business School Case Study 9-600-110, April.

Oasis. 2006. *OASIS universal business language (UBL) TC*, V2.0. Organization for the Advancement of Structured Information Standards, December.

O'Brien, J. J. 1965. *CPM in construction management: Scheduling by the critical path method*. New York: McGraw-Hill.

O'Brien, W. 2003. 4D CAD and dynamic resource planning for specialist contractors: Case study and issues. In *4D CAD and visualization in construction: Developments and applications*, eds. R. Issa, I. Flood and W. O'Brien, 101–24. Lisse, The Netherlands: A. A. Balkema.

O'Brien, W. J., and Fischer, M. A. 2000. Importance of capacity constraints to construction cost and schedule. *ASCE Journal of Construction Engineering and Management*, 125 (6), 366–73.

O'Brien, W. J., Fischer, M. A., and Jucker, J. V. 1995. An economic view of project coordination. *Construction Management and Economics*, 13 (5), 393–400.

RosettaNet. 2006. *Trading partner implementation requirements.* TPIR-PIP Design Specification Document V11.000.00, RosettaNet.

SCOR. 2006. *Supply-chain operations reference-model*, V8.0. Supply Chain Council, Washington, DC.

Shin, J., Hammer, J., and O'Brien, W. J. 2006. Distributed process integration: Experiences and opportunities for future research. *Proceedings of the 20th International Conference on Advanced Information Networking and Applications (AINA '06),* Vol. 2, 247–51.

Simchi-Levi, D., Kaminsky, P., and Simchi-Levi, E. 2000. *Designing and managing the supply chain.* New York: Irwin/McGraw-Hill.

Sprott, D., and Wilkes, L. 2004. Understanding service oriented architecture. *The Architecture Journal,* January.

Taylor, J., and Björnsson, H. 2002. Identification and classification of value drivers for a new production homebuilding supply chain. *Tenth Annual Conference of the International Group for Lean Construction (IGLC-10),* Gramado, Brazil, 171–82.

Vaidyanathan, K. 2002. Case study in application of project scheduling system for supply chain management. *Proceedings of the 10th Annual Conference of the International Group for Lean Construction,* Gramado, Brazil, 161–69.

Vaidyanathan, K. 2003. Value of visibility and planning in an engineer-to-order environment. *Proceedings of the 11th Annual Conference of the International Group for Lean Construction,* Blacksburg, VA.

Vaidyanathan, K., and Howell, G. A. 2007. Construction supply chain maturity model – conceptual framework. *Proceedings of the 15th Annual Conference of the International Group for Lean Construction,* Detroit, MI, 170–80.

Vaidyanathan, K., and O'Brien, W. J. 2003. Opportunities for IT to support the construction supply chain. *Proceedings of the Third ASCE Joint IT Symposium on IT in Civil Engineering,* Nashville, TN.

Vrijhoef, R., and Koskela, L. 1999. Roles of supply chain management in construction. *Proceedings of the 7th Annual Conference of the International Group for Lean Construction,* Berkeley, CA, 133–46.

Wandahl, S., and Bejder, E., 2003. Value based management in the supply chain of construction projects. *Proceedings of the 11th Annual Conference of the International Group for Lean Construction,* Blacksburg, VA.

16

Field Technologies and Their Impact on Management of Supply Chains

Semiha Kiziltas
Carnegie Mellon University

Burcu Akinci
Carnegie Mellon University

Esin Ergen
Istanbul Technical University

Pingbo Tang
Carnegie Mellon University

Anu Pradhan
Carnegie Mellon University

16.1   Introduction

Collection of construction field data is not only important to keep track of project performance, such as monitoring progress and productivity, but also for active management of construction supply chains. Both project and supply chain management processes require accurate, complete, and timely field data. However, more than 50% of the problems that are observed at construction sites are associated with inaccurate, missing, and delayed information [Howell and Ballard 1997; Thomas, Tucker, and Kelly 1997] coming from current manual data collection approaches that are limited in terms of timeliness, completeness, and accuracy of the data obtained [e.g., Davidson and Skibniewski 1995; Navon 2005; Reinhardt, Akinci, and Garrett 2004; Sacks et al. 2005; McCullouch 1997; Cheok et al. 2000].

Several major examples of problems observed in construction supply chains are mainly associated with: (1) tracking resources flowing from manufacturing to construction sites and within construction sites [e.g., Song et al. 2006]; (2) identifying and locating resources at manufacturing yards and at job sites [e.g., Ergen, Akinci, and Sacks 2007b]; (3) checking compliance of construction products to quality requirements [e.g., Akinci et al. 2006a; Gordon, Akinci, and Garrett 2007]; and (4) monitoring progress at job sites [e.g., Navon and Shpatnitsky 2005]. Current manual data collection and interaction approaches observed in construction supply chains have limitations in providing timely, accurate, and complete data, which contribute to late deliveries of materials, double-handling and misplacement of components, and incorrect installations [Ergen, Akinci, and Sacks 2007b; Gordon, Akinci, and Garrett 2007]. There is a need for streamlining construction supply chain processes, so that not only complete, accurate, and timely information about goods flowing in supply chains can be obtained promptly, but also compliance of constructed components with quality requirements are assured.

Field technologies play a critical role for providing possible solutions to problems currently observed in construction supply chains. Field technologies, such as smart tags, laser scanners, embedded sensors, and mobile data interaction tools provide opportunities to collect, store, and interact with field data required in construction supply chains, accurately, completely, and timely; and hence enable better supply chain management by increasing situation awareness of construction personnel. For example, laser scanners can overcome the limitations of manual or other geometric data collection methods with its peerless high and dense data collection rate, and its data accuracy. Similarly, radio frequency IDentification (RFID) tags can be used to streamline the tracking of resources flowing in the construction supply chains.

While these technologies are promising for their accuracy, timeliness, and completeness in terms of data they produce, the technical capabilities and process implications of these technologies under changing construction environments need to be assessed. Detailed field tests conducted with these technologies showed that they might behave differently as compared to the technical specifications provided by vendors, as environments within which the technologies utilized during construction differ from the ideal conditions that these specifications target [e.g., Ergen, Akinci, and Sacks 2007b; Gordon and Akinci 2005]. Another study conducted by the authors in a simulation-based environment showed that process implications of utilization of these technologies need to be incorporated for fully understanding the benefits of field technologies in streamlining management processes at job sites [Akinci et al. 2006b]. Hence, in addition to the performance evaluations of these technologies under unfavorable construction environments, the implications of utilization of these technologies within construction supply chain processes should be analyzed in detail.

In this chapter, we look at both tangible (such as material and product deliveries) and intangible flows (information flow) within construction supply chains. In this chapter, tangible and intangible flow of resources not only are bounded to the flow from suppliers to job sites, but also cover flows within storage yards of suppliers, within a job site, and from job sites to owners. The specific set of processes that will be discussed in this chapter includes: (a) identification, locating, and tracking of resources (i.e., materials, tools, and equipment) within construction supply chains; (b) improving quality of products flowing in supply chains, and defect detection processes within construction supply chains; and (c) flow of information within the supply chains from job site to field offices.

First, we will provide an overview of available field technologies and their technical capability assessments under changing supplier and construction environments. Secondly, implications of utilization of these technologies on construction supply chain processes, their potential impacts on the quality and/or timeliness of the supplied goods and final products, and accuracy of the information flows will be detailed. We will provide these discussions based on case studies, which were conducted at different parts of a construction supply chain.

Field technologies, which are utilized within construction supply chains, are grouped as technologies for: (1) resource identification (i.e., barcodes, optical character recognition [OCR], contact memory technology, RFID); (2) locating and tracking (i.e., RF-based systems, global positioning system–geographical information system [GPS–GIS], inertial sensors and gyroscopes, indoor positioning devices); (3) 2D/3D imaging for quality improvement and defect detection (i.e., 3D imaging technologies such as laser scanners and flash LADARs; 2D imaging technologies such as digital cameras); (4) embedded sensing coupled with WLAN (i.e., sensing technologies on equipment, such as equipment onboard instrumentation (OBI); on materials, such as thermocouples); and (5) data interaction (i.e., mobile and wearable devices, and applications running on them). As the naming of these groups implies, we grouped the technologies based on their roles in construction supply chains, which is aligned with the objective of this chapter, rather than based on the underlying technology. The next sections of this chapter describe these groups of technologies.

16.2 Resource Identification Technologies Within Construction Supply Chains

Various automated identification technologies that can be utilized for identifying goods within construction supply chains exist. Examples of these are barcode, OCR, contact memory, and RFID technologies [Rasdorf and Herbert 1990; McCullouch and Lueprasert 1994; Finch, Flanagan, and Marsh 1996; Jaselskis and El-Misalami 2003]. Benefits of these technologies are many. Automated identification functionality of these technologies can be used to track materials in a more accurate and reliable way than manual identification methods. In addition, some of these technologies also allow for automating the identification process. Data capture and storage functionalities of some of these technologies allow for storing information related to materials with the materials and accessing this information in the field, and hence not requiring data look-up from a remote database. In this section, an overview of some of the technologies that can be used for resource identification in construction supply chains is provided.

16.2.1 Barcodes

Barcode is the most commonly used technology, which is usually utilized for automatically identifying products in retail and manufacturing industries [Baldwin, Thorpe, and Alkaabi 1994]. A barcode consists of rectangular bars and spaces, which represents alphanumeric characters. The barcode is read by a barcode reader that sends out light across the barcode pattern and detects variations in the pattern to interpret the characters. To describe how information is encoded in the barcodes, several standards are developed and used by different organizations (e.g., American National Standards Institute [ANSI], the Automotive Industry Action Group [AIG]).

Barcodes can be grouped into three based on their data storage styles, as one-dimensional, two-dimensional (2D), and composite barcodes. One-dimensional barcode, which is more commonly used, contains one row of data; thus its storage capacity is very limited (i.e., 15–50 characters) and it can only be used for identification purposes. 2D barcodes have additional data storage capacity since data are stored in multiple rows; therefore it can contain a file of information about an object [McCullouch and Lueprasert 1994]. Composite barcodes contain both one-dimensional and 2D codes. However, 2D and composite barcodes need to be more accurately aligned during reading, and the readers are more expensive [Kärkkäinen and Ala-Risku 2003].

Barcode technology is an inexpensive and mature technology. It allows for storing some data (e.g., ID) related to the product to which it is attached, that can later be retrieved multiple times automatically. However, a limitation of barcode technology is the need that it should be very close to the component to perform the manual scanning activity, where the reader needs to be oriented to the tag at a fixed angle or vice versa. This manual scanning makes barcode technology unsuitable for multiple identification at a time and fully automated identification in semicontrolled environments. In addition, line-of-sight requirement is inconvenient for harsh construction environments. Another limitation is that the data is in a read-only format; thus, it cannot be updated. Also, typical paper-based barcodes are not durable in harsh construction environments, since a scratch or dust might result in unreadable barcodes. More durable tags that are made of other materials are also available, but are more expensive. It is recommended that plastic barcodes be used if barcodes are not subjected to much physical impact (e.g., scratching and cutting) and heat (e.g., less than 200°F). In the existence of physical impact and long periods of exposure to heat, thin metal labels are more effective [Bernold 1990].

16.2.2 OCR and Contact Memory Technologies

OCR is another automated identification technology similar to barcode. OCR is used to convert a scanned human-readable document into alphanumeric characters. It requires a scanner to capture the documents' image, and software to analyze the image and to identify the characters.

Similar to barcode, OCR requires line-of-sight and close proximity scanning; however, in some cases, it is less accurate than barcode since it analyzes human-readable characters instead of rectangular bars and spaces. Also, OCR does not perform well when the document is not clean or is damaged; therefore, it is more effective when used for data capture in a controlled environment, which is not the case for construction supply chains in general.

Contact memory is a technology that uses a microchip encapsulated in a ruggedized container to store and transfer data. Contact memory technology is preferred for use in harsh conditions that would make barcodes and OCR unreadable or impractical. Contact memory has two components: a button which has a chip enclosed in a rugged container and stores data, and a read/write device to transfer data to/from the button to a host system. The button has a unique ID number, and some types of contact memory buttons have additional memory up to 512 MB [MacSema Inc. 2007]. The data stored in the button is retrieved by contacting the button using a reader (probe).

The main advantages of using contact memory technology are the updatability of the memory in buttons and their durability. The disadvantage is the need for a direct contact during data entry, which might be problematic for locations not directly reachable, or collection that prevents multiple readings at once. This disadvantage makes the technology labor-intensive, which is also valid for barcode and OCR technologies. The direct contact requirement makes it difficult to completely automate the identification process as well.

16.2.3 RFID Technology

RFID technology is another promising technology for automated identification of resources in construction supply chains. It has two main components, a reader and a tag (Figure 16.1). A tag, which consists of an electronic chip coupled with an antenna,

FIGURE 16.1 An RFID reader and active tags.

is attached to an object, and stores data about the object. Reader, combined with an external antenna, reads/writes data from/to a tag via radio frequency (RF), and transfers data to a host computer. Some tags have light emitting diode (LED) that blinks light during communication.

RFID tags can be either active (needs a battery to operate) or passive (no battery requirement). Active tags have typically larger memory, which is currently up to 64 KB for active tags, allowing for storing additional information beyond identification information on the tag. Active tags also typically have longer reading ranges. Reading/writing ranges depend also on the operation frequency (low, high, ultra high, and microwave) in addition to tags' being active or passive. A disadvantage of the active tags is that they have a limited lifetime, requiring battery replacement periodically. Reading range and data storage capacity of passive tags are limited; however, they do not require any batteries to operate.

RFID tags are also classified according to their memory types. The information in "read-only" tags cannot be updated, whereas the information in "read–write" tags can be updated multiple times. The advantage of RFID tags with a read–write memory is that it enables data entry and access at any time throughout the life cycle of a tag.

Advantages of RFID technology are various as compared to other automated identification technologies. RFID technology does not require line-of-sight and direct contact. It can be read at longer distances, and can withstand harsh construction environments [Jaselskis and El-Misalami 2003; Song et al. 2006]. It also allows for multiple readings at once at longer distances; thus, it is suitable for automating the process of identification of multiple items at a time. Optional password mode prevents unauthorized data access and entry.

A limitation of the RFID technology is its relatively higher cost compared to the barcode technology. However, since it is an evolving technology, new RFID systems consistently have decreasing costs and larger data storage capacities. Another limitation of RFID is the performance reduction that is observed around metals and liquids when higher frequencies of communication (UHF and microwave) are used. Encapsulation of tags helps in decreasing the performance reduction. Since radio waves are absorbed or reflected by objects in the environment, technology providers strongly suggest testing the performance of RFID technology in real-life environments.

To successfully use the described identification technologies in construction supply chains, the characteristics of the technology need to meet the requirements of construction supply chains. The characteristics of each identification technology described are given in Table 16.1. These characteristics include performance of a technology indoors/outdoors, reading ranges, memory capacities, and battery requirements. As shown in Table 16.1, performances of these technologies vary; hence they should be selected for utilization in supply chains based on their characteristics. Besides, automatic identification is also important to automatically locate and track components/resources in a construction supply chain. The next section will provide an overview of field technologies for enabling locating and tracking resources within supply chains.

TABLE 16.1　Intelligent Resource Identification Technologies and their Characteristics

	Performance in Open Air	Long Distance Reading Capability	Updateable Memory	Battery Requirement
Barcode	Poor	No	No	No
OCR	Poor	No	No	No
Contact memory	Good	No	Yes	No
RFID	Good	Yes	Yes	Yes (active tags)
				No (passive tags)

16.3　Resource Locating and Tracking Technologies Within Construction Supply Chains

In a construction supply chain, materials, tools, and labor need to be tracked spatially at manufacturing sites, construction sites, and within constructed facilities. This section describes a set of technologies that enable automated locating and tracking of resources. Tracked resources can be within open environments or in closed environments, which is a significant distinction affecting the choice of field technologies for spatial locating and tracking.

16.3.1　Outdoor Location Identification and Tracking Technologies

Many commercial real-time locating systems are available to locate objects in outdoor environments (e.g., in an open air storage area). Currently, these systems mainly leverage GPS coupled with GIS technologies.

GPS technology coupled with GIS systems can provide a flexibility to track and locate resources flowing within construction supply chains. GPS is a satellite-based navigation system formed from a constellation of 24 satellites and ground stations [Leick 1990]. It consists of satellites, acting as transmitters of radio signals, and a receiver, which leverages radio signals to determine the 3D position of an attached object (i.e., latitude, longitude, and altitude). GPS receiver can also calculate velocity and bearing of an object based on its 3D position [Garmin 2006]. A GPS receiver receives time encoded radio signals from GPS satellites and then calculates its position by trilaterating GPS radio signals obtained from at least three GPS satellites. Trilateration is a method of calculating the relative positions of objects using geometry of triangles; an approach similar to triangulation used in surveying.

Based on the applications they are used in, GPS can be categorized into different types, such as surveying grade GPS, aviation GPS, automobile navigation GPS, and recreational GPS. The accuracies of these various GPS technologies vary from 50 to 1 m depending on its type. For instance, a surveying grade GPS can achieve accuracy up to 1 m, while recreational GPS is accurate up to 50 m. In construction supply chain domain, the accuracy requirement also depends on the application. For example, tracking of building components within a site [e.g., Ergen, Akinci, and Sacks 2007b] may require an accuracy of a few meters (i.e., 5–10 m), while tracking of transportation vehicles may require accuracy from 50 to 100 m depending on the desired spatial resolution.

GIS is a computer-based system used for collecting, storing, retrieving, transforming, and displaying geographical data. Such geographical data can be collected via GPS, surveying techniques, aerial photographs, and satellite images. The collected data are generally stored in two data formats as, vector format and raster format. Vector data format uses points, lines, and polygons to represent geographic objects. For instance, in a given construction supply chain domain, locations at which materials are picked up and delivered can be represented as points, routes between two locations can be represented as lines, and the area of a construction site can be represented as a polygon. In raster format, geographic area is represented as an image, consisting of columns and rows, and each cell denoting a single value generally representing elevation. For example, a construction site (250×250 m) can be represented as a collection of 250×250 m cells, each cell of size 1×1 m and each cell value represents elevation information from the mean sea level.

Many benefits can be achieved from integrating GPS data with GIS. Supply chain management domain in manufacturing industry has benefited from integrated usage of GPS and GIS in terms of intelligent tracking and locating of materials and manufactured goods. Integrated GPS–GIS systems are being widely used in scheduling and routing of material deliveries [e.g., Alexander, Nancy, and Landers 1999; Altekar 2005]. Location information obtained from GPS devices attached to transportation vehicles is integrated with GIS to monitor the location and movement of goods in real time. Such real-time information can help to assess correctly the arrival time of vehicles; hence informing interested parties accordingly [ESRI 2007]. In addition, integrated GIS and GPS systems are also used in transportation and heavy/civil projects. Thus, GPS and GIS can be instrumental in improving the construction supply chain process.

Some disadvantages with GPS and GIS utilization exist, as the respective hardware and software can be expensive depending on resolution, and require appropriate support and training [Alexander, Nancy, and Landers 1999]. Even more importantly, GPS cannot be used for tracking of goods inside a building or in dense cities with high-rise buildings, due to the unavailability of satellite signals. In such a situation, other location tracking techniques, such as inertial navigation systems (INS), which will be described in the next section, can be alternative to GPS.

16.3.2 Indoor Location Identification and Tracking Technologies

Solutions to indoor location identification and tracking problems are, in general, called indoor positioning systems (IPS) and are used to determine locations of mobile users and resources inside enclosed environments. There are a number of available IPS based on the underlying transmission medium: (a) RF-based; (b) ultrasound-based; (c) magnetic-based; and (d) infrared (IR)-based systems. In addition to IPS, there are INS that can work both indoors and outdoors.

RFID-based systems consist of active UHF RFID or Wi-Fi tags, tag exciters, and receivers. In a typical system, the tags are attached to components and tag exciters are placed at specific key locations (Figure 16.2). Whenever a tag is in the range of a tag exciter's antenna, the exciter sends a low frequency activation signal to awaken that tag and writes exciter's ID in the tag. Following the activation, the tag sends a signal including tag's ID and exciter's ID to a receiver, which can typically be communicated in the range of RFID readers. Receiver converts the signal into data, which includes the tag's ID and location, and transmits this information to a host computer or software application.

RF-based systems are suitable for tracking components in controlled environments, such as manufacturing plants, because tag exciters are needed to be placed at key locations; however, receivers are needed to be placed approximately at reading range distances. In addition, the cost of such systems is currently considered as high in construction industry. Many research studies on using RF-based systems for locating purposes have focused on locating mobile objects/people in a facility [Bahl and Padmanabhan 2000; Du, Fang, and Ning 2005] using an existing wireless LAN system. Several metrics, such as signal time of flight, angle of arrival, and signal strength are used for locating items. Due to complexities in propagation of radio signals, various location finding algorithms are used to process the metrics [Pahlavan, Li, and Makela 2002]. However, considering overall performance of wireless LAN system in locating items, there are still concerns about accuracy. Elnahrawy, Li, and Martin (2004) investigated various research studies on locating items using RF by comparing different approaches and environments, and showed that using a wireless LAN system (802.11), one can expect a localization error of 10–30 ft [Elnahrawy, Li, and Martin 2004].

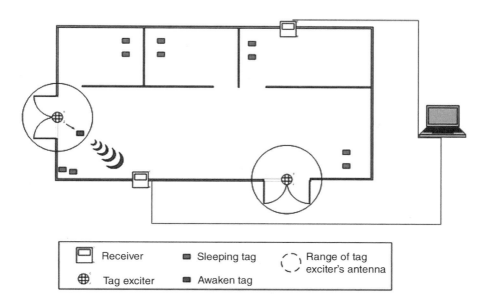

| | Receiver | | Sleeping tag | | Range of tag exciter's antenna |
| Tag exciter | | Awaken tag | | | |

FIGURE 16.2   Components of a commercial real-time locating system shown on a building plan.

To identify how RFID technology performs in real-life conditions, the authors conducted four case studies, during which technical feasibility of utilizing RFID was evaluated for identifying, locating, and tracking components and/or related information across a construction supply chain. The first case focused on automatically locating precast components (double-tees) in a large storage yard [Ergen, Akinci, and Sacks 2007b]. The second and third cases were related to tracking precast components and pipe spools as they were being shipped from a manufacturing plant to a construction site [Akinci et al. 2004]. The fourth case was conducted for tracking maintenance history of fire valves installed in a facility [Ergen et al. 2007a]. For all the tests, active UHF RFID technology was used due to its longer reading range (i.e., 30 m in open air). All the case results demonstrate that it is technically feasible to utilize active UHF RFID technology for identifying, locating, and tracking components and related information. The actual reading range of the RFID tags was reduced to 20% to 50% of the nominal reading range (i.e., 6–15 m) depending on the amount of metal and existence of obstructions in the environment. For the cases where selective identification (i.e., identification of a predetermined component) is not needed, components were successfully identified automatically by tagging the components and detecting them with a reader. For the cases where fully automated identification and selective identification are needed, a reasoning mechanism had to be developed and used to ensure that only information from the desired tag is recorded.

Ultrasound-based systems are composed of receivers and transmitters. A transmitter emits ultrasound pulses, which are collected by a receiver to determine the relative location of the receiver with respect to the transmitter. The receivers (also called tags) are attached to mobile goods within supply chains (e.g., batch of merchandise), while the transmitters are generally located at the highest points of closed environments (e.g., ceilings in facilities) [Want et al. 1992]. An ultrasound-based IPS is accurate within 9 cm of its true location for 95% of the measurements [Hightower and Borriello 2001]. From construction supply chain perspective, such a system can be deployed to track movement of goods inside a warehouse. High accuracy can help to locate goods more precisely than RF-based systems, but such accuracy may not be required for construction supply chain domain. In addition, these systems are sensitive to precise placement of sensors and require a large sensor infrastructure throughout the ceilings. Thus, utilization of this technology can be costly and suffers from the ease of deployment in large areas, such as warehouses.

Magnetic-based IPS systems are also composed of receivers and transmitters, as in the case of ultrasound-based systems. Instead of ultrasound pulse, a magnetic-based system utilizes magnetic-field pulses to determine the position and orientation of a receiver. A magnetic-based IPS has advantages in terms of high precision and accuracy up to 1 mm of the true location. However, such a system is costly and prone to increased inaccuracy with increasing distance from the transmitter [Hightower and Borriello 2001].

An IR-based system utilizes IR signal and it is also composed of a receiver and a transmitter, as in the case of ultrasound and magnetic IP systems. An IR-based indoor positioning system is also accurate up to a few meters from the true position. However, it has limited range (around 6 m) and performs poorly in the presence of sunlight due to the confusion caused by the IR band of sunlight [Hightower and Borriello 2001].

In addition, an IR system requires clear line of sight to function properly. Thus, from construction supply chains perspective, ultrasound, magnetic and IR-based positioning systems are accurate to track movement of goods inside a warehouse. However, they are costly and sensitive to increasing distance between the transmitter and the receiver.

Another technology for location identification and tracking in construction supply chains, particularly indoors, is INS. INS determines the current state (i.e., position, velocity, and orientation) of a dynamic object (e.g., moving vehicle or person) by leveraging the information collected from gyroscopes and accelerometers. An accelerometer measures linear acceleration of a dynamic object, and a gyroscope measures the angular velocity [Fraden and King 1998]. To determine the current state of a dynamic object, INS uses state-estimation algorithms, such as Kalman filter on information collected from accelerometers and gyroscopes. The Kalman filter estimates the state of a dynamic system from a series of noisy measurements. INS does not require GPS satellite signal in order to determine the position and orientation. Thus, it is particularly useful in navigating indoors (e.g., inside warehouses) and dense cities, where there is a low likelihood of receiving GPS signals. A major disadvantage of INS is that it suffers from integration drift problem where measurement errors are compounded progressively into large position and orientation errors [Thrun, Burgard, and Fox 2005]. Such a problem can be corrected by recalibrating the INS system with known reference locations and orientations. The recalibration rate of an INS system depends on the quality of the deployed sensors (e.g., accelerometers and gyroscopes).

As discussed in this section, the types of technology to be used for locating and tracking mainly depends on whether the identification, locating, and tracking will be done in an indoor or outdoor environment. Besides, it is important to know how much accuracy is needed for tracking.

16.4 2D/3D Imaging Technologies for Quality Improvement and Defect Detection Within Construction Supply Chains

Improving the quality of products by enabling capturing of more comprehensive geometric data about products is another important issue in construction supply chain management. 2D/3D imaging technologies enable capturing detailed surface features of structures in a short time, and automating the interpretation of the collected data using computer vision techniques. In this section, we will provide an overview of 2D (i.e., 2D digital cameras) and 3D (i.e., laser scanners, flash LADAR) imaging technologies.

16.4.1 2D Imaging Technologies

Traditional 2D digital cameras can collect color, texture, and reflectivity information of objects within construction sites. High-resolution digital photos can be processed by image processing techniques, such as edge detection and image filtering [Paterson, Dowling, and Chamberlain 1997]. With pattern classification and object recognition techniques, it is feasible to automatically recognize and track objects of interest using photos taken by multiple cameras on a job site [Brilakis, Soibelman, and Shinagawa 2006; Brilakis and Soibelman 2006; Brilakis and Soibelman 2005; Brilakis, Soibelman, and Shinagawa 2005]. This entity tracking method enhances automatic project progress control, as well as automatic equipment and material tracking on construction sites, and improves the construction supply chain by providing all project participants with prompt spatial information, such as position of equipment and structural components. It enhances safe and efficient construction operation and gives participants better sense in case of remote monitoring.

A series of digital photos taken from different positions of a site can be used to create mapping between detected features (such as edges and corners) in photos and 3D geometric features in the CAD model, and then can be used to reconstruct the 3D model of a building rendered with photos [Paterson, Dowling, and Chamberlain 1997; Streilein 1994]. However, since digital cameras do not directly capture 3D information, multiple photos and accurate geometric feature correspondence information between the photos are necessary to generate 3D models [Triggs 1998]. In addition, 3D reconstruction accuracy is sensitive to many other conditions, such as the illumination, accurate sensor model, and position and orientation information of the camera. Recent research studies showed that the integration of color information from digital photos and 3D point clouds is promising, since it combines the advantages of two techniques together and results in a system which can acquire accurate 3D points and their color information at the same time [Abmayr et al. 2004].

16.4.2   3D Imaging Technologies

3D imaging technologies refer to technologies that can directly position large amounts of 3D points and deliver the collected 3D data usually in 3D images, such as a range image. In a range image, each pixel contains a value, representing the distance from the 3D device to the object surface. Laser scanners and flash LADARs are among 3D imaging technologies that can be used for quality control and safety management in construction supply chains.

A laser scanner is a type of 3D imaging system that can capture a large amount of 3D points in seconds depending on its type (e.g., a few thousand points per second to hundreds of thousands per second). It is typically composed of two major modules: the laser radar system and the mechanical deflection system. The laser radar system is composed of a photon source, lenses, a receiver, and a timing circuit. The photon source emits the laser beam. Timing circuit measures the traveling time of the laser light—when the light reaches to the object of interest and returns back to the receiver—and based on that time, it determines the distance of the object from the scanner [Zoller and Fröhlich 2005]. The mechanical deflection system, composed of horizontal and vertical rotation engines, directs the laser beam onto the object of interest. It also provides the two angles (i.e., horizontal and vertical angles), which are used to specify the direction of the laser beam in polar coordinates.

There are two types of laser scanners that are mainly utilized in construction supply chains: *airborne* and *terrestrial*. Airborne scanners (LIDAR) are mainly used for large area urban modeling, since local details are usually not available in their data sets;

whereas terrestrial scanners are used for on-ground surveying and can capture local surface details. Airborne scanners can help in early stages of a construction in terms of 3D site-layout spatial modeling. Terrestrial scanners can capture detailed object surfaces; hence they are more useful for quality control of components flowing within construction supply chains.

Figure 16.3 shows what a typical terrestrial laser scanner looks like. Scanner head rests on a tripod and can rotate horizontally. A photon source continuously emits laser signals while an internal mechanical system continuously records the horizontal pose (θ) of the scanner head, the vertical pose (ϕ) of the photon source, and when each reflected laser signal is detected by the receiver. The origin of the polar coordinate system is the center of the scanner, as shown in Figure 16.3. Then, the system will get the range (ρ) from the scanner to each point on the surface of an object of interest, based on light traveling time, and determine the 3D position of each point. For calculating the traveling time of light between the scanner and the object, two major techniques are available: *time-of-flight approach*, which uses a timing circuit to record the time difference between a signal sending and receiving times; *phase-based approach*, which uses the frequency of the laser light wave and the phase difference between the phases of the sent and received laser light waves to deduce the traveling time of light. Time-of-flight scanners have higher single point accuracy in further ranges (e.g., they can collect 3D points with an accuracy of 6 mm within 100 m), as compared to a phase-based scanner, which has an accuracy around 6 mm within a range of about 50 m [Kersten, Sternberg, and Stiemer 2005]. In addition, some time-of-flight scanners have much smaller laser beam sizes compared to that of phase-based scanners, reducing the number of noisy points in a point cloud, hence improving accuracy [Kersten, Sternberg, and Stiemer 2005]. However, some phase-based scanners can achieve much higher data collection rate (typically 125,000 points per second) as compared to typical data collection rate of time-of-flight scanners (typically 5000 points per second), which results in shorter time being spent on data collection.

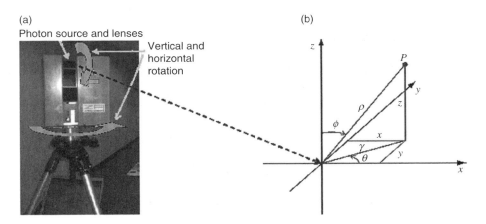

FIGURE 16.3 A typical terrestrial laser scanner and its polar coordinate system: (a) a terrestrial laser scanner, (b) polar coordinate system of a scanner.

Laser scanners can overcome the limitations of other geometric data collection methods with their peerless high data collection rate and their data accuracy comparable with that of total stations within a given range (tens of meters to hundreds of meters). Using commercial reverse engineering tools, 3D model of a structure can be reconstructed based on point clouds. From that 3D model, geometric features can be extracted to virtually inspect the structure by comparing measurements taken in 3D CAD as-planned and as-is models (generated from point clouds) for quality control and defect detection. Figure 16.4a shows a model reconstructed from point clouds for a portion of a bridge, and Figure 16.4b shows a model for a building.

Some challenges associated with the application of laser scanning to quality improvement and defect detection within construction supply chains are as follows. First, in order to acquire the geometric information with required level of detail and accuracy, users need to analyze the spatial occlusions to ensure surfaces of objects of interest are visible to the scanner. In addition, they need to consider the data accuracy and data resolution issues of the scanner while deciding on scanner positions on the job sites. Second, laser-scanned data are large data files containing noisy data. Removal of noise from the data and manipulation of large number of points in a point cloud are still time-consuming and require a large amount of human involvement. Third, laser-scanned data only contains coordinate (some scanners also provide color and surface reflectivity) information of scanned points and carry no semantic information, such as which points are associated with which component. Working with point clouds makes the geometric reasoning based on laser-scanned data tedious and error-prone.

In order to understand technical capabilities of laser-scanning technology for quality improvement and defect detection in construction supply chains, four representative case studies were conducted using a phase-based laser scanner. The results showed that a laser

FIGURE 16.4 Example 3D models generated from laser scanned data: (a) a bridge model, (b) a building model.

scanner can capture surfaces of building objects of interest with a very small sampling step in a short time, such as a few minutes and generate dense point clouds. From those dense point clouds, features smaller than 1 cm can be extracted given the proper hardware and software configuration. Hence, dense point clouds are promising resources for geometric feature extraction and reasoning for quality control. Second, 3D point clouds directly capture the geometric relationship between different components of a given facility or infrastructure, making it possible to directly analyze the geometric relationship between components. Also, during construction, as-built information provided by the laser scanner can be utilized by suppliers to manufacture components, which can accurately fit in the existing parts of the building, hence assuring quality and reducing costly reworks.

Process implications of utilization of laser scanners in construction supply chains are also various. Preparation steps for using laser-scanning technology include site preparation (i.e., site investigation, surface cleaning, removal of interfering objects such as equipment), equipment setup (i.e., assembling the laser scanner, leveling of equipment, preview scanning, and several trial scans to confirm correctness of data collection plan), and target placement (i.e., setting up fiducial markers). Also, accuracy requirements of the scanned data should be considered; hence, this might necessitate steps to ensure that the scanner is in the right location and with the right resolution. Majority of the process implications and requirements of laser-scanning technology are observed during postprocessing stage. These processes are data registration, noise removal, and manipulation of point clouds. Depending on the scanned facility, more than one scanning might be required to create the entire 3D model. Hence, data registration process, which is required to merge each scan file to a cohesive whole, is required. In addition, laser-scanned data are large data files containing various types of noise. Removal of noise from the data and manipulation of those large 3D data sets are still time-consuming and require a large amount of human involvement. In addition, for simple virtual inspection tasks, users have to manually select certain parts of the point clouds and extract geometric features for generating a certain piece of geometric information.

Another technology within the 3D family is flash laser radar (LADAR), also known as 3D camera [Bosche et al. 2006]. A flash LADAR can capture simultaneously many 3D points just like a camera can capture color and reflectivity information. Hence, a flash LADAR can be described as a digital camera providing arrays of range distances, which correspond to the distance from the object surface to the camera, instead of brightness [Bosche et al. 2006]. Therefore, like LADARs, this technology provides dense range point clouds. Contrary to a regular LADAR, however, "photos" can be acquired with a frequency up to 30 Hz [Bosche et al. 2006], which results in a series of snapshots of a scene. A flash LADAR is composed of two major components: a broad field illumination source and a focal plane array (FPA) detector [Stone et al. 2004]. The broad field illumination source emits a laser to the space of the field of view. The FPA detector is a 2D "chip" in which individually addressable photosensitive detectors can be accessed. In this detection, individual 3D points in the field of view of the LADAR are mapped to those laser sensitive detectors on the 2D chip [Stone et al. 2004]. Each single detector on the FPA records the distance of its corresponding 3D point to the camera, just like a pixel on a digital photo records the color information of its corresponding point.

A flash LADAR can capture a snapshot of a construction site with a flash, and it can generate tens of snapshots in one second. The resolution of each snapshot is much lower than the range images created by a laser scanner. Compared with laser scanners, flash LADARs can also collect many range images in a short time (30 per second). However, each range image collected by it will not achieve the high resolution of scanned range images. In temporal dimension, a flash LADAR is more powerful, even though its data is sparse in spatial dimension, making them useful for capturing rough 3D information on a dynamic construction site for analyzing the on-site spatial changes using a sequence of range images. It can enable detecting people and equipment moving around a job site with the captured snapshots [Teizer et al. 2005; Teizer, Caldas, and Haas 2006].

2D and 3D imaging techniques improve the quality control within construction supply chains. Laser scanners make it technically feasible to reconstruct a 3D model of scanned environments in high-level detail and in millimeter-level accuracy in relatively short time. This generated 3D model is available for multiple uses, such as geometric information retrieval, or storage of geometric information for keeping historical heritage for possible future referral. Digital cameras can capture color and texture information. Flash LADAR can capture sparse point clouds within a very short time, which is important for object tracking and dynamic construction site monitoring. As explained, each technology has its own strengths. Hence, combinations of these imaging techniques are promising to improve quality control in construction supply chains.

16.5 Embedded Sensing Technologies in Construction Supply Chains

Embedded sensors, coupled with wireless networks, are designed to sense and communicate data about materials, components, or equipment within which they are embedded. They are designed for sensing specific signals to convert to data (e.g., temperature, humidity). Embedded sensors provide opportunities to sense data to improve the quality of construction products and to streamline processes in construction supply chains. The following subsections provide an overview of different sensing technologies embedded in construction equipment, materials, and components.

16.5.1 Embedded Sensors on Construction Equipment

Commercial construction equipment vendors, such as Caterpillar™ and Komatsu™, have embedded a suite of OBI for various types of construction equipment (e.g., dump trucks, graders, excavators, and hoes). OBI is an electronic hardware that contains a set of sensors (e.g., GPS, inertial sensors, temperature sensor, and pressure sensor) for monitoring the state and utilization of a piece of equipment. In addition to sensors, it has an electronic circuitry board that processes and stores the collected data by sensors, respectively. For instance, OBI installed on a dump truck can monitor various characteristics related to payload productivity: time and date of the payload cycle, idle times, loading times, empty travel times, loaded travel times, empty travel distances, loaded travel distances, and payload weights. Such information can be used to monitor payload productivity per truck, hence enabling assessment of resource utilization. For instance, payload weight for each cycle can indicate whether a given payload truck is filled to its capacity or not. Similarly, a truck's idle times within a day can point out whether it is being utilized fully or not.

In addition, OBI can also constantly monitor the health of a payload truck by sensing temperature and pressure. Temperature sensors measure engine and coolant temperatures. Such temperature information can give diagnostic feedback on the condition of an engine. Similarly, pressure sensors can provide air pressure information about the truck's tires. Such critical information can help to reduce costly maintenance of equipment by taking preventive actions on time. Thus, OBI can be used for both intelligent productivity monitoring and structural health monitoring of construction equipment. Within the context of construction supply chains, understanding the productivity and health status of equipment can effectively help manage equipment in use, resulting in low operational and maintenance costs.

In order to assess the technical capabilities of embedded sensors on equipment for improving the productivity monitoring processes at job sites, we conducted a case study on a highway construction project. The case study was conducted on the entire bulk excavation activity on the route of the highway, where we investigated data collected via OBIs installed on dump trucks. The bulk excavation activity was conducted in two shifts: morning shift and day shift. The collected OBI data was rich in content as it contained time and date, idle time, loading time, empty travel time, loaded travel time, empty travel distance, loaded travel distance, and payload weight for each payload cycle. The analysis of the data collected from OBI suggested certain truck loading patterns, which were significantly different than what the project team was perceiving as happening on the job site. The case study demonstrated that human situation assessment can sometimes be biased due to the lack of factual information, as detailed in Pradhan and Akinci (2007). Thus, OBI installed in construction equipment can provide valuable information related to intelligent productivity monitoring. In the construction supply chain domain, OBIs, installed in pieces of equipment (e.g., transportation vehicles), can aid in better understanding of a given process and make objective decisions based on correct information.

Process implications of utilization of OBI for productivity monitoring were also identified during the case study. It was observed that, before the OBI is used for data collection, it has to be properly embedded to the equipment, connected, and made sure that it records data. A couple of trial recordings should be downloaded and observed before the equipment and sensors are in action for collecting productivity monitoring data. During utilization of the equipment and OBI, the most important thing is related to the recording of data. When sensors are activated as the payload data is loaded, the equipment on which the OBI is embedded should be stationary. Another important task is downloading the collected data to computers for productivity analyses. For this purpose, software applications are provided by OBI vendors.

16.5.2   Embedded Sensors on Materials and Components

Embedded sensors on materials and components are used to sense and communicate specific data items, such as temperature, material strength, and humidity of the object to which they are attached. In construction supply chains, temperature sensors, such as thermocouples, thermistors, and resistance temperature detectors (RTD) started to be utilized for monitoring the strength of curing concrete using the maturity method.

Among these sensors, thermocouples are simple self-measuring temperature sensors composed of two conductors separated from each other with insulation along their lengths and welded at the measuring point. Ranges that a thermocouple can measure vary based on the application temperature (i.e., the environment for which the measurements are taken) and sensor type; however, this range may cover from –250°C to 2500°C [Peak Sensors Ltd. 2007]. Due to their characteristics as being durable, reliable, robust, and small, thermocouples are widely preferred for applications requiring embedding into materials [WP 2007].

In addition to thermocouples, there are thermistors and RTDs utilized for measuring temperature. Thermistors, which are constructed mainly with a bead attached to two wires, are also temperature sensors with a temperature range around –100°C to 300°C. They are not preferred in applications requiring measurement of high temperatures and working under mechanical stresses, since they are not as durable as thermocouples. RTDs are usually platinum and nickel metals utilizing the basic knowledge of changing in the resistance of metals with temperature. They can measure larger ranges of temperatures (–250°C to 650°C), and are more expensive than thermistors, and perform slower than thermocouples due to their sizes being ten times bigger than the smallest thermocouples.

Utilization of emerging embedded sensing technologies within construction supply chains is becoming more common to improve delivery time to owners, improve quality of products and productivity. For example, traditional methods of strength measurement of concrete (such as field-cured cylinders) sometimes do not provide timely (e.g., without necessity to wait for strength gaining of cylinders) and accurate measurements. However, concrete strength information is required to be timely and needs to be accurate, since many of the activities' start times at job sites are dependent on this information. Such activities vary from removal of formworks to removal of winter protections around building elements, which directly affect total project completions and quality of products delivered. Embedded sensors used in the concrete maturity method, which is based on the correlation between concrete temperature and maturity, provide opportunities to get accurate timing information about when the concrete will gain its expected strength based on time-series temperature measures of concrete during the period of concrete curing [Goodrum et al. 2004]. In addition, they provide more accurate information since they are in situ and report back the actual behavior of concrete at a job site. Hence, accurate and timely assessment is possible with embedded temperature sensors at job sites.

On construction sites, durable and wireless sensors are preferred due to the harsh conditions to which these sensors get exposed. Experiments with temperature sensors at job sites were conducted to assess technical capabilities of these sensors under harsh construction conditions. Experiment results highlighted that sensor failures might be observed under extreme weather conditions, silent failures can be expected due to technological imperfections, and users might experience loss of data before analyses, if data loggers are overfilled and not transferred timely by the construction personnel [Gordon, Akinci, and Garrett 2007].

In addition to embedded sensors, wireless sensor networks are used for monitoring the environment or physical conditions of components within construction supply chains during construction and maintenance operations. Wireless sensor network consists of a large number of small sensors that can sense the physical conditions (e.g., temperature, humidity, tension, and motion), process data, and communicate wirelessly with each other. The sensed data can be used for quality control purposes in the construction phase, and for increasing facility efficiency and occupant comfort, as well as continuous commissioning, monitoring and verification of a facility, and monitoring of deformation and safety of a structure (e.g., a bridge) during the service lives of facilities and infrastructures.

This technology enables a project manager or a facility manager to place sensors without wiring and disrupting the existing facility, and allows sensors to be placed in spaces that may see changing configurations. For example, in a modular office environment, wireless sensors can easily be relocated whenever the office modules are reorganized. The cost of wiring for sensors ranges from 20 to 80% of the cost of a sensor, depending on the circumstances, such as whether the sensor is being installed in a new construction or is a retrofit, or the length of wiring [Brambley et al. 2005]. Wireless sensors are especially advantageous over traditional wired solutions for retrofits and fast track projects where wiring is either too expensive or physically impossible. In retrofits, wiring might damage the existing structure, especially if it is a historic building.

There are commercially available wireless sensors for collecting various types of data (e.g., zone temperature, occupancy, outdoor air, lighting, air quality). However, relatively narrow communication bandwidth can reduce download rates (e.g., to one kilobyte per second) and this might result in slow data transfer, which is not suitable for transferring large amounts of data, such as data generated by a bridge in operation. Research studies are being conducted to manufacture smaller and less costly sensors and to integrate all the available data from sensors.

16.6 Overview of Technologies for Data Interaction

Field technologies suggested for process improvements in supply chains are not bounded with data capturing. In addition to data capturing, there are technologies to improve data interaction in the field. Mobile (e.g., personal digital assistants [PDAs], tablet, portable data terminal and other pocket PCs) or wearable computing devices (e.g., wrist attachable keyboards, hand-keys, finger trackballs, digital hardhats) enable access to data on sites (Figure 16.5). They are utilized in support of field tasks, such as progress monitoring or punch list administration [Reinhardt, Akinci, and Garrett 2004].

Mobile and wearable devices are usually designed to satisfy physical constraints and requirements of users. These physical constraints and requirements vary based on the environment in which these tools will be utilized. For instance, for data access and modification during bridge inspection tasks, a physical constraint might be inspectors' necessity to use both of their hands, and a relevant requirement can be a hands-free, light and small device to improve mobility or visibility of screen in bright sunlight or near darkness [Elzarka, Bell, and Floyd 1997; Bowden 2005]. In addition, applications running on these devices need to satisfy physical constraints and requirements

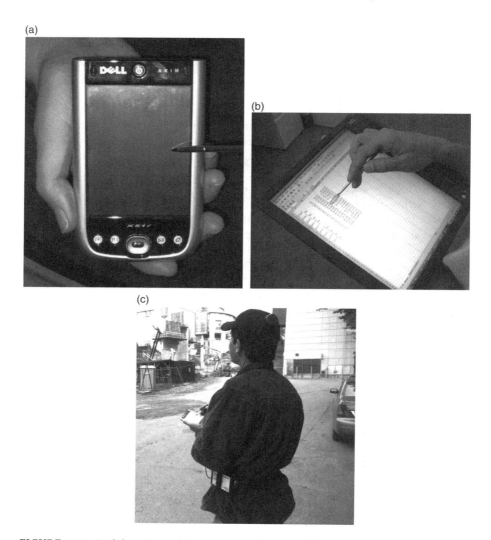

FIGURE 16.5   Mobile and wearable computers that are usable in construction supply chains: (a) a mobile computer PDA Dell™ AXIM, (b) an IBM™ X-60 tablet PC, (c) a wearable computer.

(e.g., user interfaces and ergonomics) to reach the data at a right level of detail, as well. Studies were conducted to help both system designers and end-users to enable better user interaction [Burgy 2002], and improve navigation within large data sets to reach the right level of detail as required by a given task and a user at a job site [Reinhardt, Akinci, and Garrett 2004]. Currently, these systems vary from direct data interaction (data access and modification) [Reinhardt, Akinci, and Garrett 2004] to augmented reality-based systems (e.g., applications for bridge inspection) [Hammad et al. 2005]. Both mobile devices and applications running on them are promising for timely information access for prompt actions in construction supply chains.

16.7   Technology Comparison

In this section, we provide a comparison of the field technologies discussed in this chapter based on a set of criteria. This set of criteria is listed in the first column of Table 16.2 and includes the technique on which the corresponding technology is based; beneficiary groups from the technologies utilized within construction supply chains; relative costs of technology acquisition and operation; scale of technologies showing coverage range for a given technology set; data collection rates; data accuracies, levels of details; and technical limitations and ideal working requirements of the technologies.

Many parties along a construction supply chain, such as suppliers, contractors, owners, and facility managers, can benefit from field technologies, as shown in Table 16.2. Suppliers and contractors can utilize them to streamline the flow of status information about deliveries to job sites, and track and locate them in supply chains; project managers and engineers can use them to detect defects early during construction, get accurate and timely data about components being constructed and temporary components to be removed (e.g., formworks), and productivity monitoring of activities. Similarly, facility managers can locate components, and get component information in a timely and accurate way.

Considering RFID, it is expected that whenever a party within a supply chain needs to access status or historical information about a component tagged with RFID, he or she will be able to do so by interrogating the RFID tag attached to that component. RFID technology coupled with advanced tracking technologies, such as inertial sensors and GPS, will enable: (a) tracking and locating components in an automated way in manufacturing plants, storage yards, construction sites, and in facilities; (b) streamlining the flow of status information from upstream parties (e.g., material suppliers) to downstream parties (e.g., contractors) through shared databases and automated alerts (e.g., shipping notification sent to contractor by manufacturer); (c) storage of related component information on components and instant access to this readily available information; and (d) integration of information that flows with components with downstream parties' databases at various milestones (e.g., at the gate of a construction site, during turnover of the building to the owner) as components move in the supply chain, as discussed in detail in the previous sections of this chapter.

Similarly, laser scanners, on board instrumentations (OBI), and embedded sensors are beneficiary to parties, such as construction project managers and engineers during different phases of construction (e.g., site survey and progress monitoring). For example, a laser scanner can be used to survey the construction site during the initial phase of a construction and later to detect defects during construction. Early defect detection can lead to significant cost savings and avoid construction delays, which can hamper supply of construction materials at the job site. OBI data can also assist project engineers in progress monitoring of construction activities (e.g., excavation) and operating conditions (i.e., health monitoring) of construction equipment. Such progress monitoring ensures that the project is on schedule and in compliance with plans and specifications. Similarly, embedded temperature sensors can assist project engineers to have a better understanding of concrete curing and strength, and such understanding can help to decide when to remove the formwork without having to wait for the 28 days concrete strength gaining period.

TABLE 16.2   Comparison of Technologies

	RFID	Laser Scanner	OBI	Inertial Sensors	Embedded Sensors (Temperature, Humidity)
Technique	Radio signal-based identification	Time of flight-based or phased-based 3D positioning	Pressure-induced measurement of payloads	– Piezoelectric induced measurement of acceleration – Conservation of angular momentum principle for gyroscopes	– Thermoelectric effect (Seebeck) based measurement of temperature – Absorption-based humidity sensing
Beneficiary groups	– Material providers – Construction group – Owners/facility managers	– Construction group	– Project engineers – Project managers – Owners	– Facility managers – First responders	– Project engineers – Project managers – Facility managers
Cost	Major cost item is associated with the middleware (to filter and transfer data) for an RFID technology Another major cost item is associated with acquiring the basic components (reader + antenna)	Major cost item is associated with acquiring a laser scanner Software required for data processing is also another major cost item for a laser scanner technology to be fully functional	Major cost item is associated with acquiring the basic components (sensor system + data loggers + software to download data)	Major cost item is associated with purchasing the software to process raw sensor data Acquiring the basic components (sensor system + data loggers) is the second major cost item	Major cost item is associated with purchasing basic components (sensors + loggers)
Scale	Passive tags cover 2–4 m, and active tags cover 6–30 m indoors Passive tags cover 4–7 m and active tags cover 30–100 m outdoors	A short–medium range scanner can capture a space with dimensions 50–70 m A long-range scanner can capture a space up to 1000 m, and a typical open space with diameter of 500 m	One OBI per construction equipment will capture payload data for all cycles of the equipment	One inertial sensor per object tracked will capture acceleration and orientation of that user within the tracked timeframe	One temperature sensor per measurement Temperature sensors can communicate with loggers 70 m to 20 miles line of site

Data collection rate	100–1000 tag/sec for UHF frequency, lower rates for low and high frequency	~1000s of points/second –625,000 points/second	Typically 1–2 Hz (i.e., measurements/second)	Typically 4–5 Hz	Temperature measurement rate: 10 seconds–1 h (resolution for min/max)
Level of detail of data	Object IDs	Dense point cloud's surface sampling step can be less than 1 cm within a range of 100 m	Cycle time information can be captured in minutes and seconds	Object acceleration or orientation/reading	Temperature resolution: 1°C
Data accuracy	Accurate (~100%)	Changes with laser scanner's range. Considering ideal working conditions, for time-of-flight scanners: 6–12 mm for 50–100 m range For phase-based scanners: 3–10 mm for 25–50 m range	Within 3%–5% of its difference between the min and max sensor reading	Within 3%–5% difference between the min and max sensor reading	Temperature accuracy: +/−1°C
Technical limitations and requirements	Performance reduction around metal, water, and indoors due to reflection of signals. Line-of-sight is not required; hence the user does not need to scan the tag, data can be collected without interfering with construction. Operating temperature: −40°C to +85°C	No scanned data of an object if object is occluded from the scanner. As long as objects are visible from the scanning position, scanning does not interfere with traffic or construction activities, and users do not need to access risky spaces to put tags due to nontouch nature of the scanning technique. Calibrated temperature range: 0°C to +40°C	Construction equipment needs to be parked on neutral during loading for accurate measurement of payloads. Temperature: −20°C to +70°C	The sensing device needs to be firmly attached to the object (e.g., vehicle). Temperature: 0°C to +70°C	Problems can occur due to data logger size, wire crimping, water intrusion, sensing range and sensor accuracy, and ability of supporting devices to provide power to the sensor system. Sensors typically must be in contact with the material or building element being measured

Current manual data capture processes are labor-intensive and time-consuming; therefore, timely, complete and accurate field data is limited in construction supply chains. Utilization of field technologies will streamline data capture from the field and thus, improve information flow and its quality in terms of completeness and accuracy across a supply chain. This would decrease waste and control over project in supply chains. Resource identification, locating and tracking technologies integrated with back-end information systems enable real-time material tracking through a construction supply chain. Using material tracking information, it would be possible to generate reliable plans and streamline lean supply chain management. For example, project engineers at construction sites could effectively schedule just-in-time deliveries to construction sites when up to date status of materials/components that are produced or prefabricated at an off-site location is available. To decrease the waste in construction, 2D/3D imaging technologies could be used for quality improvement and early defect detection. Embedded sensing technologies, utilized on materials and components, improve project delivery time to owners, and improve quality of products and productivity. Resource identification, locating and tracking technologies could reduce the number of materials or components that are misplaced in a supply chain. In addition, productivity monitoring and structural health monitoring of construction equipment decrease the operational and maintenance costs of equipment.

16.8   Conclusions

This chapter provided an overview of field technologies that are applicable to construction supply chains. It included discussion of technologies that enable the flow of materials from suppliers to construction sites, within construction sites, and from construction sites to the final users of facilities. From case studies conducted over a decade on many construction job sites via utilizing several of these technologies, discussions on technological assessments and their process implications were provided in details at the end of each relevant technology overview.

Detailed investigation of these technologies showed that, while these technologies are capable of streamlining the associated processes, their performances can differ from the manufacturers' specifications when they are utilized at the field on construction supply chains. For example, laser scanners are promising in improving the defect detection at job sites for quality assurance and quality control, RFID systems are feasible in not only tracking the components from suppliers to job sites and within job sites, but also locating other resources (small tools, labor, equipment) during construction for progress and productivity improvement. In addition to those, OBIs are capable of providing accurate and timely data for information flow within the supply chains for quality and progress improvements; whereas embedded sensors, GPS and GIS applications are providing timely and complete data.

In terms of technological assessment, these technologies can have weaknesses to be considered while utilized at job sites. Silent failures of thermocouples, phantom points in laser scanners, reading problems of RFID around obstacles, and unstable measurements in OBIs are some of the examples to be considered while such technologies are utilized on job sites.

While these technologies eliminate some of the nonvalue adding activities, they also add new tasks that need to be performed. For example, data collected by laser scanners need to be registered and processed further to glean relevant information for a given project management task, such as progress monitoring and quality control. Sometimes, such additional tasks can be significant in comparison to the nonvalue adding activities eliminated, limiting the process improvements to be achieved by using advanced field technologies. Hence, a thorough understanding and assessment of both the technological capabilities and process implications of these technologies are needed to be able to utilize them effectively on construction supply chains.

References

Abmayr, T., Hartl, F., Mettenleiter, M., Heinz, I., Heinz, A., Neumann, B., and Frohlich, C. 2004. Realistic 3D reconstruction – combining laser scan data with RGB color information. *XXth International Society for Photogrammetry and Remote Sensing (ISPRS) Congress*, Istanbul, Turkey.

Akinci, B., Boukamp, F., Gordon, C., Huber, D., Lyons, C., and Park, K. 2006a. A formalism for utilization of sensor systems and integrated project models for active construction quality control. *Automation in Construction*, 15 (2), 124–38.

Akinci, B., Ergen, E., Haas, C., Caldas, C., Song, J., Wood, C. R., and Wadephul, J. 2004. Field trials of RFID technology for tracking fabricated pipe. *The Fully Integrated and Automated Technologies Consortium (FIATECH) Smart Chips Report*, FIATECH, Austin, TX.

Akinci, B., Kiziltas, S., Ergen, E., Karaesmen, I. Z., and Keceli, F. 2006b. Modeling and analyzing the impact of technology on data capture and transfer processes at construction sites: A case study. *Journal of Construction Engineering and Management*, 132 (11), 1148–1157.

Alexander, B., Nancy, S., and Landers, T. L. 1999. Intelligent tracking in manufacturing. *Journal of Intelligent Manufacturing*, 10 (3), 245–50.

Altekar, R. V. 2005. *Supply chain management: Concepts and cases*. India: Prentice Hall.

Bahl, P., and Padmanabhan, V. N. 2000. RADAR: An in-building RF-based user location and tracking system. In *Proceedings of the Nineteenth Annual Joint Conference of the IEEE Computer and Communications Societies*, Vol. 2, 775–84. IEEE (Institute of Electrical and Electronics Engineers, Inc.), Tel Aviv, Israel.

Baldwin, A. N., Thorpe, A., and Alkaabi, J. A. 1994. Improved materials management through bar-coding: Results and implications from a feasibility study. In *Civil engineering, Proceedings of the Institution of Civil Engineers*, ICE, London, November, 102 (4), 156–162, ISSN 0965-089X.

Bernold, L. 1990. Bar code driven equipment and material tracking for construction. *ASCE, Journal of Computing in Civil Engineering*, 4 (4), 381–95.

Bosche, F., Teizer, J., Haas, C. T., and Caldas, C. H. 2006. Integrating data from 3D CAD and 3D cameras for real-time modeling. *Joint International Conference on Computing and Decision Making in Civil and Building Engineering*, Montreal, Canada.

Bowden, S. L. 2005. Applications of mobile IT in construction. PhD thesis, Loughborough University, UK.

Brambley, M. R., Kintner-Meyer, M., Katipamula, S., and O'Neill, P. J. 2005. Wireless sensor applications for building operation and management. Chapter 27 in *Web-based energy information and control systems: Case studies and applications*, eds. B. L. Capehart and L. C. Capehart, 341–67. Lilburn, GA: The Fairmont Press/CRC Press.

Brilakis, I., and Soibelman, L. 2005. Content-based search engines for construction image databases. *Automation in Construction*, 14 (4), 537–50.

――――. 2006. Multi-modal image retrieval from construction databases and model-based systems. *Journal of Construction Engineering and Management*, 132 (7), 777–85.

Brilakis, I., Soibelman, L., and Shinagawa, Y. 2005. Material-based construction site image retrieval. *Journal of Computing in Civil Engineering*, 19 (4), 341–55.

――――. 2006. Construction site image retrieval based on material cluster recognition. *Advanced Engineering Informatics*, 20 (4), 443–52.

Burgy, C. 2002. An interaction constraints model for mobile and wearable computer-aided engineering systems in industrial applications. MS thesis, Carnegie Mellon University, Pittsburgh, PA.

Cheok, G. S., Lipman, R. R., Witzgall, C., Bernal, J., and Stone, W. C. 2000. NIST construction automation program report No. 4 non-intrusive scanning technology for construction status determination. *Building and Fire Research Laboratory National Institute of Standards and Technology*, Gaithersburg, MD.

Davidson, I. N., and Skibniewski, M. J. 1995. Simulation of automated data collection in buildings. *Journal of Computing in Civil Engineering*, 9 (1), 9–20.

Du, W., Fang, L., and Ning, P. 2005. LAD: Localization anomaly detection for wireless sensor networks. *19th IEEE International Parallel and Distributed Processing Symposium*. April 3–8, 2005 Denver, Colorado.

Elnahrawy, E., Li, X., and Martin, R. P. 2004. The limits of localization using signal strength: A comparative study. *Proceedings of the First IEEE International Conference on Sensor and Ad hoc Communications and Networks (SECON 2004)*. IEEE, Santa Clara, CA, October.

Elzarka, H. M., Bell, L. C., and Floyd, R. L. 1997. Applications of pen based computing in bridge inspection. *Proceedings of the Fourth Congress in Computing in Civil Engineering*, ASCE, Philadelphia, PA, 327–34.

Ergen, E., Akinci, B., East, B., and Kirby, J. 2007a. Tracking components and maintenance history within a facility utilizing radio frequency identification technology. *Journal of Computing in Civil Engineering*, 21 (1), 11–20.

Ergen, E., Akinci, B., and Sacks, R. 2007b. Tracking and locating components in a precast storage yard utilizing radio frequency identification technology and GPS. *Automation in Construction*, 16 (3), 354–67.

ESRI: GIS and Mapping Software 2007. GIS logistics: Routing and scheduling. http://www.esri.com/industries/logistics/business/routing.html (accessed April 20, 2007).

Finch, E., Flanagan, R., and Marsh, L. 1996. Auto-ID application in construction. *Construction Management and Economics*, 14 (2), 121–29.

Fraden, J., and King, J. G. 1998. Handbook of modern sensors: Physics, designs, and applications. *American Journal of Physics,* 66 (4), 357–59.

Garmin. 2006. What is GPS? Garmin Ltd. http://www8.garmin.com/aboutGPS (accessed May 1, 2007).

Goodrum, P., Dai, J., Wood, C., and King, M. 2004. The use of the concrete maturity method in the construction of industrial facilities: A case study. *FIATECH Report,* January.

Gordon, C., and Akinci, B. 2005. Technology and process assessment of the use of Ladar and embedded sensing for construction quality control. *Proceedings of Construction Research Congress,* ASCE, San Diego, CA.

Gordon, C., Akinci, B., and Garrett, J. H. 2007. Formalism for construction inspection planning: Requirements and process concept. *Journal of Computing in Civil Engineering,* 21 (1), 29–38.

Hammad, A., Khabeer, B., Mozaffari, E., Devarakonda, P., and Bauchkar, P. 2005. Augmented reality interaction model for mobile infrastructure management systems. *First Canadian Society for Civil Engineering (CSCE) Specialty Conference on Infrastructure Technologies, Management and Policy.* CSCE, Toronto, Canada.

Hightower, J., and Borriello, G. 2001. Location systems for ubiquitous computing. *IEEE Computer,* 34 (8), 57–66.

Howell, G., and Ballard, G. 1997. Factors affecting project success in the piping function. In *Lean construction,* ed. L. Alarcon, 161–85. Rotterdam, The Netherlands: A. A. Balkema.

Jaselskis, E. J., and El-Misalami, T. 2003. Implementing radio frequency identification in the construction process. *Journal of Construction Engineering and Management,* 129 (6), 680–88.

Kärkkäinen, M., and Ala-Risku, T. 2003. Automatic identification – applications and technologies. *Proceedings of Logistics Research Network (LRN) Eighth Annual Conference,* London, UK.

Kersten, T., Sternberg, H., and Stiemer, E. 2005. First experience with terrestrial laser scanning for indoor cultural heritage applications using two different scanning systems. *Panoramic Photogrammetry Workshop,* International Society of Photogrammetry and Remote Sensing (ISPRS), Berlin, Germany.

Leick, A. 1990. *GPS satellite surveying.* New York: Wiley.

MacSema, Inc. 2007. http://www.macsema.com (accessed March 12, 2007).

McCullouch, B. 1997. Automating field data collection on construction organizations. *Proceedings of Construction Congress,* ASCE, Minneapolis, MN, 957–63.

McCullouch, B., and Lueprasert, G. K. 1994. 2D bar-code applications in construction. *Journal of Construction Engineering and Management,* 120 (4), 739–52.

Navon, R. 2005. Automated project performance control of construction projects. *Automation in Construction,* 14, 467–76.

Navon, R., and Shpatnitsky, Y. 2005. Field experiments in automated monitoring of road construction. *Journal of Construction Engineering and Management,* 131 (4), 487–93.

Pahlavan, K., Li, X., and Makela, J. P. 2002. Indoor geo-location science and technology. *Communications Magazine,* 40 (2), 112–18.

Paterson, A. M., Dowling, G. R., and Chamberlain, D. A. 1997. Building inspection: Can computer vision help? *Automation in Construction*, 7, 13–20.

Peak Sensors, Ltd. 2007. Thermocouple technical information. http://www.peaksensors. co.uk/thermocoupledatasheets.html (accessed May 12, 2007).

Pradhan, A., and Akinci, B. 2007. A planning based approach for fusing data from multiple sources for productivity monitoring. *Proceedings of Construction Research Congress*, American Society of Civil Engineers (ASCE), Grand Bahama Island, Bahamas, 5–7 May.

Rasdorf, W. J., and Herbert, M. J. 1990. Automated identification systems: Focus on bar coding. *Journal of Computing in Civil Engineering*, 4 (3), 279–96.

Reinhardt, J., Akinci, B., and Garrett, J. H. 2004. Navigational models for computer supported project management tasks on construction sites. *Journal of Computing in Civil Engineering*, 18 (4), 281–90.

Sacks, R., Navon, R., Brodetskaia, I., and Shapira, A. 2005. Feasibility of automated monitoring of lifting equipment in support of project control. *Journal of Construction Engineering and Management*, 131 (5), 604–14.

Song, J., Haas, C. T., Caldas, C., Ergen, E., and Akinci, B. 2006. Automating the task of tracking the delivery and receipt of fabricated pipe spools in industrial projects. *Automation in Construction*, 15 (2), 166–77.

Stone, W. C., Juberts, M., Dagalakis, N., Stone, J., and Gorman, J. 2004. Performance analysis of next generation LADAR for manufacturing, construction, and mobility. *NISTIR 7117*, National Institute of Standards and Technology, Gaithersburg, MD.

Streilein, A. 1994. Towards automation in architectural photogrammetry: CAD-based 3D feature extraction. *Journal of Photogrammetry and Remote Sensing*, 49 (5), 4–15.

Teizer, J., Caldas, C. H., Haas, C. T. 2006. Real-time three-dimensional modeling for the detection and tracking of construction resources. *ASCE Journal of Construction Engineering and Management*, 133 (11), 880–88.

Teizer, J., Kim, C., Haas, C. T., Liapi, K. A., and Caldas, C. H. 2005. A framework for real time 3D modeling of infrastructure. *Transportation Research Record*, Washington DC, No. 1913, 177–86.

Thomas, S. R., Tucker, R. L., and Kelly, R. W. 1997. An assessment tool for improving team communications. *Research Report, RR105-11*. Construction Industry Institute (CII), Austin, TX.

Thrun, S., Burgard, W., and Fox, D. 2005. *Probabilistic robotics*. Boston, MA: MIT Press.

Triggs, B. 1998. *3D reconstruction from multiple images*. http://homepages.inf.ed.ac.uk/ rbf/CVonline/LOCAL_COPIES/MOHR_TRIGGS/node51.html (accessed May 30, 2007).

Want, R., Hopper, A., Falcao, V., and Gibbons, J. 1992. The active badge location system. *ACM Transactions on Information Systems*, 10 (1), 91–102.

WP. 2007. How to choose the right temperature sensor. White Paper WP05-001. http:// www.singleiteration.com/library/document.cfm?id=51 (accessed April 2007).

Zoller and Fröhlich GmbH. 2005. *Understanding imager 5003 accuracy specifications*. Technical data explanation. Z+F Inc., Wangen im Allgäu, Germany.

17

Benefits of Using E-Marketplace in Construction Companies: A Case Study

Luis F. Alarcón
Pontificia Universidad
Católica de Chile

Sergio Maturana
Pontificia Universidad
Católica de Chile

Ignacio Schonherr
Pontificia Universidad
Católica de Chile

17.1 Introduction

The construction industry is one of the main movers behind the economic development of any country. In the United States, in 1997, the construction sector represented 10% of the gross domestic product (GDP) [Allmon et al. 2000]. In the European Union, the construction sector is the largest sector of the economy, contributing approximately 11% of the GDP [Bourdeau 2000]. In Chile this sector has contributed, on average, 8.4% of the GDP for the past seven years. Furthermore, the construction sector requires a large quantity of materials, machinery, and services from other productive sectors, producing a multiplying effect in the economy.

Supply is one of the main processes of construction companies because they need a wide range of materials and services in order to execute their projects. The aim of supply is to deliver the correct quantity of a quality product, at the correct time, in the correct place, and at the best price. The vital nature of supplies in a construction project is summarized in the following points [Alarcón and Rivas 1998]:

Supply must support the construction processes, providing the necessary supplies and resources when they are required, so as not to affect the project schedule. The high relative value of supplies with respect to the total cost of the project (between 50% and 70% of the total cost of the project) makes it imperative to have strict and constant control of what is being purchased.

Any type of supplies can be vital, due to the relationships of precedence and interaction between the different areas of the project. This paper analyzes the benefits that have accrued as a result of using e-marketplace to carry out the procurement process, based on a study of a representative sample of construction companies that have adopted this means of making their purchases.

17.2   E-Marketplace in the Construction Industry

In economics, a market is defined as a virtual meeting-place of supply and demand. E-marketplaces fulfill the same purpose. E-marketplaces are a business to business relationship model (B2B) in which multiple organizations, both sellers and buyers, can communicate, collaborate, and perform commercial transactions by means of a Web platform which is common to all those participating in the market. The main advantage of the e-marketplace compared with other B2B models is that it allows a purchasing company access to multiple selling companies, and selling company's access to multiple purchasing companies. Some examples of e-marketplaces include farms.com, specializing in agricultural products in the USA; convisint.com, which specializes in the automotive and healthcare industries; and Aeroxchange.com, which specializes in aviation supplies.

E-marketplaces may be horizontal or vertical [Issa, Flood, and Caglasin 2003]. Horizontal markets are those that provide a common service to companies of many industries. By contrast, vertical markets operate in a particular industry, developing functions, transactions, and contents in a specific sector of the economy.

In order to be successful, an e-marketplace must meet the following conditions:

- It must have a critical mass of participants, both buyers and sellers.
- It needs a large volume of transactions to achieve economies of scale that allow diminishing charges.
- It must be neutral to ensure transparency and confidentiality in commercial transactions.
- For a vertical market, it is necessary for the operator to be familiar with the industry in which it is going to work.
- It must have tools that are easy to use and that meet the clients' requirements.
- It must have the ability to provide personalized solutions for each business need.
- It needs trained personnel who are capable of providing adequate training for new users of the e-marketplace and for solving clients' problems as quickly as possible.

Implementing an e-marketplace is much more than simply implementing a new system. It implies a new way of working, which calls for profound changes in processes and in the way of thinking and acting [Neef 2001]. However, the construction industry does not favor the implementation of these initiatives, characterized by a strong resistance to change, attributable to the strong, rigid nature of construction culture [Aouad and Hassan 2002]. Some factors that may explain this resistance to change in construction companies are:

- The high degree of fragmentation in the industry, which produces interruptions that make communication/coordination difficult between the various phases of construction projects and their disciplines and subsystems [Rivard et al. 2004].
- Strong price competition within the sector, which prevents many companies from making investments to improve their processes.
- Every project is different, so that each is basically a prototype [Alarcón and Campero 2003], and they are of relatively short duration, meaning that there is high turnover in the work force [De Solminihac and Thenoux 2005].
- The work force is not highly trained and this leads to fear of new technologies.
- Each job is carried out in a different place, meaning that it is difficult for all of them to have the necessary infrastructure (a broadband connection and computers) to use an e-marketplace.

Nevertheless, there has been progress on construction e-marketplaces. Ekström and Björnsson (2002) argue that the slow adoption of electronic commerce in the Architecture Engineering Construction (AEC) industry is largely due to the lack of trust in the participants in electronic marketplaces. They investigated on how different theoretical frameworks can be used as a basis for an AEC rating system. In Ekström, Björnsson, and Nass (2003), a prototype rating tool called TrustBuilder, is described which facilitates information sharing between peer industry practitioners by calculating a weighted rating based on source credibility theory. In Ekström, García, and Björnsson (2005), an incentive mechanism called HYRIWYG (How You Rate Influences What You Get) is presented that encourages evaluators to volunteer their true opinions.

Li et al. (2002) present the concept of e-union, which integrates the services provided by different e-trading construction sites to provide an open e-trading service. They also describe the design of a mobile agent-enabled framework for building such an open e-trading marketplace environment. Castro-Lacouture and Skibniewski (2003) discuss the applicability of e-work models in the automation of construction materials management systems, using knowledge of the past in addition to observations of projects in progress or recently completed. Kong et al. (2004) present the system architecture and functions of a prototype e-union system, which utilizes Web services technology to provide information sharing between construction material e-commerce systems. Kong et al. (2005) discuss construction products information standardization and technologies for communicating and exchanging this information. Also, Castro-Lacouture, Medaglia, and Skibniewski (2007) address the purchase of construction materials as the last component in the supply chain and present a tool for optimizing purchasing decisions in B2B construction marketplaces.

Finally, Hadikusumo, Petchpong, and Charoenngam (2005) propose a decentralized database system equipped with electronic agents for material procurement. They show that their system can be used to assist human purchasers without some of the problems that the closed systems used by e-trading portals can produce.

17.3   Case Study: Iconstruye

The Chilean construction industry is highly fragmented, constituted by many small companies (about three thousand). The preferred procurement method is traditional design-bid-build, even though engineering-procurement-construction (EPC) contracts are becoming increasingly popular in the mining sector. The annual volume of the construction industry is around 15 billion dollars. Most supplies are highly concentrated in only a few distributors and suppliers.

Despite the difficulties of introducing new technologies in the construction industry, there is a vertical e-marketplace in Chile called Iconstruye, owned by the Chilean Chamber of Construction, which has over 600 clients. There is also another e-marketplace, called Seconstruye, which is owned by the main producers of steel, cement, and concrete. However, Seconstruye is focused on those products (steel, cement, and concrete), while Iconstruye offers a complete variety of construction supplies.

Iconstruye acts as a neutral facilitator in business relations between companies, providing technology and services that make the business cycle more efficient, quick, and transparent. Iconstruye has experienced steady growth in the number of its clients, as shown in Figure 17.1, and the amounts traded since the year it was set up, as shown in Figure 17.2, which currently corresponds to 20% of the total purchases of the

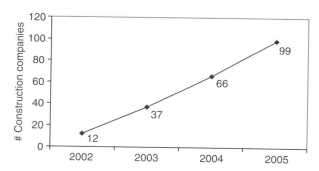

FIGURE 17.1   Enrollment of purchasing companies—accumulated.

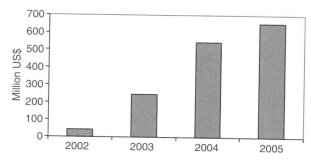

FIGURE 17.2   Amounts traded annually.

construction sector in Chile. Seconstruye, on the other hand, has only a small fraction of the market.

17.4 Aims and Methods

The aim of this study was to measure the benefits of using some e-marketplace to support the supply process of construction companies [Alarcón, Schonherr, and Maturana 2007]. More than 70 professionals and executives from 25 companies that use the Iconstruye e-marketplace were randomly selected to complete a questionnaire that would allow estimating the benefits of using the e-marketplace. Since the total number of construction companies that worked with Iconstruye when the survey was conducted was 89, a sample size of 25 gave us a standard error of 6% at a confidence level of 85%. Furthermore, the construction companies in the sample specialized in different types of construction projects and were of different sizes. Table 17.1 shows the size segments that were considered, and Figure 17.3 shows the distribution of the construction companies in the sample according to this segmentation.

TABLE 17.1   Construction Companies Size Segments According to Annual Sales

Company Size	Annual Sales (US$ Millions)
Extra large (XL)	More than 70
Large (L)	Between 35 and 70
Medium (M)	Between 14 and 35
Small (S)	Less than 14

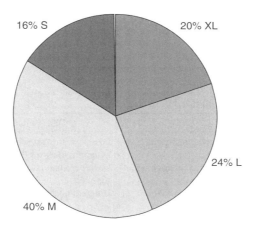

FIGURE 17.3   Distribution of the construction companies in the sample according to annual sales.

Each company answered the questionnaire individually in the presence of a researcher. The data compiled with the questionnaire was complemented with information obtained from the Iconstruye database, to verify some of the data collected with the questionnaire.

17.5 Traditional Supply Process

The supply process may be understood as all the activities required to obtain the appropriate supplies and services for a project, on the requested date, with the quality and in the quantities needed, in the appropriate place, and at a reasonable price [Al-Khalil et al. 2004]. The supply process may be centralized or decentralized. Centralized supply means that a central office makes all the purchases for a company, while in the decentralized procurement process, each project is responsible for all its purchases, depending on its own requirements. Each of these methods has advantages and disadvantages, as shown in Table 17.2 and Table 17.3.

The traditional approach in which a company makes its purchases has several drawbacks:

- The existence of a large number of manual processes, a considerable amount of paperwork, and a large number of validations and signatures, which are carried out physically and are based on physical information

TABLE 17.2 Centralized Purchases

Advantages	Disadvantages
• Better prices due to volume discounts. • Greater bargaining power with suppliers. • Avoids price anomalies and competition for scarce commodities between jobs [Serpell 2002]. • Standardized procedures [Serpell 2002].	• Increased level of inventories. • For orders delivered to warehouse, the company has to pay for transport to the building sites. • Greater administrative difficulties (errors in the type of product sent, mistakes in quantity delivered, etc.). • The sites have to compete to ensure that the head of procurement in the central office listens to their requests.

TABLE 17.3 Decentralized Purchases

Advantages	Disadvantages
• Greater speed in purchasing due to reduced bureaucracy. • Greater ability to react in emergencies (e.g., urgent orders) than the centralized model [Pooler and Pooler 1997]. • Better knowledge of the needs of the particular job.	• Loss of savings on volume purchases. • Different prices for the same product when purchased from different suppliers. • Different quality for similar products.

- The information on the stages of the process are completed unintegrated, with no repository in which one can have access to all the information and the relationships between the different data and sources of information. This results in disconnection between the central office and the building sites.
- Lack of control and transparency, due to the fact that the stages of the process are scarcely visible to the other members of the organization.
- A reactive process, in other words, the need to purchase arises as a result of the lack of supplies.
- Lack of materials is one of the main reasons for noncompletion of planned activities in companies.
- Nearly 80% of the order of materials are labeled urgent in most companies.
- The generation of reports is a long, arduous process with relatively unreliable results, susceptible to typing errors, and manipulation of the information.

17.6 Procurement Process Using an E-Marketplace

E-marketplaces can significantly improve the procurement process. In general, there were two main approaches to construction project supply in the companies studied: centralized process where the supply process is managed at company level for all the company projects at the same time, and decentralized process where each individual project has its own supply process. Figure 17.4 shows the procurement process of a company with centralized supply. The figure describes the sequence of activities carried out on-site, at the central office, and at the supplier office, to complete the supply process. Figure 17.5 shows how the use of an e-marketplace allows automating a series of activities in the centralized supply process, where the colored rectangles are the activities that are carried out via the e-marketplace. The case of decentralized purchases is similar, with the main difference being that all the activities take place on-site.

The automation of the supply process produces an important reduction in the problems presented by traditional supplying, and a series of benefits, as shown below:

- Access to a larger number of suppliers [Neef 2001].
- Reduction in administrative expenses in the procurement process [Issa, Flood, and Caglasin 2003].
- Reduction in prices of goods and services [Subramaniam, Qualls, and Shaw 2003].
- Establishment of long-term relationships with suppliers [Issa, Flood, and Caglasin 2003].
- Time savings [Hitech 2002].
- Reduction in last-minute purchases [Neef 2001].
- Reduction in inefficiencies and errors, as a result of less paperwork and repetitive steps involved in the procurement process [Neef 2001].
- Reduction in the costs of searching for the best price for a product [Dai and Kauffman 2002].
- Decrease in inventories [Issa, Flood, and Caglasin 2003].

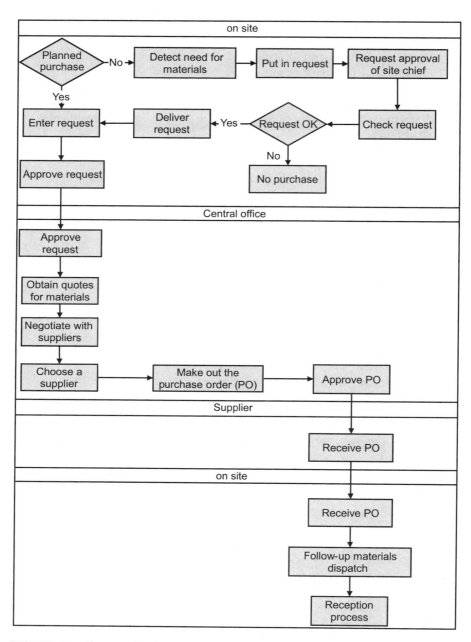

FIGURE 17.4 The centralized supply process.

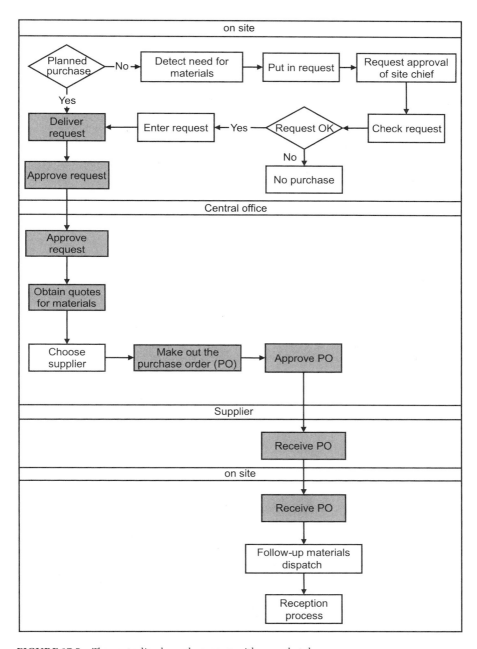

FIGURE 17.5 The centralized supply process with e-marketplace.

In the case of Iconstruye, requests are made in the orders for materials (OM), which are presented in a single format. Once the OM has been generated, it goes through an approval process to obtain the go-ahead from the supervisor or administrator, after which the purchase is negotiated.

Quotes may be made online in the case of integrated suppliers (IS); that is, those that belong to the Iconstruye business community. For the other suppliers, quotes must be made in the traditional way, but they can be entered in the system in order to generate a comparative table of offers.

Iconstruye's purchases module formalizes and controls the procurement process by submitting each purchase order (PO) generated in the system to an approval process, where it is checked and signed by each responsible person defined by the company. In this way, every PO sent to a supplier, whether integrated or not, has the approval of the company. For the IS, the POs are sent via the e-marketplace.

17.7   Results

One of the first questions posed to the professionals and executives of the companies that were surveyed was what were the main difficulties they had using Iconstruye. The results are shown in Figure 17.6.

In order to quantify the benefits of using the e-marketplace in construction companies, the performance indicators shown in Table 17.4 were defined for the supply process to determine whether companies that used Iconstruye showed an improvement.

Figure 17.7 shows the reduction of cycle time achieved by using the e-marketplace reported by the companies that were surveyed. As can be seen in this figure, around 75% of the companies reduced their cycle times by more than 20%.

Figure 17.8 shows the reduction in OM errors. We can see that 63% of the companies surveyed reported reductions of 20% or more in the OM errors with a significant number (25%) reporting reductions between 40 and 50%.

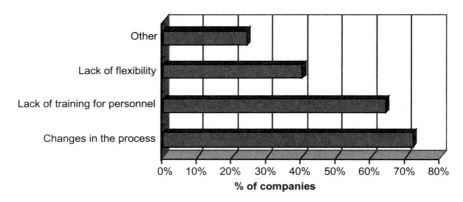

FIGURE 17.6   Main difficulties in using Iconstruye.

TABLE 17.4   Performance Indicators

Name of Indicator	Definition
Cycle time (CT)	Defined as the time from when a request is entered until the purchase order (PO) for that order is sent to the supplier.
Errors in OM	$$\text{OM with errors (\%)} = \frac{\text{Number of OM with errors} \times 100\%}{\text{Total quantity of OM}}$$
Errors in PO	$$\text{PO with errors (\%)} = \frac{\text{Number of PO with errors} \times 100\%}{\text{Total quantity of PO}}$$
Urgent orders (UO)	A UO is an order that exceeds the minimum cost and requires an additional effort to buy it, so that it will arrive in time. $$\text{UO (\%)} = \frac{\text{Quantity of UO} \times 100\%}{\text{Total quantity of OM}}$$
Irregular purchases	Irregular purchases are those made without following the internal procedures of the company. $$\text{Regularizations (\%)} = \frac{\text{Quantity of regularizations} \times 100\%}{\text{Total quantity of PO}}$$

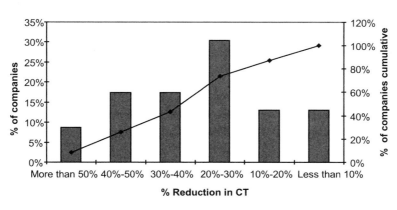

FIGURE 17.7   Distribution of cycle time reduction.

Figure 17.9 shows the reduction in PO errors. We can see that 52% of the companies surveyed reported reductions of 20% or more in the PO errors. In this case, the reduction was less than in the OM errors, with more than 20% of the companies surveyed reporting reductions of less than 10%.

Figure 17.10 shows the reduction in urgent orders (UO). We can see that 70% of the companies surveyed reported reductions of 20% or more in the PO errors. Another

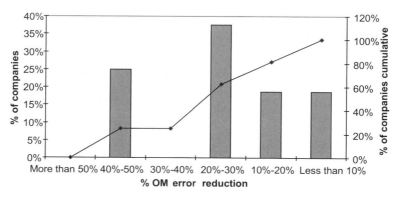

FIGURE 17.8   Distribution of reduction of orders for material (OM) errors.

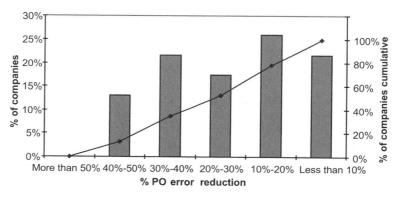

FIGURE 17.9   Distribution of reduction of purchase order (PO) errors.

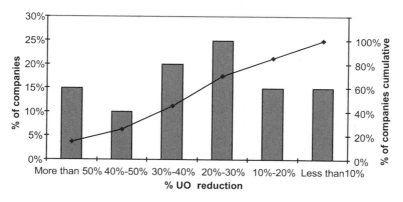

FIGURE 17.10   Distribution of reduction of urgent orders (UO).

study showed that UO cost an average of 20% more than normal orders, so reducing them can produce significant savings.

Figure 17.11 shows the reduction in irregular purchases. We can see that 61% of the companies surveyed reported reductions of 20% or more in irregular purchases. The companies that experienced a greater reduction were medium-sized companies, with an average reduction of 36%. However, even the largest companies reported reductions of 20% on average.

Figure 17.12 summarizes the improvements in the performance indicators. This figure compares the value of the indicators with a base value of 100 and shows how much each of them improved.

The perception of the companies using Iconstruye regarding possible benefits that may be obtained by using the e-marketplace was also measured on a Likert scale of (1) very negative to (5) very positive.

- *Reduction in administrative expenses:* Subramaniam, Qualls, and Shaw (2003) and Neef (2001), among others, maintain that the use of the e-marketplace reduces costs, due to the greater speed with which commercial transactions take place and the reduction in paperwork.

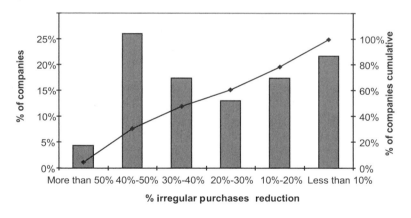

FIGURE 17.11 Distribution of reduction of irregular purchases.

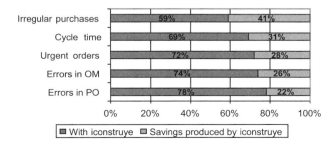

FIGURE 17.12 Performance indicators improvement due to use of an e-marketplace.

- *Increase in productivity:* good supply management contributes to the elimination of inefficiencies in the workforce [Thomas, Sanvido, and Sanders 1989]. On the other hand, some authors, such as Brynjolfsson and Hitt (2000), Teicholz (2004), and Samuelson (2002) maintain that the use of information technology to support processes produces an increase in the productivity of companies.
- *Increase in transparency:* Alarcón and Campero (2003) maintain that the transparency of a process seeks to make the main flow of the operations or activities involved in the process visible and comprehensible to all employees. The use of an e-marketplace should produce an increase in the transparency of the purchasing process, because with this tool any member of the organization with access to the Internet can review all the information related to the supply process, thus reducing possible points of corruption.
- *Increase in control:* due to the formalization of the purchasing process in organizations, as a result of the approval process for documents online (requests for supplies, POs, reception of products, and invoices for payment). Also, the related information is stored and can be consulted by the different players in the company.
- *Reduction in prices of products:* Subramaniam, Qualls, and Shaw (2003) maintain that the organizations using some system of e-procurement can obtain price reductions in two ways: (1) by consolidating orders, thanks to the centralization of the information about purchases; and (2) by increasing competition among their suppliers.

Figure 17.13 shows the impact of the e-marketplace on the reduction in administrative expenses reported by the companies that were surveyed. In general, about 50% of the companies reported a positive or very positive impact, and 60% of the largest companies reported a positive impact. Note that the area of the circle is proportional to the percentage of companies of each size that have a certain perception.

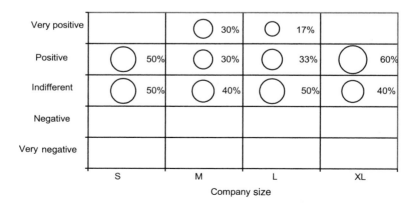

FIGURE 17.13   Perceived impact on reduction in administrative expenses by company size.

Figure 17.14 shows the perceived impact of the e-marketplace on the increase in productivity reported by the companies that were surveyed. As can be seen by the figure, a very large percentage reported a positive or very positive impact.

Figure 17.15 shows the perceived impact of the e-marketplace on the increase of control reported by the companies that were surveyed. As can be seen in the figure, almost all of the companies reported a positive or very positive impact.

Figure 17.16 shows the perceived impact of the e-marketplace on the increase in transparency reported by the companies that were surveyed. As can be seen by the figure, a very large number of the companies reported a positive or very positive impact.

Figure 17.17 shows the perceived impact of the e-marketplace on price reduction reported by the companies that were surveyed. As can be seen by the figure, smaller companies reported a more significant impact than larger companies.

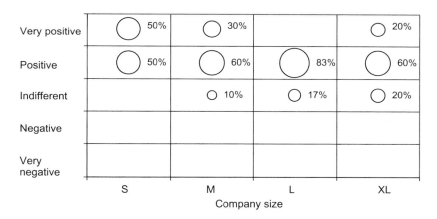

FIGURE 17.14 Perceived impact on productivity increase by company size.

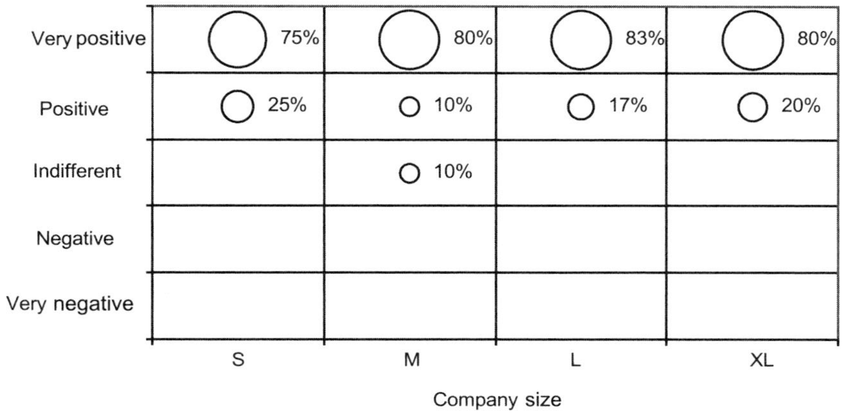

FIGURE 17.15 Perceived impact on increase of control by company size.

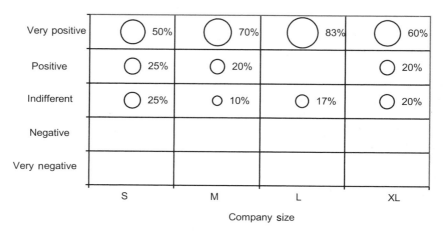

FIGURE 17.16 Perceived impact on increase in transparency by company size.

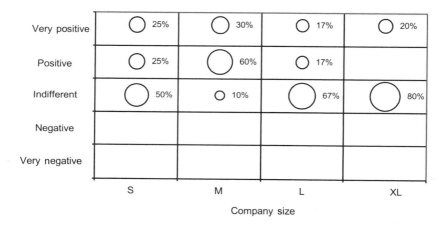

FIGURE 17.17 Perceived impact on price reduction by company size.

Finally, Figure 17.18 summarizes the distribution of replies from the companies and the average perception per indicator.

17.8 Conclusions

The present study has shown that, despite the characteristics of the construction industry that hinder the implementation of e-marketplaces, there are many companies that have adopted this new technology and have experienced important benefits. Figure 17.12 shows that the companies that were studied have significantly reduced the proposed indicator, which means that they have improved the performance of their procurement process.

	Reduction adm. expenses	Increase in productivity	Increase in control	Increase in transparency	Reduction in price
Very positive	◯ 16%	◯ 24%	◯ 80%	◯ 68%	◯ 24%
Positive	◯ 36%	◯ 64%	◯ 16%	◯ 16%	◯ 32%
Indifferent	◯ 48%	◯ 12%	◦ 4%	◯ 16%	◯ 44%
Negative					
Very negative					
Average perception	**3.7**	**4.1**	**4.8**	**4.5**	**3.8**

FIGURE 17.18 Distribution for perception of global indicators.

It is important to note that for all the global indicators considered, the average mark was over 3, as shown in Figure 17.18, meaning that the e-marketplace has not had negative effects in any company. The perception of the clients of the impact that the e-marketplace has had on the global indicators is consistent with that shown in Figure 17.12. Even though the study did not measure the impacts on-site of the implementation of the e-marketplace, the authors believe that directly contributes to the implementation of lean construction principles such as: increased process transparency, reduced process cycle time, reduced upstream variability, reduced inventories and waste of materials. Considering data available on the impact of procurement on overall construction performance [Alarcón et al. 2005], it is reasonable to expect that indirect impacts of improvements in the procurement process can be even more significant than the impacts on the process itself.

It is also worth noting that the companies involved in the study might have greater benefits if they made better use of the opportunities offered by Iconstruye, because not all of the organizations use all the functions of the e-marketplace. For example, there are still some companies that continue seeking quotes by telephone or fax, and practically none use the tools for reception of materials, which would allow them to keep better control of these.

Finally, although the conclusions of this study are based on the experience of Iconstruye and the Chilean construction industry, we believe our main conclusions are also applicable to other countries. However, the role of the Chilean Chamber of Construction was important in Iconstruye's success, and not all countries have such a strong association.

References

Al-Khalil, M., Assaf, S., Al Faraj, T., and Al-Darweesh, A. 2004. Measuring effectiveness of materials management for industrial projects. *Journal of Management in Engineering*, 20 (3), 82–87.

Alarcón, L., and Campero, M. 2003. *Administración de proyectos civiles*, 2nd ed. Santiago, Chile: Ediciones Universidad Católica de Chile.

Alarcón, L., Diethelm, S., Rojo, O., and Calderon, R. 2005. Assessing the impact of implementing lean construction. *Proceedings XIII Conference on Lean Construction*, IGLC-13, Sydney, Australia, July, 387–93.

Alarcón, L., and Rivas, R. 1998. Diagnóstico, evaluación y mejoramiento del proceso de abastecimiento en proyectos de inversión. *Revista Ingeniería de Construcción*, No. 18, Departamento de Ingeniería y Gestión de la Construcción, Pontificia Universidad Católica de Chile, Santiago, Chile.

Alarcón, L., Schonherr, I., and Maturana, S. 2007. Quantifying the benefits of using e-marketplace in construction. *Proceedings XV Conference on Lean Construction*, IGLC-15, East Lansing, MI, July, 560–70.

Allmon, E., Hass, C., Borcheding, J., and Goodrum, P. 2000. US construction labor productivity trends, 1970–1998. *Journal of Construction Engineering and Management*, 126 (2), 97–104.

Aouad, G., and Hassan, M. 2002. *Implementation of information technology in the construction industry: The conceptual methodology approach*. Salford, UK: University of Salford.

Bourdeau, L. 2000. The agenda 21 on sustainable construction. *CIB Symposium-Construction and Environment: From theory into practice*, Sao Paulo, Brazil.

Brynjolfsson, E., and Hitt, L. 2000. Beyond computation: Information technology, organizational transformation and business performance. *Journal of Economics Perspectives*, 14 (4), 23–48.

Castro-Lacouture, D., and Skibniewski, M. 2003. Applicability of e-work models for the automation of construction materials management systems. *Production Planning & Control*, 14 (8), 789–97.

Castro-Lacouture, D., Medaglia, A., and Skibniewski, M. 2007. Supply chain optimization tool for purchasing decisions in B2B construction marketplaces. *Automation in Construction*, 16, 569–75.

Dai, Q., and Kauffman, R. 2002. Business models for internet-based B2B electronic markets. *International Journal on Electronic Commerce*, 6 (4), 41–72.

De Solminihac, H., and Thenoux, G. 2005. *Procesos y técnicas de construcción*, 4th ed. Santiago, Chile: Ediciones Universidad Católica de Chile.

Ekström, M., and Björnsson, H. 2002. A rating system for AEC e-bidding that accounts for rater credibility. *CIB W65 Symposium*, Cincinnati, OH, September, 1001–1006.

Ekström, M., Björnsson, H., and Nass, C. 2003. Accounting for rater credibility when evaluating AEC subcontractors. *Construction Management and Economics*, 21 (2), 197–208.

Ekström, M., García, A., and Björnsson, H. 2005. Rewarding honest ratings through personalised recommendations in electronic commerce. *International Journal of Electronic Business*, 1 (3–4), 392–410.

Hadikusumo, B., Petchpong, S., and Charoenngam, C. 2005. Construction material procurement using internet-based agent system. *Automation in Construction*, 14 (6), 736–49.

Hitech Dimensions, Inc. 2002. *From procurement to e-procurement and e-sourcing.* Hitech Dimensions Inc. Publication. http://www.amazon.com/Procurement-E-procurement-E-sourcing–Hitech-Dimensions/dp/B00006AS5V/ref=sr_1_6?ie= UTF8&s=books&qid=1213210979&sr=1-6.

Issa, R., Flood, I., and Caglasin, G. 2003. A survey of e-business implementation in the US construction industry. *Electronic Journal of Information Technology in Construction*, 8, 15–28.

Kong, S., Li, H., Hung, T., Shi, J., Castro-Lacouture, D., and Skibniewski, M. 2004. Enabling information sharing between e-commerce systems for construction material procurement. *Automation in Construction*, 13, 261–76.

Kong, S., Li, H., Liang, Y., Hung, T., Anumba, C., and Chen, Z. 2005. Web services enhanced interoperable construction products catalogue. *Automation in Construction*, 14, 343–52.

Li, H., Cao, J., Castro-Lacouture, D., and Skibniewski, M. 2002. A framework for developing a unified B2B e-trading construction marketplace. *Automation in Construction*, 12, 201–11.

Neef, D. 2001. *E-procurement: From strategy to implementation.* Upper Saddle River, NJ: Prentice-Hall.

Pooler, V., and Pooler, D. 1997. *Purchasing and supply management: Creating the vision.* New York: Chapman & Hall.

Rivard, H., Froese, T., Waugh, L., El-Diraby, T., Mora, R., Torres, H., Gill, S., and O'Reilly, T. 2004. Case studies on the use of information technology in the Canadian construction industry. *Electronic Journal of Information Technology in Construction*, 9, 267–79.

Samuelson, O. 2002. IT-barometer 2000: The use of IT in the Nordic construction industry. *Electronic Journal of Information Technology in Construction*, 7, 1–26.

Serpell, A. 2002. *Administración de operaciones de construcción*, 2nd ed. Santiago, Chile: Ediciones Universidad Católica de Chile.

Subramaniam, C., Qualls, W., and Shaw, M. 2003. *Impact of B2B e-procurement systems: A summary report.* CITeBM and Industrial Distribution Management, University of Illinois, IL.

Teicholz, P. 2004. Labor productivity declines in the construction industry: Cause and remedies. *AECbytes* Viewpoint No. 4. http://www.aecbytes.com/viewpoint/ 2004/issue-4.html.

Thomas, R., Sanvido, V., and Sanders, R. 1989. Impact of material management on productivity – a case study. *Journal of Construction and Management*, 115 (3), 370–84.

18

Standards-Based Approaches to Interoperability in Supply Chain Management: Overview and Case Study Using the ISO 18629 PSL Standard

A.F. Cutting-Decelle
Ecole Centrale de Lille

R.I. Young
Loughborough University

B.P. Das
Loughborough University

C.J. Anumba
The Pennsylvania State University

J. Stal-Le Cardinal
Ecole Centrale de Paris

18.1   Introduction

Many construction and manufacturing organizations do business either directly or indirectly with other industrial sectors. Interoperability problems are aggravated by different cultures and disciplines as well as by the broad range of suppliers and subcontractors having different business functions. These problems are an important source of dysfunctioning of the supply chain (SC) communications, and the way they will be solved will have a strong impact on the quality of the exchanges [Stal-Le Cardinal 2001].

All of this has prompted the authors to undertake this research study within a cross-disciplinary SC environment.

The term "interoperability" refers to the ability to share technical and business data, information, and knowledge seamlessly across two or more software tools or application systems in an error-free manner with minimum manual interventions [Ray and Jones 2003]. For example, a weekly production scheduling package would require customer order details, production capacity details of machinery at the shop floor and its availability, production process details which would eventually generate weekly production plans, as well as product delivery details for various customers. Customer order details data may come from sales/marketing, and production capacity details may come from manufacturing, such as part prefabrication. Under normal circumstances for an interoperable system, the scheduling package should be able to capture that data seamlessly from the system and generate the necessary output. This does not usually happen, except where large integrated custom-made database application systems are used. In the worst case, the customer's system cannot understand the output data relayed by the supplier as such, and cannot use this information in his/her system without reinputting the data manually.

The ability to capture and share information seamlessly amongst a suite of software systems is very important, as it reduces data handling errors, facilitates concurrent business activities, and improves the responsiveness of an organization. However, this feature is not always available amongst the commonly used software applications. This lack of interoperability is costly to many globally distributed industries [see Gallaher, O'Connor, and Phelps 2002]; all of this has encouraged the research community to explore ways to reduce this cost.

There are many reasons for this lack of interoperability: different software operating systems; different software development approaches; different high-level software languages for interfacing data/information, etc. The most common reason is due to incompatibility between the syntaxes of the languages and the semantics of the terms used by the languages of software application systems. This is mainly due to arbitrary definitions provided by users to the developers of the proposed systems. Therefore, there is a strong need for the development of an approach which would overcome these incompatibilities.

There are three principal approaches to handle these issues [Gallaher, O'Connor, and Phelps 2002; Cutting-Decelle et al. 2004]. The first is a point-to-point customized solution, which can be achieved by contracting the services of systems integrators. This approach is expensive since each pair of systems needs a dedicated solution. A second approach, adopted in some large SCs, obliges all SC partners to conform to a particular solution. This approach does not solve the interoperability problem since the first or sub-tier suppliers are forced to purchase and maintain multiple, redundant systems. The third approach involves neutral, open, published standards. By adopting open standards, the combinatorial problem is reduced from n^2 to n, with bidirectional translators. Published standards also offer some stability in the representation they propose of the information models, an essential property for long-term data archiving. The third approach appears to be promising, suggesting development of a common, shared communication language understandable to each participating software application. It is important to note that many standard approaches to integration and interoperability provide a syntactic standard, but do not provide standard, interoperable semantics. Some commercial approaches are starting to provide a level of semantic support for SC communication, such as ebXML (2007) and Rosettanet (2007). However, a critical issue is the level of rigor involved in the semantic definition. If interoperability is to be checked and confirmed by computer analysis, it is essential that sufficient mathematical rigor underlies the semantic definitions being used. It is for this reason that this paper has focused on the application of the process specification language (PSL), which is based on first order logic, to the SC interoperability problem.

PSL [Schlenoff, Gruninger, and Ciocoiu 1999] has been developed to provide a common shared language to support process interoperability and is now an ISO standard—ISO 18629. PSL is a formal language based on first order logic and on mathematical set theory [Barwise and Etchemendy 1992]. The work reported here has focused on exploring the potential of PSL as a shared communication language within the context of a cross-disciplinary SC. The particular aim of the work was to explore the use of PSL as a formal route to the comparison of potentially interoperable processes in order to identify their level of compatibility.

We have pursued a scenario study approach to developing and understanding the requirements of a cross-disciplinary SC system in terms of likely communication processes. This scenario is introduced in Section 18.2. Within the context of this scenario, the chapter proposes an analysis of a particular cross-disciplinary transaction, the "purchase order transaction," in order to explore the detailed functions, processes, and tasks that may occur in the chain.

The third section of this chapter focuses on the information exchanges and sharing problems as they are dealt with in terms of standardization by the ISO TC 184 committee, particularly through the presentation of some standards commonly used for product data management: ISO 10303 Standard for the Exchange of Product model data (STEP), ISO 13584 Parts Library (P-LIB), and ISO 18629 PSL, but also within the International Alliance for Interoperability (IAI) industry foundation classes (IFCs).

A practical example showing how the ISO 18629 PSL standard has been used in the SC scenario is then presented, and analyzed from the point of view of ambiguity problems caused by the use of similar terms, although with different meanings, thus leading to interoperability conflicts.

The last section of the chapter then provides a general discussion about our work, about the use of standards to improve the management of the information handled in SCs, and the conclusions that can be drawn from it.

18.2   Example of Interoperability Problems: The Case of a Cross-Disciplinary SC Scenario

The cross-disciplinary SC scenario proposed here provides a set of processes and information flows, which are required to support a typical set of business functions. Given the number of interrelations and functions to be achieved between the partners, this example can be considered as a good way of highlighting the communication problems, notably those related to a lack of interoperability among the software tools used. The scenario encompasses a construction company, a construction site, a manufacturer, a retailer, and a transporter. The scenario is then used as the basis for the exploration of specific SC processes and the applicability of PSL to process interoperability analysis.

Figure 18.1 shows the SC network as devised for this research study. The main actor of this network is the construction company since it generally initiates most of the activities.

18.2.1   Functions of the SC Nodes

Various functions (tasks, activities, actions, processes, operations) normally performed by each node of the network within the context of processing a client's order by the construction company are briefly described below.

- *Construction company:* first receives an order from its client for implementing a construction project. The project manager in charge of this project asks his/her various departments to design, cost estimate, and prepare a construction schedule for the project.

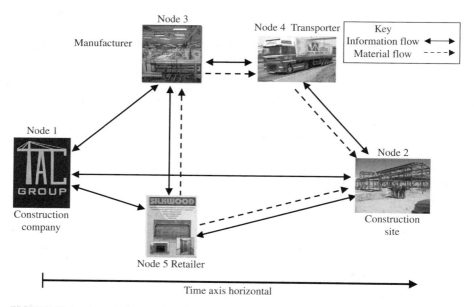

FIGURE 18.1   A typical cross-disciplinary supply chain network.

Based on the large list of items which needs to be ordered for the construction schedule to be completed in time, the project manager places orders through the materials management department to the manufacturer and the retailer for the goods to be delivered at the construction site on a particular date. Once the orders are accepted and confirmed by the supplier, the project manager requests its accounting department to prepare payment to the supplier for the set of goods he/she has ordered. However, the payment will not be released until the project manager receives confirmation of delivery of the goods at the construction site from the construction site manager and the satisfactory quality report from the quality department. Many other functions take place within the company, but they are out of the scope of this scenario. The main functions involved within the construction company in this business scenario are: *order processing, project planning and management, designing of building, cost-estimating, construction scheduling, procuring, quality management*, and *accounting*.

- **Manufacturer:** based on the order received from the construction company, the manufacturer processes the order; if feasible, the goods will be produced and shipped to customer's construction site via the transporter; accounting department prepares payments to suppliers and obtains remittance from the customer. The main functions involved within the manufacturer's business scenario are: *order processing, manufacturing, procuring, goods handling*, and *accounting*.

- **Transporter:** handles all the tasks related to transport and delivery of finished goods to customer's selected location. It receives the shipment order from the manufacturer, collects the goods, and delivers them to the construction site; the accounting department obtains remittance from the customer for the service. The main functions involved within the transporter's business scenario are: *order processing, distribution planning and scheduling, goods handling*, and *accounting*.

- **Retailer:** based on the order received from the customer (manufacturer and construction company), the retailer analyzes the order; if feasible to supply, goods will be delivered; if items are not in stock, the retailer places an order to a distributor, and delivers them to the customer when they are received from the distributor; the accounting department prepares payment to the distributor and releases it when goods are recveived, and also obtains remittance from the customer for the goods supplied. The main functions involved within the retailer's business scenario are: *order processing, distribution planning and scheduling, procuring, goods handling*, and *accounting*.

- **Construction site:** is a part of the construction company's business until it is handed over to the client. The construction site starts building work according to project plans and the building work schedule, which will be prepared by the construction company. The construction site receives notification from the construction company, manufacturer, transporter, and the retailer; the construction site notifies, appropriately, the construction company, manufacturer, transporter, and the retailer when goods are received. All damaged goods will be sent back to relevant suppliers and undamaged goods are released to the relevant users. The main functions involved within the construction site's business scenario are: *project management and goods handling, including quality management.*

It is essential to mention here that it is assumed that each actor uses separate workflow engines, as each represents an autonomous company, and there is, therefore, bound to be a problem with interoperability. Detailed organizational structure and function analysis of all the five nodes have been carried out within the scope of the project/research work. The information presented in this subsection represents a brief overview. The functions described here for each node are basically interbusiness processes, and it is essential to understand how they are related to each other within the node, as well as across the SC nodes, for supporting process interoperation in cross-disciplinary SCs. These issues are addressed in the next subsection.

18.2.2 Processes and Flows Supporting the SC Node Functions

This subsection introduces a set of processes, which are assumed to be needed to support the various functions of the SC nodes. For this research work, the term "process" means a structured collection of *activities/tasks* that have sequential relationships, while an "activity/task" means the transformation of a set of inputs into a set of outputs.

We mention here a short excerpt of the overall list of processes that may occur within the *construction company* node to facilitate the functions described earlier; for a more complete description, see Das et al. (2007):

- Construction company:
  Function name: Order processing
  Process name: 1. Receive client's order
                  2. Process client's order
  Function name: Procuring
  Process name: 3. Raise purchase order
                  4. Send purchase order
                  5. Receive acceptance of purchase order
                  6. Receive rejection of purchase order

Similarly, four other sets of processes supporting the functions of the other nodes have been identified and developed within the scope of the project work.

An overall information flow diagram across the SC to support an order processing transaction ("client's order") as initiated by the construction company is shown in Figure 18.2. The numbering procedure used on the information flow line of Figure 18.2 is as follows: the first digit indicates the node number, which is followed by a dot, and then the next two digits indicate information list number of that particular node. The diagram shows the nature of interactions of information that may occur across the SC network. Some of this information will be *going out* from the node, and some will be *coming into*

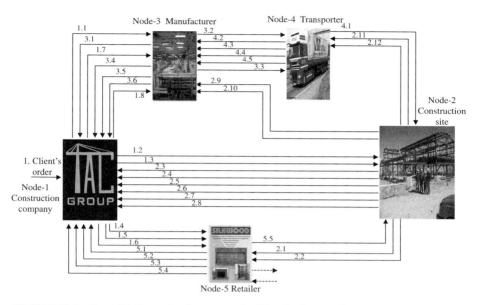

FIGURE 18.2   Overall information flow across the SC nodes for an order processing transaction.

the node from another external node. Visualization of the flows of information within and across the system boundaries is important, particularly from the point of view of software development, as well as in understanding the requirements for better communication/interoperability. It also highlights the variations of system requirements. This diagram clearly shows the nature and the complexity of the system and emphasizes that a clear understanding of the system is essential for managing the SC efficiently [Cutting-Decelle et al. 2004; Young et al. 2004; Cutting-Decelle et al. 2007].

18.2.3   Example of the "Purchase Order Transaction"

The "purchase order transaction" has been developed within the framework of the project to test the scenario, since it enables a detailed exploration of the communication processes across the SC, thus facilitating a good understanding of the related information structure and the process connectivity. As such, it has provided a good platform on which it has been possible to test the "applicability" of the ISO 18629 PSL language as a tool to improving SC interoperability. In this work, the term "communication process" means the flow of information from the sender to the receiver and vice versa in order to execute an activity/task. In this scenario, as shown in Figure 18.1, the purchase department of the construction company of the SC places an order for supplying a set of doors to a manufacturer: doors should be delivered to its construction site on a particular date. The order may contain the following information: "Supply 10 pieces of 6.5×3.5-ft (2-inch thick) plane, varnished plywood door complete with lock and handle to our Loughborough construction site on 10 October 2004." The manufacturer will supply the goods to the construction site through a transporter.

The analysis of the "purchase order transaction scenario" suggests that the processes within the system are the main sources in facilitating the communication across the SC. The next phase of the work explored the interoperability characteristics of those processes before attempting to develop mechanisms to share information amongst the nodes of the chain. This had been done through the use of the ISO 18629 PSL language presented in the next sections of this chapter.

18.3   ISO TC 184 Approach to Interoperability

In this section, we explore the potential of standards to increase interoperability. Then we focus on some of the standards developed by the ISO TC 184 standardization committee: ISO 10303 STEP, ISO 13584 P-LIB and ISO 18629 PSL.

18.3.1   Potential of Standards to Increase Interoperability

There are three principal approaches to compensate for the lack of interoperability: The first is a point-to-point customized solution, which can be achieved by contracting the services of systems integrators. This approach is expensive since each pair of systems needs a dedicated solution. A second approach, adopted in some large SCs, requires all partners to conform to a particular solution. This approach does not solve the interoperability problem since the first or subtier suppliers are forced to purchase and maintain multiple, redundant systems. It can also be costly to the smaller organizations in a SC since they are rarely in a position to influence the choice of infrastructure, and may not have enough resources to comply. The third approach involves neutral, open, published standards. By adopting open standards, the combinatorial problem is reduced from n^2 to n, with bidirectional translators [Pouchard and Cutting-Decelle 2007].

Published standards also offer some stability in the representation they propose of the information models, an essential property for long-term data archiving. This chapter highlights some of the standards developed within the ISO TC184 "Industrial Automation Systems and Integration" Committee. However, the problem is far from solved. Interoperability standards are used in layers, from the cables and connectors, through networking standards, to the application or content standards such as those mentioned here, that is STEP, P-LIB, and PSL. All of these layers must function correctly for interoperability to be achieved. The greatest challenges remain at the top of this stack of standards, in order to make them interoperable. Due to the capability of the PSL language to be extended (through its ontology) for accommodating concepts in other standards, this language can be considered as a powerful tool of this interoperability, enabling, for a near future, the consideration of a "universal interfacing."

We will briefly present some of the main (de jure and de facto) standards that can be used in construction SCs—references are mentioned for further information about them. Since this approach of the sector with an interoperability based on standards is rather new, we describe the most known in the domain of product data modeling (ISO 10303 STEP, ISO 13584 P-LIB, and IAI/IFCs) and the ISO 18629 PSL standard used for the specification of process-related information. This language brings an important contribution to the problem of the semantic ambiguity met in information exchanges.

18.3.2 ISO TC 184 Standardization Committee

The ISO TC184 is one of the two hundred committees managed by the ISO (International Standardization Organization, Geneva, Switzerland). Its scope is: "Standardization in the field of industrial automation and integration concerning discrete part manufacturing and encompassing the applications of multiple technologies; that is, information systems, machines and equipments, and telecommunications." This means that the standards developed are applicable to manufacturing and process industries, applicable to all sizes of business, applicable to extending exchanges across the globe through e-business. Are excluded from the scope the following areas: electrical and electronic equipment (dealt with by the IEC/TC44, International Electrotechnical Commission, Technical Committee 44), and programmable logical controllers for general applications (IEC/TC65). The standards developed by the ISO TC184 and its different subcommittees (particularly the subcommittee 4, SC4 "Industrial Data" for the standards mentioned here) cover various domains related to industrial automation and integration, among which: enterprise modeling, enterprise architecture, communications and processes, integration of industrial data for exchange, access and sharing, life-cycle data for process plants, manufacturing management, mechanical interfaces and programming methods, part libraries, physical device control, PSL, product data, and robots for manufacturing environment.

18.3.3 SC4 Standards: ISO 10303 STEP, ISO 13584 P-LIB, ISO 18629 PSL

18.3.3.1 ISO 10303 STEP

Each part of ISO 10303 contains the following introductory paragraph that summarizes the significant challenges undertaken in this standardization effort [Kemmerer 1999]: "ISO 10303 is an International Standard for the computer-interpretable representation and exchange of product data. The objective is to provide a neutral mechanism capable of describing product data throughout the life cycle of a product, independent from any particular system. The nature of this description makes STEP suitable not only for neutral file exchange, but also as a basis for implementing, sharing product databases, and archiving [ISO IS 10303-1 1994]."

STEP is intended to support data sharing and data archiving in the following way:

- Product data exchange: transfer of product data between a pair of applications. STEP defines the form of the product data that is to be transferred between a pair of applications. Each application holds its own copy of the product data in its own preferred form. The data conforming to STEP is transitory and defined only for the purposes of exchange.
- Product data sharing: the access of and operation on a single copy of the same product data by more than one application, potentially simultaneously. STEP is designed to support the interfaces between the single copy of the product data and the applications that share it. The applications do not hold the data in their own preferred forms. The architectural elements of STEP may be used to support the realization of the shared product data itself.

- Product data archiving: storage of product data, usually long term. STEP is suitable to support the interface to the archive. As in product data sharing, the architectural elements of STEP may be used to support the development of the archived product data itself. Archiving requires that the data conforming to STEP for exchange purposes are kept for use at some other time.

One important feature of the ISO 10303 standard is the use of the EXPRESS language, defined within the standard [ISO IS 10303-11 2003] throughout the standard, and even throughout the whole SC4 standards. The use of this language is an important stage towards the interoperability of the software applications targeted by the scope of the standard.

18.3.3.2   ISO 13584 P-LIB

ISO 13584 [ISO IS 13584-1 1999] specifies the structure of a library system, which provides an unambiguous representation and exchange of computer-interpretable parts library information. The data held in the library provide a description that enables the library system to generate various representations of the parts held in the library. The structure is independent of any particular computer system and permits any kind of part representation. The structure will enable consistent implementations to be made across multiple applications and systems. ISO 13584 does not specify the content of a supplier library. The content of a supplier library is the responsibility of the library data supplier. The library management system used in the implementation of the structure defined in ISO 13584, and any interface between this system and a user of the system is the responsibility of the library management system vendor and is not specified in ISO 13584.

ISO 13584 separates the representation of information held in a parts library from the implementation methods used in data exchange. The standard makes use of the EXPRESS language to specify information about the structure of a library. ISO 13584 separates information about the structure of a parts library from the information about different representations of each part or family of parts in the library. ISO 13584 permits information about part representation to be specified by different standards, and includes mechanisms that enable references to such descriptions [Pierra et al. 1998].

ISO 13584 is divided into a series of parts, each with a unique function. Each series may have one or more parts. The series are: Conceptual Description, Logical Resources, Implementation Resources, Description Methodology, Conformance Testing, View Exchange Protocol, and Standardized Content.

18.3.3.3   ISO 18629 PSL

ISO 18629 is the newest in the family of standards aimed at facilitating interoperability for industrial data integration (of products and processes) in industrial applications in TC 184. Standardized within a joint committee, ISO TC 184 SC4/SC5, PSL provides a generic language for process specifications applicable to a broad range of specific process representations in manufacturing and other applications. PSL is an ontology for discrete processes written in the knowledge interchange format (KIF) [Genesereth and Fikes 1992]. The scope of KIF enlarged and the language became a candidate to standardization at the international level, under the auspices of the ISO/IEC Joint

Technical Committee (JTC1) and the name of Common Logic (see the Common Logic Web site 2007) [Delugach 2007]. Each concept in the PSL ontology is specified with a set of definitions, relations, and axioms all formally expressed in KIF. Relations specify types of links between definitions or elements of definitions; axioms constrain the use of these elements. In addition, the PSL ontology is based on set theory, first order logic, and situation calculus [Barwise and Etchemendy 1992]. Because of this reliance on theories, every element in the PSL language can be proven for consistency and completeness [Gruninger 2004]. At the time of this writing, approximately half of the PSL definitions, relations, and axioms have been proven to be consistent with the base theories.

PSL is an international standard for providing semantics to the computer-interpretable exchange of information related to manufacturing and other discrete processes [ISO IS 18629-1 2004]. Taken together, all the parts contained in PSL provide a language for describing processes throughout the entire production within the same industrial company or across several industrial sectors or companies, independently from any particular representation model. The nature of this language makes it suitable for sharing process information during all the stages of production. The process representations used by engineering and business software applications are influenced by the specific needs and objectives of the applications. The use of these representation models varies from one application to another, and is often implicit in the implementation of a particular application.

A major purpose of PSL is to enable the interoperability of processes between software applications that utilize different process models and process representations. As a result of implementing process interoperability, economies of scale are made in the integration of manufacturing applications.

All parts in ISO 18629 are independent of any specific process representation or model used in a given application. Collectively, they provide a structural framework for interoperability. PSL describes what elements should constitute interoperable systems, but not how a specific application implements these elements. The purpose is not to enforce uniformity in process representations. As objectives and design of software applications vary, the implementation of interoperability in an application must necessarily be influenced by the particular objectives and processes of each specific application.

PSL [ISO IS 18629-1 2004] is organized in a series of parts using a numbering system consistent with that adopted for the other standards developed within ISO TC184/SC4. PSL contains core theories (Parts 1x), external mappings (Parts 2x), and definitional extensions (Parts 4x). Parts 1x and 4x contain the bulk of ISO 18629, including formal theories and the extensions that model concepts found in applications. Parts 1x are the foundation of the ontology, and Parts 4x contain the concepts useful for modeling applications and their implementation. Table 18.1 presents the organization of ISO 18629. Except noted otherwise, PSL version 2.2 is presented.

In practice, ISO 18629 Parts 4x are what software applications will utilize to specify and exchange their processes using PSL. In order to facilitate specification, the National Institute of Standards and Technology (NIST) has implemented the 20-Question-Wizard, a utility that points to the appropriate definitions [NIST 2003]. A user specifies a process in detail by answering questions and checking boxes for its process. The wizard returns the PSL definition for this process written in the KIF syntax.

TABLE 18.1   Organization of ISO 18629

Series	ISO Numbers	Names
Core theories	ISO IS 18629-1	Overview and basic principles
	ISO IS 18629-11	PSL core
	ISO IS 18629-12	Outer core
	ISO IS 18629-13	Duration and ordering theories
	ISO IS 18629-14	Resource theories
	ISO WD 18629-15	Actor and agent theories
External mappings	ISO 18629-2x	Mappings to EXPRESS, UML, XML
Definitional extensions	ISO IS 18629-41	Activity extensions
	ISO IS 18629-42	Temporal and state extensions
	ISO IS 18629-43	Activity ordering and duration extensions
	ISO IS 18629-44	Resource extensions
	ISO WD 18629-45	Process intent extensions

The use of PSL for developing a high-level translation between the source and target applications is now provided below: two applications do not necessarily exchange all their processes for interoperability. Only one or a set of processes may be translated. After identifying the concepts to be exchanged, the translation is performed in three steps, beginning with a syntactic translation, where the native syntax of an application is parsed to PSL syntax (KIF). This parser keeps the terminology of the application. A semantic translation to PSL then follows, keeping the KIF syntax for the terminology of the application of interest, KIF definitions are written for that application using PSL definitions. These definitions are found within the concepts of the PSL extensions. The question wizard facilitates the attribution of definitions to the terminology and concepts of an application to PSL definitions. The third stage is made of a semantic translation from one application to another. At this point, the processes of the source and target applications have been expressed using PSL terms and KIF syntax. Each should have a one-to-one correspondence between each process definition and a PSL definition. The concepts of the source application are mapped to concepts of the target application using PSL as the intermediate language. On this basis, data for the relevant process can be exchanged.

Several research work have been done or are currently on-going, showing examples of interoperability amongst software tools using PSL, notably at the University of Stanford, at the Center for Integrated Facility Engineering (CIFE) [Law 2001; Cheng et al. 2003], and at the University of Loughborough [Cutting-Decelle, Michel, and Schlenoff 2000; Cutting-Decelle et al. 2002, 2004; Tesfagaber et al. 2002].

18.3.4   De Facto Standard Developed by the IAI: IFCs

The IAI is an international consortium of regional chapters registered and listed as not-for-profit organizations in North America, United Kingdom, Germany, France, Scandinavia, Japan, Singapore, Korea, and Australia. Currently, the IAI has about 650 membership organizations worldwide, being construction companies, engineering firms, building owners and operators, software companies, and academic institutions.

The vision of the IAI is "to provide a universal basis for process improvement and information sharing in the construction and facilities management industries." More information about the IAI is available at IAI (2007).

The IFC are a data sharing specification, written in EXPRESS (10303-11, 1994). Content according to IFC is currently exchanged between IFC compliant software applications using the clear text encoding of the exchange structure, the STEP physical file, ISO 10303 part 21.

The scope of the IFC specification is the project life cycle of construction facilities, including all phases as identified by generic process protocols for the construction and facilities management industries. Development of IFC is guided by versions and releases which do extend the scope successively. The processes supported by the current IFC2x specifications are: outline conceptual design, full conceptual design, coordinated design, procurement and full financial authority, production information, construction, operation, and maintenance.

The target applications to exchange and share information according to IFC2x are: CAD systems, heating ventilating and air conditioning (HVAC) design systems, electrical design systems, formwork design and scheduling systems, structural analysis systems, energy simulation systems, quantity take-off systems, cost estimation systems, production scheduling systems, clash-detection systems, product information providers, steel and timber frame construction systems, prefabricated systems, stand-alone visualization tools, and others.

Traditionally, IFC model view definitions have been understood as subsets of the IFC model specification, and have been defined primarily for IFC implementation purposes. The format defined in this document covers the same scope. For definition of IFC model views, the goal was set to "finding a useful balance between the wishes of users/customers and the possibilities of software developers, and documenting the outcome clearly." The IFC model view definition format is used for documenting this outcome. The format must be well-defined and unambiguous, but the format is only one part of what is needed.

The IFC model architecture has been developed using a set of principles governing its organization and structure [Hietanen 2006]. These principles focus on basic requirements and can be summarized as: to provide a modular structure to the model, to provide a framework for sharing information between different disciplines within the architecture, engineering and construction/facility management (AEC/FM) industry, to ease the continued maintenance and development of the model, to enable information modelers to reuse model components, to enable software authors to reuse software components, and to facilitate the provision of better upward compatibility between model releases.

The IFC model architecture provides a modular structure for the development of model components, the "model schemata." There are four conceptual layers within the architecture, which use a strict referencing principle, and within each conceptual layer a set of model schemata are defined:

- The first conceptual layer provides resource classes used by classes in the higher levels
- The second conceptual layer provides a core project model. This core contains the kernel and several core extensions

- The third conceptual layer provides a set of modules defining concepts or objects common across multiple application types or AEC industry domains. This is the interoperability layer
- The fourth and highest layer in the IFC model is the domain layer. It provides a set of modules tailored for specific AEC industry domain or application type

IFCs have been endorsed by the International Standardization Organisation as a publicly available specification (PAS) under the ISO label ISO/PAS 16739.

18.4 Practical Example: Use of the ISO 18629 PSL Standard in the SC Scenario

The scenario proposed within the framework of our project focused on the use of the ISO 18629 PSL standard, since the objective of the project was to test the feasibility of the use of the language to facilitate the information exchanges in a multidisciplinary SC.

For the scenario, we used the "PSL Wizard" developed at NIST: this tool is aimed at providing help in the selection of the PSL concepts most suited to the translations between the ontologies of the different software applications to be interfaced [NIST 2003]. This method helped us to get the PSL representation of the "send_purchase_order" process as mentioned in Figure 18.3.

This translation definition highlights 10 properties of the process in terms of 10 PSL concepts such as: simple, nondet_folded, variegated, conditional, etc. Repeating the PSL mapping method with other SC processes allows us to identify where semantically similar processes exist. If the semantic translation of two processes is very similar, then there is likelihood that these two processes are shareable or interoperable. Based on this argument, the wizard has been applied to the set of 10 processes that are identified in our scenario. Some of the translation results are proposed below.

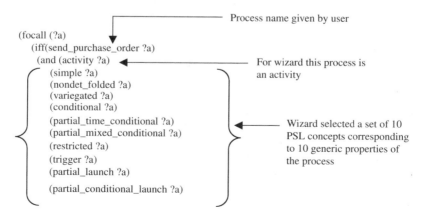

FIGURE 18.3   Answer proposed by the PSL_wizard for the "send_purchase_order" process.

The analysis of the scenario and the list of processes identified prompted us to classify these processes into four generic process types. These types are:

- **SEND**: formulated around activities involving monetary transactions
- **NOTIFY**: formulated around activities involving response by supplier to customer's requirements
- **HANDOVER**: formulated around handling activities of goods, which are finished, semifinished or raw materials, and responded to by supplier based on customer's requirements
- **DELIVER**: formulated around logistics/transporting activities of finished goods, semifinished goods or raw materials, and responded to by supplier based on customer's requirements.

The specific processes have been allocated under the relevant generic process class as shown in Table 18.2.

We might reasonably expect that all of the processes which are allocated under any of these four generic processes, as shown in Table 18.2, should be semantically similar. Also, the processes belonging to different generic process classes would not be semantically similar. This is not always the case, which will be evident from the following analysis. The translation results of the processes in our scenario are analyzed for semantic similarities by comparing the generic PSL properties of each process with the corresponding properties of the other processes. This leads to four groupings of these processes: Table 18.3 and Table 18.4 show the result obtained for the "SEND" and "NOTIFY" processes.

These groupings can be used to explore the potential for "shareability" of the processes. For example, Table 18.3 shows the translation results and the grouping of the Process 1 and the Process 3 under the generic process class "SEND." Both processes are semantically similar in terms of the 10 PSL concepts. We can conclude that there is a scope of sharing between these two processes which is in line with our expectations.

When we analyze the "NOTIFY" processes in Table 18.4 (Table 18.4 only shows an excerpt of the processes involved in "NOTIFY"), we find that there is a range of different PSL properties: if we examine Process 2 and Process 4, it may be seen that these two processes

TABLE 18.2 Allocation of Processes of the Scenario under the Generic Processes

Generic Process Class	Processes of the Scenario
Send	Process 1: send purchase order to manufacturer
	Process 3: send goods collection order to transporter
Notify	Process 2: notify goods supply date to customer
	Process 4: notify goods collection date to customer
	Process 5: notify goods arrival instruction to site
	Process 6: notify goods delivery date to site
	Process 9: notify receipt of undamaged goods to construction company
	Process 10: notify payment to supplier
Handover	Process 7: handover goods to transporter
Deliver	Process 8: deliver goods to customer's chosen site

TABLE 18.3   *SEND* Processes Against their PSL Concepts

Property No.	Process 1: Send Purchase Order to Manufacturer	Process 3: Send Goods Collection Order to Transporter
1	Simple	Simple
2	Nondet_folded	Nondet_folded
3	Variegated	Variegated
4	Conditional	Conditional
5	Partial_time_conditional	Partial_time_conditional
6	Partial_mixed_conditional	Partial_mixed_conditional
7	Restricted	Restricted
8	Trigger	Trigger
9	Partial_launch	Partial_launch
10	Partial_conditional_launch	Partial_conditional_launch

TABLE 18.4   *NOTIFY* Processes Showing Partially Different PSL Concepts (Excerpt)

Property No.	Process 2: Notify Goods Supply Date to Customer	Process 4: Notify Goods Collection Date to Customer	Process 10: Notify Payment to Supplier
1	Partial_permuted	Partial_permuted	Nondet_permuted
2	Partial_folded	Partial_folded	Partial_folded
3	Variegated	Variegated	Variegated
4	Conditional	Conditional	Conditional
5	Partial_time_conditional	Partial_time_conditional	Rigid_time_conditional
6	Partial_mixed_conditional	Mixed_conditional	Rigid_mixed_conditional
7	Restricted	Restricted	Local
8	Trigger	Partial_trigger	Trigger
9	Partial_launch	Partial_launch	Launch
10	Partial_conditional_launch	Partial_conditional_launch	Unconditional_launch

are semantically different, since property numbers 6 and 8 of these two processes are different. Similarly, if we consider Process 4 and Process 10 of the same table, they are also semantically different because their properties—1, 5, 6, 7, 8, 9, and 10—are all different.

There are many reasons for these differences which can be better understood by further investigations. For example, if we look more closely at property number 8 for Process numbers 4 and 10: for Process 10, the value assigned to property 8 is "trigger." The reason for allocating this property to this process is because this process is triggered once all the relevant information is received from all of the four participating nodes of the SC. On the other hand, the "Partial_trigger" chosen by the PSL_wizard for Process 4 is due to the fact that it waits for information only from within the node itself, in this case the transporter. In this manner, analytical reasons for selecting the properties of the process by the wizard can be developed.

From this evaluation, we can say that the PSL tool has enabled us to sort out the different processes into categories based on semantic identity, which was not easy (or not possible) to do by simply observing their contents. Such analyses facilitate the understanding of the transactions made of similar processes. This, in turn, provides a critical contribution to identifying the interoperability problems between SC processes.

18.5 Discussion and Perspectives

Standards-based approaches can be helpful to facilitate information sharing and interoperability among software applications commonly used in manufacturing and construction. Most of the time, technical terms handled by those applications look similar, or, even worse, are exactly the same—however their meanings are different. Given its properties and its structure, ISO 18629 PSL can be considered as a powerful interoperability "tool" for the information systems of the enterprises. It introduces economy of scales—each application only needs to provide interoperability to PSL once for information exchange. If an application changes, it is up to the developers of this application to provide new translations to PSL. Thus, only one application in the chain of interoperation is affected and not the others.

However, implementation of standards is nontrivial and costly. It is not until the use and implementation of these standards in a particular industry has reached critical mass that costs will decrease. STEP already has made great strides in this direction. The construction industry may benefit from the lessons learned in other domains.

In the domain of SC management, the identification of processes is very important for the development of information systems. Many processes occur within SC nodes and the names of those processes can be very similar across them. However, their requirements and function may be different, which creates the potential for interoperability problems amongst the nodes and leads to cross-disciplinary communication problems.

This chapter has shown an approach to the evaluation of processes in a SC in terms of the semantic compatibility. This has been achieved by identifying the nature of information flows, business nodes, and processes. For a particular node, a part of the information is generated within the organization, another part is received from external sources. Sometimes the information processes have similar names, sometimes the content can be similar. This necessitates a formal analysis of the information contained within the processes, in order to identify where semantic similarities or differences exist.

It is thus possible to set up a set of requirements expected from standards to be used for improving the communications within the SC and the interoperability of the software applications that are used. Among them, let us mention:

- The use of a common representation language (what is achieved by the standards developed within the ISO TC 184 committee: use of the ISO 10303-11 EXPRESS language)
- The use of some kinds of "common resources," expressed in terms of time management, resource management, and other common features always met in SC projects and related to product definition
- The development of a common approach, focusing on the exchange, storage, and archiving of data, and expressed in terms of product information, process information, resource management, materials, and components information

related management (this is the core approach of the ISO TC 184 way of managing the development of new standards and the maintenance of current ones)

- The consideration of traceability information, thus enabling it to manage some kind of history of the production/operation process of a given product
- The integration of the whole set of information related to products into a sustainable environment

One of the standards mentioned in this chapter, ISO 18629 PSL, defines a communication language which can support process interoperation. The language has the following features:

- A sufficient mathematical rigor to enable computational comparisons of the meaning of the terms
- A formal definition of process concepts as well as the more generally accepted entity information concepts
- Process concepts have to include a broad range of constraints which may be applicable, such as sequencing, state, time, and resource

PSL, although initially developed for discrete manufacturing processes, appears to provide a sufficiently generic and rigorous foundation ontology from which to compare the process semantics between SC processes. There is, however, a need to develop more effective methods for mapping system specific terminology to PSL, particularly in the domain of the analysis of dysfunctioning problems and their origins. PSL can still be seen as a powerful tool to aid communication between processes in the SC.

In summary, while there is great potential for improved interoperability in the approach taken, there is still substantial research needed before effective commercial solutions are available. The authors of this chapter are currently working on various issues in order to propose more effective solutions to these communication problems in SC management.

Acknowledgments

The work reported in this paper has been conducted as part of the X-Slang project, led by the University of Loughborough, and whose objective was to explore the communication requirements of a language to support interoperation across a multidisciplinary supply chain (including manufacturing and construction actors). The authors would like to acknowledge the support of the Engineering and Physical Sciences Research Council (EPSRC), which funded this project through Loughborough University's "Innovative Manufacturing and Construction Research Centre" under the Grant Number GR/N22526.

References

Barwise, J., and Etchemendy, J. 1992. *The language of first-order logic (Windows program, Tarski's world)*, 3rd ed., revised & expanded. CSLI Lecture Notes Number 34, Center for the Study of Language and Information—CSLI Stanford, Ventura Hall, Stanford University, Stanford, CA94305. Chicago, IL: The University of Chicago Press.

Cheng, J., Gruninger, M., Sriram, R. D., and Law, K. H. 2003. Process specification language for project scheduling information exchange. *International Journal of IT in Architecture, Engineering, and Construction*, 1 (4), 307–28.

Cutting-Decelle, A. F., Anumba, C. J., Baldwin, A. N., Bouchlaghem, N. M., and Tesfagaber, G. 2002. Exchanges of process information between software tools in construction: The PSL language. *Proceedings of the International Conference ECPPM02*, Portoroz, Slovenia, September 2002.

Cutting-Decelle, A. F., Das, B., Young, R. I. M., Case, K., Rahimifard, S., Anumba, C. J., and Bouchlaghem, N. 2007. A review of approaches to supply chain communications: From manufacturing to construction. *Journal of Information Technology in Construction (ITCON)*, 12, 73–102.

Cutting-Decelle, A. F., Michel, J. J., and Schlenoff, C. 2000. Manufacturing and construction common process representation: The PSL approach. *Proceedings of the CE2000: Seventh ISPE International Conference on Concurrent Engineering*. International Society for Productivity Enhancement (ISPE), Lyon, France, July 17–20.

Cutting-Decelle, A. F., Pouchard, L. C., Das, B. P., Young, R. I., and Michel, J. J. 2004. Utilising standards based approaches to information sharing and interoperability in manufacturing decision support. *Proceedings of the Flexible Automation and Intelligent Manufacturing Conference, 14th International Conference (FAIM04)*, National Research Council (NRC), Toronto, Canada, July 12–14.

Das, B. P., Cutting-Decelle, A. F., Young, R. I., Case, K., Rahimifard, S., Anumba, C. J., and Bouchlaghem, N. M. 2007. Towards the understanding of the requirements of a communication language to support process interoperation in cross-disciplinary supply chains. *International Journal of Computer Integrated Manufacturing*, 20 (4), 396–410.

Delugach, H. 2007. Common logic standard. http://cl.tamu.edu/ (accessed 2007).

ebXML. 2007. Electronic business using eXtensible markup language home page. http://www.ebxml.org/ (accessed 2007).

Gallaher, M. P., O'Connor, A. C., and Phelps, T. 2002. Economic impact assessment of the international standard for the exchange of product model data (STEP) in transportation equipment industries—final report. Research Triangle Institute (RTI) Project Number 07007.016, December 2002.

Genesereth, M., and Fikes, R. E. 1992. Knowledge interchange format, version 3.0. Reference manual. *Technical Report Logic-92-1*. Computer Science Department, Stanford University, Stanford, CA. www-ksl.stanford.edu/knowledge-sharing/papers/kif.ps (accessed 2008).

Gruninger, M. 2004. Ontology of the process specification language. In *Handbook on ontologies in information systems*, eds. R. Studer and S. Staab, 575–92. Heidelberg: Springer-Verlag.

Hietanen, J. 2006. IFC model view definition format. *International Alliance for Interoperability*. http://www.blis-project.org/IAI-MVD/ (accessed 2008).

International Alliance for Interoperability. 2007. http://www.iai-international.org (accessed 2008).

ISO IS 10303-1. 1994. Industrial automation systems and integration—product data representation and exchange—part 1: Overview and fundamental principles.

ISO IS 10303-11. 2003. Industrial automation systems and integration—product data representation and exchange—part 11: The EXPRESS language reference manual.

ISO IS 13584-1. 1999. Industrial automation systems and integration—Parts Library: Overview and fundamental principles.

ISO IS 18629. 2004. Industrial automation systems and integration—Process specification language.

Kemmerer, S. J. 1999. *STEP: The grand experience,* ed. Sharon J. Kemmerer. United States Of America Department Of Commerce, Manufacturing Engineering Laboratory, National Institute of Standards and Technology (NIST), Special Publication 939, 194 p., Gaithersburg, MD.

Law, K. H. 2001. Process specification and simulation. *PSL quarterly progress report.* Center for Integrated Facility Engineering (CIFE) Stanford, CA, June 1999.

NIST. 2003. 20 Questions: Classes of atomic activities, classes of complex activities. *Process Specification Language (PSL).* http://www.mel.nist.gov/psl/20questions.html (accessed February 2008).

Pierra, G., Sardet, E., Potier, J. C., Battier, G., Derouet, J. C., Willmann, N., and Mahir, A. 1998. Exchange of component data: The PLIB (ISO 13584) model, standard and tools. *Proceedings of the CALS EUROPE '98 Conference,* 160–6, Paris, France, September 16–18.

Pouchard, L. C., and Cutting-Decelle, A. F. 2007. Ontologies and standards-based approaches to interoperability for concurrent engineering. In *Concurrent engineering in construction projects,* eds. C. J. Anumba, J. M. Kamara and A. F. Cutting-Decelle, 118–61. London: Taylor & Francis Group.

Ray, S. R., and Jones, A. T. 2003. Manufacturing interoperability. Concurrent engineering: Enhanced interoperable system. *Proceedings of the 10th ISPE International Conference on Concurrent Engineering: Research and Applications,* 535–40, UNINOVA, Funchal, Portugal, July 26–30.

RosettaNet. 2007. http://www.rosettanet.org/cms/sites/RosettaNet/ (accessed 2007).

Schlenoff, C., Gruninger, M., and Ciocoiu, M. 1999. The essence of the process specification language. *Transactions of the Society for Computer Simulation,* 16 (4), 204–16.

Stal-Le Cardinal, J. 2001. Decision-making: How to improve an organisation by its dysfunctions? *Proceedings of the ICED01: 13th International Conference on Engineering Design,* 617–24, Glasgow, Scotland, August 21–23.

Tesfagaber, G., Cutting-Decelle, A. F., Anumba, C. J., Baldwin, A. N., and Bouchlaghem, N. M. 2002. Semantic process modelling for applications integration in AEC. *International Workshop on Information Technology in Civil Engineering,* ASCE, Washington, DC, November 2–3.

Young, R. I. M., Guerra, D., Gunendran, G., Das, B., Cochrane, S., Cutting-Decelle, A. F. 2005. Sharing manufacturing information and knowledge in design decision support. In *Advances in Integrated Design and Manufacturing in Mechanical Engineering,* eds. A. Bramley, D. Brissaud, D. Coutellier, and C. McMahon. 173–88. ISBN 1-4020-3481-4, Berlin: Springer.

19

Lean Enterprise Web-Based Information System for Supply Chain Integration: Design and Prototyping

Nashwan Dawood
University of Teesside

19.1 Overview

This chapter focuses on design and development of a lean enterprise Web-based information system (LEWIS). It is based on the overall system framework for multiconstraint planning and control. The aim is to develop a system that would serve as a backbone information infrastructure for supply chain integration, where information regarding construction products, processes, resources, and documentation could be integrated. The system can accommodate the statuses and problems from project supply chain. In turn, the system has the ability to perform look-ahead analysis, query constraint information, and generate a workable backlog for weekly work planning at the work face.

This chapter is concerned with a new methodology to integrate supply chain through reliable construction planning and control. In complex and fast-track construction projects, planning and control is very complicated due to the tremendous pressure to complete projects under conditions of uncertainty, in less time, and without sacrifice to cost and quality. Without reliable planning and control, there is a strong tendency for weak coordination across upstream supply chains and downstream operations at the work face. In many cases, activities will be executed even if not all the prerequisite works are completed and not all required resources and information are available. This tendency, known as "separation of execution from planning," frequently results in variability of activities' duration and obsoleteness of the plan.

To better integrate the construction processes and relieve the problem of separation of execution from planning, two solution paradigms have been offered. Firstly, computer integrated construction (CIC) has been globally highlighted as a technology-led paradigm aiming to reduce fragmentation, improve coordination and communication throughout the supply chain, and allow what-if analyses of various construction plans. Secondly, a process-led paradigm, which focuses on developing a solution to the well-defined business needs and problems, has gained much acceptance in the construction management discipline. Major concepts such as lean construction and theory of constraints have recently been promoted. These concepts additionally introduce management of flow and constraints, such as physical, information, execution space, etc., to the traditional project management body of knowledge (PMBOK). It is argued that the view of seeing production as a conversion of inputs into outputs and the idea of breaking up the total conversion into smaller, more manageable conversions (analytical reductionism) in the PMBOK is inadequate. To enhance ability to predict and handle variability and uncertainty in construction projects, the focus should be changed from managing activities or contracts, to managing physical material, information, and work flows.

This chapter proposes a new technique, and a set of developed tools for construction planning and control are derived from a synergy of technology-led and process-led paradigms. More specifically, it incorporates flow and constraint management model into the design and development of an integrated decision support system for construction planning and control.

The evolutionary prototype approach has been used as the main guideline for the system development. Two main system development steps, including system requirements identification and system design and development, are elaborated in this chapter. For the first step, rationale for implementing project Web-based information system is devised, and limitations of existing research prototypes and commercial systems are identified. Various aspects of requirement specifications including functionality, information, and technology needs are then analyzed. Based on these requirement specifications, the system design and development in the second step involves a number of processes, which include structuring and classifying data, conceptual data modelling, development of database structure, data population, and development and utilization of Web interfaces. Each of these processes is elaborated in the chapter. Finally, a real case study has been used to evaluate LEWIS.

19.2 Integrated Decision Support System for Multiconstraints Construction Planning and Control

It is concluded from previous literature in the areas of decision support systems and construction planning that more research and development should be conducted so as to seek more applications that better suit the construction industry needs and, in turn, promote wider benefits realization by the industry. This section is focused on the identification of technical requirements and suitable architecture that will enable development of a compatible integrated decision support system for the multiconstraints construction planning in the industry.

19.2.1 Requirements Identification for System Development

As a main guideline for development of an integrated decision support system for multiconstraint construction planning and control, achievable requirements are identified and categorized into seven aspects as follows:

- *Aim and improvement strategy*: following the aim of CIC research, this chapter will extend the application level of integrated information model to be an integrated decision support system that provides intelligence and visualization for evaluation of construction plans. Unlike previous research projects, technological development will be based on a rigorous planning process improvement from state-of-the-art planning techniques, such as last planner, critical chain scheduling, 4D/nD visualization, and multicriteria optimization. The system must support five superior characteristics including (1) collaborative and multilevel planning, (2) multiconstraint consideration, (3) effective uncertainty handling, (4) appropriate visual representation, and (5) practicable optimization
- *Phases and applications coverage*: the system development presented here aims to focus only on the planning and control during construction stage. However, the development will utilize the concept of integrated database in which project information can be shared among upstream supportive organizations (i.e., designers, suppliers, subcontractors), project planners, and work-face personnel
- *Integration and modelling approach*: the Web-based project database integration approach will be employed in the system development. To protect the database from becoming very large and unwieldy with consequent maintenance and information retrieval difficulties, the database will be designed as

a "semicentralized" architecture. This means that only information required for executing multiconstraint visualization and optimization at run time will be stored in the database. Other information produced by each discipline can be kept separately, given regular update of related information to the central database. Regarding the modelling approach, the relational model will be a preference. This is because the majority of the systems used and developed by the industry are based upon this model, hence the cost of changes would be minimized. Furthermore, it is relatively easier for prototype development due to high maturity of implementation of the relational model. Most commercial software packages provide features to import and export data between relational database management systems (RDBMS) and their systems

- *Data standardization*: subject to the limitation of current IFC version (IFC2x2) regarding its insufficient coverage for construction process domain, and especially multiconstraint planning and control, it is considered that the system development in this chapter will not be able to fully support the implementation of IFC. However, the system will be developed as an added-in to IFC-compliant software wherever possible. For instance, the multiconstraint visualization feature can be developed on top of the Autodesk Architectural Desktop 1.5.1 that supports product model based on IFC version 1.5.1. In addition, other standards, such as BS1192-5 (British Standard) for structuring CAD layers, and Uniclass (A standard universal system for CAD layering) for classification of product and process data, will be implemented where appropriate
- *System platform*: the system will be developed on the PC due to the fact that PCs represent the dominant platform for construction applications. Customized applications will be developed as plug-ins to standardized software packages including AutoCAD, Microsoft suite, and Internet browsers
- *DSS class and level of integration*: this chapter will consider the integration at the departments' level in which multiple applications supporting planning and control in construction phase will be developed. Subject to this focus, it is intended that the developed system will have characteristics of being: (1) analysis information system for generating special reports, (2) representation models for estimating consequences of particular actions, and (3) optimization models for calculating an optimal solution to a multi-criteria problem
- *Usability*: the developed system will be a prototype system that will be verified and validated through a real-life construction project

19.2.2 System Architecture

The design of DSS is based on a compound applications concept in which data extensibility and making use of the existing capabilities in off-the-shelf application packages can be beneficial [Heindel and Kasten 1997]. The core of the architecture is a central RDBMS where product model (CAD), process model (schedule), upstream information (i.e., drawings, specifications, method statements, resources information, etc.) and downstream information (i.e., weekly work plan and feedback) are integrated. The database system is named LEWIS. Microsoft SQL Server 2000 is chosen for the database implementation because of its wide availability, scalability, and multiusers supportability.

However, it is important to note that the design of this DSS is neither meant to be a fully integrated system nor to replace enterprise resource planning (ERP) systems and proprietary applications. Instead, it is designed as a system that gathers processed information from upstream supportive organizations for the benefits of planning and control at multiple levels. Figure 19.1 represents an overall DSS architecture for multiconstraint planning and control. The architecture is organized into three main layers and each of them is described in the following subsections.

19.2.2.1 Upstream Supportive Functions Layer

Upstream supportive functions may include design, engineering, contract management, accounting and cost control, procurement, inventory control, quality assurance, safety, and risk management. These functions are usually performed by multiple organizations (i.e., designer, consultant, contractor, suppliers, and subcontractors) from multiple locations both off-site and on-site and during multiple periods throughout the construction stage using heterogeneous applications and databases.

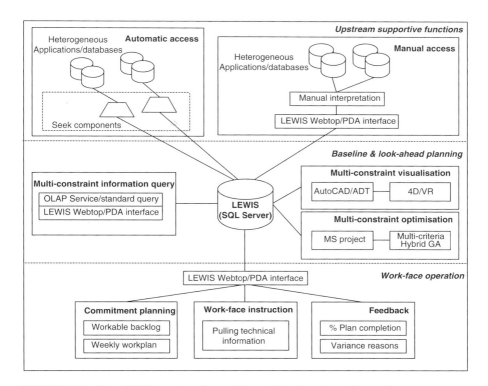

FIGURE 19.1 Overall DSS structure for multiconstraint planning and control.

As seen in Figure 19.1, there are two possible ways to access the LEWIS database:

1. *Semiautomatic access*: utilization of the scalable extraction of enterprise knowledge (SEEK) components [O'Brien et al. 2002] as a translator between heterogeneous applications/databases and the LEWIS is seen to be appropriate. As far as data standardization and system interoperability are concerned, our assumption is similar to those of O'Brien et al. (2002), Amor and Faraj (2001), Turk (2001), and Zamanian and Pittman (1999) in that hundreds or thousands of firms composing a project will not subscribe to a common data standard for process-related data. It seems more likely that firms will maintain the use of legacy applications/databases for process data, selectively transitioning to new applications [O'Brien and Hammer 2002]. Furthermore, the work on IAI/IFC has focussed on product models while limited process extensions have been developed.

2. *Manual access*: upstream supportive participants can currently access the LEWIS via Web interface (ordinary PC and Web browser) and wireless PDA interface (Pocket PC). Nevertheless, the users have to manually interpret the data from their legacy systems and reinput in the LEWIS. Examples of data input are estimated readiness time and delivery details of resources and information.

19.2.2.2 Baseline and Look-Ahead Planning Layer

Given contract finished date and allowable budget, an optimum baseline plan that has accounted for unforeseen circumstances is generally prepared at the very beginning of construction stage. Once the construction progresses, and deliveries of resources and information are confirmed, look-ahead planning should be performed so that constraints that are associated with scheduled activities can be proactively satisfied [Ballard 1997]. The main contractor's planning department with collaborative supports from upstream and downstream personnel is normally in charge of these two functions. To facilitate this collaborative process, three major components were developed and are described briefly as follows:

1. *Multiconstraint information query*: utilizes the online analysis processing (OLAP) service to create interactive browser-based queries for multiconstraint information. This technology provides a powerful tool helping project planners in understanding constraints more clearly and making decisions more effectively [Chau et al. 2002]. Furthermore, the system provides users with the ability to generate on-demand queries using standard search forms

2. *Multiconstraint visualization*: extends the capability of 4D (3D+time) and virtual reality (VR) visualization technologies to evaluate not only physical constraints, but also contract, resource, and information constraints. The 4D prototype has been developed using visual basic for application (VBA) embedded in the Autodesk Architectural Desktop 3.3 (IFC 1.5.1 supported) environment (Dawood et al. 2002)

3. *Multiconstraint optimization*: employs a hybrid optimization approach to reschedule project plan. Given multiple constraints, such as activity dependency, limited working area, and resource and information readiness, the algorithm alters tasks' priorities and construction methods so as to arrive at an optimum or near optimum set of project duration, cost, and smooth resources profiles. This

feature has been developed and embedded into macro in the project management software such as MS Project

19.2.2.3 Work-Face Operation Layer

Information systems at the work face can be classified into three main areas including commitment planning, work-face instruction, and feedback. Each is explained as follows:

1. *Commitment planning*: importance of the work in this area has not been raised until the introduction of lean construction concept and development of Work Plan tool [Choo et al. 1999]. Unlike the Work Plan tool that was developed as a stand-alone application in MS Access, this research employs a system architecture that enables an integration of commitment planning with upstream management systems and higher-level planning

2. *Work-face instruction*: an extensive literature review in the area of IT applications in construction reveals that research and development of systems that can provide technical information and instruction to the work-face personnel during construction stage has been largely ignored [Sriprasert and Dawood 2001]. A few prototype systems for delivering information to operators during facility management stage [Pakanen et al. 2001; Song, Clayton, and Johnson 2002] may be adaptable

3. *Feedback*: this area involves the process of monitoring project progress and providing feedback to planning department and upstream supply chain. The focus of the development is to gather feedback about weekly percent plan completion (PPC) and reasons why the committed tasks were not completed as planned

19.3 Prototype Development Approach

Most software development for the past 20 years has been based on a life cycle (waterfall model) that consists of distinct stages for requirements analysis, requirements specification, design, implementation, validation, verification, operation, and maintenance. However, this linear and phase-oriented process has led to dramatic cost increase in correcting errors as the life cycle progresses [Isensee and Rudd 1996]. In many cases, it was found that the delivered systems do not satisfactorily meet the user's needs [Green 1996].

In this chapter, an evolutionary prototyping approach (spiral model) has been used to develop each component in the whole suite of the integrated decision support system for multiconstraint construction planning and control. This approach involves the production of a basic implementation, on which incremental changes can be made to the system before its delivery [McMonnies 2001]. The prime function of the prototype is to aid in the analysis and validation of requirements, and as such needs only to be a mock-up of the proposed system, concentrating on the main system functions and the more visible features such as the user interface [Green 1996]. Many advantages of employing such a prototyping approach have been identified by many researchers [Alavi 1984; Boar 1984; Connell and Shafer 1989; Isensee and Rudd 1996; Lantz 1986]. These advantages include cost saving, better collection of customer requirements, increased quality, early demonstration of feasibility test, and improved user satisfaction.

According to the outlined processes shown in Figure 19.1, the development firstly begins with an identification of initial requirements in construction planning and control. This requirements identification determines important features that are missing in the existing research and commercialized systems. In addition, the process analyzes detailed specifications regarding system functionality, information coverage, and suitable technology. Once these have been achieved, the second process involves design and implementation of initial prototype. As the development progresses, the early versions of the prototype can be used to ascertain the true needs of the system. The third process involves repetitive testing and refining system's design and implementation until it is acceptable. Finally, the fourth process involves the process of releasing the prototype and its satisfied specifications to the development of final system.

19.4 System Requirements

System requirements identification consists of two stages including requirements acquisition and requirements analysis.

19.4.1 Requirements Acquisition

The requirements acquisition stage involves a process of obtaining background understanding about a problem domain as well as state-of-the-art processes and technologies that have been offered to remedy the problem [Davis and Olson 1985]. There are four typical means for acquiring system requirements, which are observation, literature review, interview, and questionnaire. In the context of this chapter, the problem domain considered is information management in construction. A review of literature in this domain provides general background and reveals important features that are missing from existing research prototypes and commercialized systems.

19.4.1.1 Information Management in Construction

Functional and geographical fragmentations are distinct characteristics of the construction industry [Howard et al. 1989; O'Brien and Al-Soufi 1993]. Most of the construction projects are based on temporary collaborations of owners, architects, contractors, subcontractors, suppliers, and so on. These partners take different roles and may join the team at different construction stages. In addition, the locations of projects and the locations of these partners are usually geographically different, sometimes thousands of miles apart. Many communication and information transfer problems in construction projects have been publicized.

Tam (1999) considers that enormous amounts and diversity of forms of information, which include detailed drawings and photos, cost analysis sheets, budget reports, risk analysis charts, contract documents, planning schedules, etc., are generated and exchanged during a project life-time. This problem can be referred to as information overloading in which a large part of the information is pushed to construction professionals and workers on a "just-in-case-they-need-it" basis [Thorpe and Mead 2001]. This scenario may hinder productivity because each individual must spend time sorting through piles of irrelevant data in multiple formats to find it.

A study reported that communications of design and technical information are heavily relying on informal methods, such as face-to-face discussions and paper-based documents. This typical practice originates from inadequate, inaccurate, inconsistent, and delayed information, which ultimately leads to poor coordination and reworks.

In addition to the above problems, traditional reporting systems and the lack of standard systems for data interchange can create other communication barriers. For instance, individuals may withhold information in order to gain a competitive advantage [Thorpe and Mead 2001], or may be unable to exchange information due to technological limitations [Rojas and Songer 1999]. Also, construction information can be distorted as it passes linearly from person to person [Thorpe and Mead 2001].

19.4.1.2 Project Web-Enabled Systems

The Internet is a global network that is not restricted by locations, time, or different computer operating systems [Deng et al. 2001].

Recent surveys [Johnson and Clayton 1998; Mohamed and Stewart 2003] indicate that the Internet-based information technology solutions are very useful at varying degrees in the construction industry. These solutions include e-mail, sharing files with e-mail attachment, shared databases, Intranet, and World Wide Web (WWW). One of the advantages of using the Internet as the communication platform is that data can be exchanged much more quickly and efficiently without geographical boundary [Tam 1999]. Thorpe and Mead (2001) also support the use of the Internet for just-in-time information management. Unlike the traditional push or just-in-case information model, this pull model allows users to access information when they need it, eliminating information that is out of phase, and reducing the transmission of irrelevant data.

After almost a decade of technological evolvement, the construction industry has witnessed the emergence of a number of Web-enabled research prototypes that allow participants to monitor, control, manipulate, and store project information [i.e., Abdelsayed and Navon 1999; Chan, Chua, and Lang 1999; Deng et al. 2001; Hajjar and AbouRizk 2000; Rojas and Songer 1999; Song, Clayton, and Johnson 2002; Tam 1999]. Furthermore, a number of software packages are currently available in the market. Many of these Web-enabled packages cover a wide range of functionalities in different stages, including tender stage, design and construction stage, and procurement stage (e-commerce). A sample list of the commercialized software can be found in Alshawi and Ingirige (2003).

By reviewing a number of research prototypes and commercialized systems (i.e., BIW Information Channel and Primavera Expedition) common functions that have been implemented are identified as follows:

1. *Communication management*: this function allows users to communicate and interact with other project members speedily and economically. General features for communication consist of information exchange through e-mail, Internet chat enhanced with on-screen images/pictures/drawings, and video conferencing. Furthermore, the systems provide standard forms for meeting minutes, tele-

phone records, correspondence sent and received, and team member's address information

2. *Data and file management*: the major data exchange tools provided by the systems include Telnet and File Transfer Protocol (FTP). First, the Telnet allows users to directly connect to the remote host. The system can provide a cheap and efficient method to get information compared with fax, phone or post. Second, FTP enables file transfer from one computer to another, even if each computer has a different operating system and file storage format. Once logged into the Internet FTP system, files such as CAD drawings and project documents can be transferred to the directory of the user's computer in a relatively short time. Users can only access the directory in which a password has been registered

3. *Contract and change management*: this function plays a major role for improving contractor, vendor, and supplier management. The systems enable contract managers to verify contractor's insurance, delivery of submittals, purchase orders, payment requisitions, and the impact of changes to cost or schedule. Any changes required can be broadcasted to relevant project members for rapid response or approval

4. *Job cost management*: as subcontracts are awarded, purchase orders are issued, requisitions are paid, and changes are requested, the systems can instantly update the cost worksheet. Project managers can track budgets, review commitments, record actual cost, and analyze cost variances for a complete financial picture. The systems can be set to notify managers when critical events occur, such as change requests or budget adjustments that exceed a set monetary threshold

5. *Site management*: reporting events on the job site is critical in maintaining an accurate record of daily occurrences. The systems provide standard forms for contractors to easily record weather conditions, work completed, field force, visitors, and materials delivered on site. A system like Primavera Expedition provides a link to its Primavera Project Planner (P3e/c™), thus allowing users to update actual progress on construction schedule activities from remote PCs or PDAs. In addition, a live videocam can be set up for transmitting snapshots (i.e., at every 10–20 s) of site activities to the head offices. These snapshots are useful for remote monitoring and analysis of site productivity

6. *Online office and library*: the objective of this feature is to provide the facility to project participants to prepare project-related documents directly in the server computer. Normally, the project participants prepare their project documents in their own computers, and then upload them to the server computer. The online office is an alternative means of communication between the client and server computers so that no data file transfer is necessary from the client computer to the server. In addition, standard-working procedures can be posted for references

7. *Auxiliary services*: construction firms can use their Web sites for auxiliary purposes, such as recruitment of new staff, questionnaire survey, and promoting their company's image to potential customers around the world

Despite the realized benefits and wide availability of the project Web-enabled systems, several challenging improvements remain. Limitations of the existing systems can be identified as follows:

1. *Lack of proper decision-making tools for project planning*: planning is a lengthy process and needs contributions from the entire project team. It is also context dependent. This process can be significantly improved if appropriate decision-making tools are incorporated into their structure. Unfortunately, the review of major CIC research points out that only a few research projects have just started pursuing their developments in this direction. One of the most sophisticated commercial packages, SAP R/3, an ERP system, also has not offered appropriate features for construction planning and analysis. A survey by O'Connor and Dodd (2000) reveal that users have low–moderate satisfaction in this aspect due to missing functionality for handling project work breakdown structure (WBS) and short interval planning

2. *Lack of integration within the supply chain*: the current ordering, purchasing, and invoicing practices have a lot of shortcomings in terms of delays in suppliers being received, less collaboration with manufacturers and supplies, and low integration of purchasing with accounts and planning software. For example, many delays are caused by the lack of integration between current material procurement systems and project plans. This deficiency tends to impact on stock control policies (e.g., carrying a high quantity of stock) of construction firms due to the inability to make accurate predictions of resource requirements for the project. The main reason for this is the poor communication and coordination among the supply chain partners and the overall lack of an integrated system to cater for this need [Alshawi and Ingirige 2003]

3. *Lack of standardized platform for information exchange*: incompatibility between hardware and software has raised a serious technical problem that has prevented users easily accessing project information. Zhu, Issa, and Cox (2001) mention that the current Web-based approaches do not effectively solve the problem of data interoperability. Since different software vendors may use different data formats, automation of document processing is still impossible even if everybody were on the Web

4. *Lack of real time analysis and collaborative support*: in civil engineering projects, substantial work is done on-site requiring the workforce to be mobile. Most of the problems encountered in civil engineering projects need the assistance of both on-site and in-office personnel to be solved. However, current project Web-enabled systems allow information dissemination for a limited segment of desktop users and with no effective collaborative support. There needs to be a system that enables access to information and application from anywhere with minimum device specification requirements [Penã-Mora and Dwivedi 2002]. Examples of such devices that allow Web content to be displayed on the screen are Web-enabled phones (WAP) and personal digital assistants (PDA and PALM)

19.4.2   Requirements Analysis

The purpose of this requirements analysis stage is to organize the facts that result from the requirements acquisition into a more structural form and with a scope that is feasible. The requirements analysis can be classified into three aspects including required functionality, required information, and required technology. Analysis of each of these aspects is elaborated in the following subsections.

19.4.2.1   Required Functionalities

The main objective is to develop an integrated decision support system that will enable implementation of the multiconstraint planning and control. The development of LEWIS is focused on the delivery of this objective rather than reinventing the common features that have already been implemented in the existing systems. Overall, LEWIS will act as a hub of project information and as a platform that allows integration among different applications, such as multiconstraint visualization and multiconstraint optimization. Innovative functions, such as integrating supply chain with project planning, look-ahead planning through multiple constraints analysis, and weekly work planning based on the last planner approach will be implemented. The details of each of these functions are provided as follows:

1. *Integrating supply chain with project planning*: the implementation of this function has been largely ignored in the development of the existing systems. Based on the concept of multiconstraint planning and control system, the ability to capture and analyze the readiness status of information and resources for each scheduled activity is crucial for project success. To avoid delays and cost overruns, related information and resources should be in place just before execution of any activity. Late delivery or imperfection of these items can substantially cause idle time or reworks. On the other hand, if these items (particularly large bunches of materials) are delivered too soon, additional time and cost must be spent for their storage and rehandling. Therefore, there is a need for a system that is capable of not only informing suppliers about the actual status of construction on-site, but also allowing them to prioritize deliverables for each activity in the just-in-time manner. In the development of LEWIS, a tracking system for drawings and materials is included. The system will allow users to allocate drawings and materials to a number of related activities in the project schedule. With this relationship, the users will be able to check the readiness status and estimated ready time (ERT) of information and resources for each activity. If the ERT in any item is not appropriate, project planners and suppliers will be able to proactively negotiate and solve the problems

2. *Analysis of multiple constraints and look-ahead planning*: this function will cover the analysis of four types of constraints, which include physical, contract, resources, and information constraints. Once the schedule and supply chain information is updated, a set of queries that list potential constraints of each scheduled activity will be generated. With the assistance of multiconstraint visualization, the planners can then perform look-ahead planning (i.e., once a week or once a

month) by reviewing all the constraints that will prevent upcoming scheduled activities from being started as planned. This look-ahead planning process will notify and allow planners to remove the constraints in advance. The activities that are free of constraints will be registered in a workable backlog and will be released for weekly work planning at the work-face level. It should be noted that safety and environmental constraints are not included in the prototype development

3. *Commitment planning*: commitment planning involves processes of weekly work planning, monitoring PPC, and identification of reasons for variance. In this case, the system will allow users to generate weekly work plans (what will be done) by selecting constraint-free activities from the workable backlog (what can be done). The system will then ask users to break down these sub-activities into that match available labor capacity in each particular week. At the end of each week, the system should also provide interface for monitoring progress of work and reasons for variance. Once the monitoring data is inputted, the system will automatically calculate PPC (a percentage ratio of number of completed tasks to total number of planned tasks in that week), and summarize reasons for variance. It should be noted that a similar commitment planning system has been developed earlier by Choo et al. (1999). However, their stand-alone application is not integrated with higher-level planning system, visualization system, and optimization system as explained earlier in this chapter

19.4.2.2 Required Information

To achieve the system functionalities identified, required information input and output can be classified as follows:

1. *Required information input*: this input information is referred to as information that must be preliminarily prepared and inputted into LEWIS. Four major groups of the information input are:
   - General project information including project details, contact details of project teams and staff, and standard data dictionary for organizing project information, such as document and material status, project areas, project phases, etc.
   - Product information including product classification code, geometrical data of 3D CAD model, product groups, and construction spaces or zones
   - Process information including process classification code based on WBS, list of scheduled activities, activity duration, activity start and finish dates, predecessors, successors, resource and space allocations, and relationship between activities and groups of CAD product
   - Documents and resources information including documents and resources details (i.e., ID, description, amount, and unit cost), delivery sheets, readiness status, ERT, and relationship with activities

2. *Required information output*: this information is referred to information that is analyzed and generated by LEWIS. The information may be reported directly to

the users or may be used by visualization and optimization applications. Three major groups of the information output are:

- Constraint information including status of prerequisite works, space utilization (% overload), readiness of information and resources
- Workable backlog including list of constraint-free activities at particular period of time
- Feedback including PPC and reasons for variance

19.5   System Design and Development

Based on the system requirements identified in the previous section, this section elaborates on detailed design and development of LEWIS. The section starts with an explanation on how product and process data in LEWIS database were structured using the Unified Classification for Construction (Uniclass). Given the appropriate data structure, a conceptual data model that presents relationships between different data types is designed using EXPRESS-G. This conceptual data model is then translated into a database physical diagram using Structure Query Language (SQL) server. The section further describes the methodology to populate product and process data into LEWIS databases. Finally, the design and utilization of the Web-based interfaces for capturing readiness status of information and resources, multiconstraint analysis, and commitment planning are demonstrated.

19.5.1   Data Classification

19.5.1.1   Classification System

In the construction industry, classification plays a major role for structuring information in specifications, organizing drawings, classifying product data and technical operation literature, and indexing historical cost data, etc. The need for general classification systems grows with the increased internationalization of the construction market and the rapid development towards a CIC process based on computer-aided product data modelling [Ekholm 1996]. Substantial efforts have been exerted in developing a standard classification system. Examples of these systems are Master-format (North America), CI/SfB (Europe), BSAB (Sweden), DIN (Germany), Lexicon (Netherlands), and Uniclass (UK). However, none of them has been universally recognized or widely employed as an international standard. Recently, the Construction Specifications Institute (CSI) and the International Alliance for Interoperability (IAI) have provided a co-chairmanship support for development of a new system called "Overall Construction Classification System (OCCS)." Although the development of the OCCS would initially be focused on the needs of the North American construction industry, the hope is that a truly universal classification system will eventually emerge from this effort [OCCS Development Committee 2001].

19.5.1.2   Uniclass Implementation

Uniclass is the first classification scheme for the construction industry that follows the international work set out by ISO technical report 14177 [Crawford, Cann, and O'Leary 1997]. One of the benefits of using the Uniclass, apart from providing standards for structuring building information, is that it provides a systematic standard for classifying and integrating product breakdown structure (PBS) with WBS. This is an important aspect for delivering a meaningful 4D visualization model. Other major reasons for selecting the Uniclass are:

1. The IFC standard is still being developed and has a long-term vision for standardizing information exchange in the industry. Amor and Faraj (2001) argue that IFC standard will not be able to contain the multiple views and needs of all projects participants. They predict that rather than a single standard, multiple protocols will evolve over time. Consequently, it is anticipated that contractors and designers will still maintain the use of legacy application for process data for a foreseen period in the future

2. Almost all of the data currently generated in the construction industry are non-IFC compatible. Only in a few projects where substantial collaborations between research teams and industrial partners are initiated, certain parts of product models are constructed based on the IFC standard

3. The OCCS is still in the developing stage and being built on Uniclass's hierarchical multifacets concept. If the OCCS becomes a widely acceptable standard, the main structure of the LEWIS database will not be obsolete

According to Kang and Paulson (2000), there may be two methods to apply the Uniclass in building projects. One is the partial classification for specific subjects, such as a work section or an element in a WBS, and the other is the integrated information classification through the life cycle of a project. In this chapter, the four facets of the Uniclass, including D (facilities), F (spaces), G (elements for buildings), and J (work sections for buildings) are used for representing physical products and work items in a building project. Furthermore, facet A, M, and P are used for classifying information, construction aids, and materials, respectively. Figure 19.2 shows an example of an organization of product-based WBS based on the Uniclass.

Depending upon the degree of project complexity, the structure in this example is organized into six levels including facility, summary, product, subproduct, process, and related information and resources levels. As seen in the structure, every item is represented by a Uniclass code. The classification code of each item can then be used with the link to the other items higher than their levels. For example, the code of [D72111 – G26 – F134:G261: G311 – JG10z1 – M4114] means a mobile crane [M4114] that is allocated for structural steel framing works group 1 (Grid-line A-C/1-5) [JG10z1] of ground floor column [F134:G261: G311] for superstructure frame [G26] in the school of health building facilities [D72111].

This structure allows planners to systematically define process, information, and resources required to construct each type of product, which, in turn, facilitates multiconstraint analysis. For example, planners will be able to monitor whether required information and resources have been made ready for processes of constructing each type of product.

19.5.2 Conceptual Data Modelling

Data modelling provides a set of powerful techniques that allow confident transitions from analysis to design, and from design to implementation. It is an understanding of

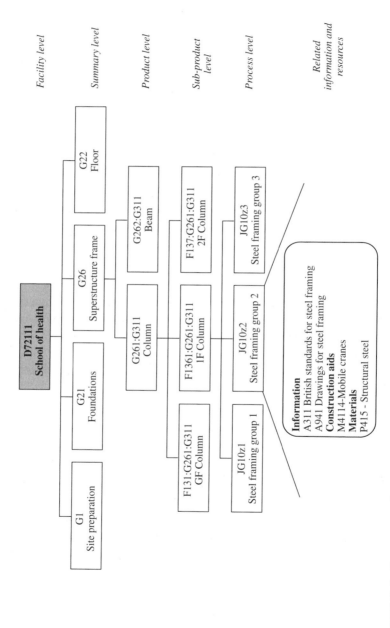

FIGURE 19.2   Organization of product-based work breakdown structure based on Uniclass.

the data structures of an information system, which is the key to an effective information systems design [Green 1996]. A data model represents entities, attributes, and relationships that will be stored in a database. There are several graphical data modelling methods, such as entity-relationship (E-R models), Nijssen's information analysis method (NIAM), unified modelling language (UML), EXPRESS-G, and integration definition for information modelling (IDEF1x).

In this chapter, EXPRESS-G is used to model an overall data structure of LEWIS database. EXPRESS-G is a conceptual diagrammatic modelling notation developed by the International Organization for Standardization [ISO/DIS 10303-11 1994]. Its unique concept of entity generalization (supertype) and specialization (subtype) allows users to hierarchically and clearly create an abstract parent model with its detailed child models [Schenck and Wilson 1994]. Other reasons for choosing the EXPRESS-G are familiarity of the author with this method, and availability of suitable modelling tools.

An abstract of LEWIS high level relationship data model (EXPRESS-G) is presented in Figure 19.3. The database structure shown in Figure 19.4 illustrates how product, information, resource, space, process, and work-face information are interrelated and integrated. Some explanations of the model are provided as follows:

1. *Products*: the term (ABS) is used with products to indicate that it is an abstract supertype. This means that it cannot exist in itself, only by virtue of its subtypes. For example, beams, columns, and walls are discussed on building sites, but the products never are. The abstract supertype is useful since it enables attributes (i.e., width, length, height, and coordinates (x, y, z)) to be collected at a higher level within the data model and then inherited. Additionally, the relationship "PartOf-Project S [1:?]" means that a project has one-to-many relationship with products. In other words, many products constitute a part of the project

2. *Project*: the project entity is used to represent general project profile. Examples of project's attributes are project name, address, duration, cost, contracted start date, and contracted finish date, etc. The relationship "Managed By S[1:?]" means that a project is managed by many participating companies and many responsible persons

3. *Company/person*: this entity is used to represent the contact information of companies and their staff that are involved in the project. These companies may include clients, designers, consultants, general contractors, subcontractors, suppliers, and so on. Examples of related attributes for this entity are name, abbreviation, role, address, contact number, and email, etc. The relationship "Supply/Check S [1:?]" means that each company and its staff have responsibility to supply or check a variety of information, resources, and site spaces. For instance, a structural designer has a duty to supply drawings for roof trusses, while a steel supplier must supply the structural steel, and a contractor must supply storage and working spaces on site

4. *Information*: similar to the products, the term (ABS) is used to indicate that the information is an abstract subtype. Subtypes of information include drawings, specifications, method statements, and those that are required for work-face personnel to construct a product. General attributes for each type of information are information/document code, title, responsible person, discipline, phase, and ERT. However, additional attributes can be specified for each type of information.

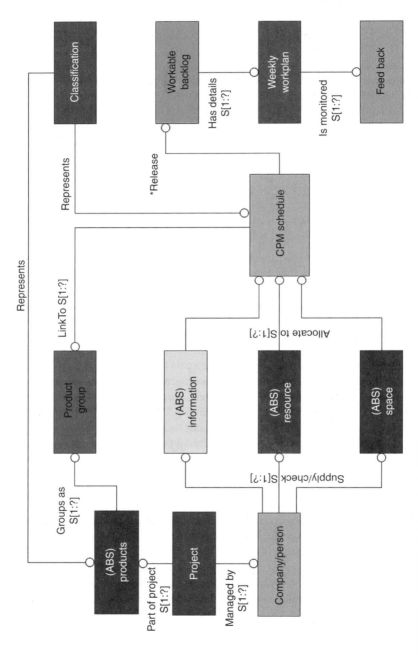

FIGURE 19.3   EXPRESS-G data model for LEWIS.

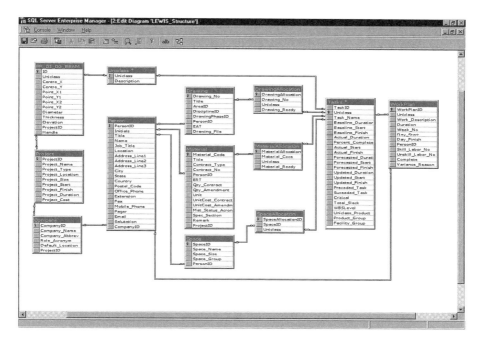

FIGURE 19.4   Abstract level of LEWIS database physical diagram.

The relationship "Allocate To S[1:?]" means that each piece of information can be allocated to one or more activities in the project schedule

5. *Resource*: the resource entity is modelled as an abstract supertype. Subtypes of resource include materials, equipments, tools, temporary supports, and labor. General attributes for each type of resource are resource code, title, responsible person, quantity, unit, unit cost, resource status, and ERT. The relationship "Allocate To S[1:?]" means that each resource can be allocated to one or more activities in the project schedule

6. *Space*: the space entity is modelled as an abstract supertype, which may include execution spaces, product spaces, and storage spaces and resources paths. General attributes for each type of spaces are space code, space name, size, group (a group/zone of spaces identified in product model), and responsible person. The relationship "Allocate To S[1:?]" means that each space can be allocated to one or more activities in the project schedule

7. *CPM schedule*: this entity represents all information that is related to project schedule. Examples of attributes for CPM schedule are WBS code, activity name, duration, start date, finish date, percent completion, predecessors, successors, criticality, and total float. It should be noted that information of various schedules; that is, baseline schedule, actual schedule, forecasted schedule, and updated schedule can be included. The relationship "*Release" means that each scheduled activity can be released and registered in the workable backlog where the activity is free from constraints. The presence of an asterisk * against a relation name indicates that an activity can be released

8. *Weekly work plan*: unlike CPM scheduling, weekly work planning is usually a simple list of tasks that are to be completed each week. The last planner (i.e., foreman) may select a number of activities from the workable backlog and break them down into a series of small manageable tasks that he or she is committed to perform in a particular week (represented by the relationship "Has Details S[1:?]"). Attributes of the weekly work plan may include task ID, task description, week number, day start, day finish, responsible person, and numbers of skilled and unskilled labor required

9. *Feedback*: this entity represents feedback of each weekly work plan. The feedback's attributes include completion status and reasons for not completed as planned. The relationship "is Monitored S[1:?]" means that many feedbacks can be given to a monitored weekly work plan

10. *Product group*: this entity represents groups of CAD products, such as a group of GF columns in gridline A–D. For the purpose of product-process integration and generation of 4D visualization, the relationship "Link To S[1:?]" is created to allow one or more products to be allocated to a scheduled activity

11. *Classification*: the Uniclass can be used to consistently classify products and processes. The relationship "Represents" means that either a product or a process must be represented by a unique Uniclass code.

19.5.3 Population of Product and Process Data Using Real Life Case Study

A real life case study has been used to populate LEWIS product and process database. The case study is composed of two wings primary school at value of £1.6 million. Figure 19.6 and Figure 19.7 show 3D of the building.

The two following subsections individually elaborate on preparation and population of product and process data.

19.5.3.1 Product Data

Product data refers to data that are related to the CAD model of a construction project. These data include product classification, dimensions, and coordinates. Once these data are populated into a database, they can be used to assist preparation of bill of quantities, purchase orders, and progress payments. Furthermore, the product data were recently used to automatically generate Virtual Reality Modelling Language (VRML) models and space objects; that is, execution spaces surrounding the product [Heesom and Mahdjoubi 2003; Mallasi and Dawood 2003].

In this chapter, an intelligent VBA macro called "DataExtractMan" was developed to automatically extract and populate 2D or 3D product data from CAD software (i.e., AutoCAD 2000 or Architectural Desktop 3.3). To utilize this macro, the CAD model must, however, be prepared according to a certain standard. The British Standard on structuring and exchange of CAD data [BS 1192-5 1998] was chosen for consistent organization of CAD layers. In addition, the CAD model must be drawn using objects or polylines. Broken lines and arcs are not acceptable formats that the macro can understand. Figure 19.5 provides an example of layering convention based on the BS 1192-5.

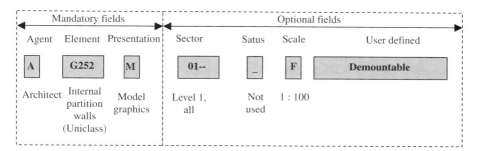

FIGURE 19.5 Example of layering convention based on the BS 1192-5.

The DataExtractMan interface allows users to extract and populate all the product data in a single click.

19.5.3.2 Process Data

Process data is referred to as data that are related to schedule and utilization of resource and execution space in a construction project. Execution spaces are regarded as resources in LEWIS and allocated to activities.

More elaboration on execution space is given later in this chapter. Examples of data attributes are process classification, activity name, duration, start and finish dates, predecessors, successors, criticality, total float, resource and space name, resource and space allocation units, and resource and space allocation period. Relationships among the process data, product data, and information and resource readiness data can be created and used for multiconstraint analysis and visualization.

The schedule data from project planning and scheduling software (i.e., MS Project or Primavera) is simply extracted to the database using Open Database Connectivity (ODBC) and a built-in import/export template feature.

19.5.4 LEWIS Interfaces and Data Input

Continuing from the development of database structure and initial population of product and process data, this subsection describes the design of menus and forms for viewing, inputting, and updating information through the Web browser. The subsection starts with the development of a main interface, which contains a number of menus for accessing forms and project information. Examples of forms and interfaces for accessing each type of information are then presented.

19.5.4.1 Main Interface

To achieve Web-based functionality, several Web programming languages including HTML (Hyper-Text Markup Language), DHTML (Dynamic HTML), ASP (Active Server Pages), Visual Basic Script, Java Script, and Microsoft Data Access Pages were employed. Figure 19.6 illustrates the main interface of LEWIS system.

As presented in Figure 19.6, LEWIS main interface contains a set of multilevel pulldown menus that enable direct access to different categories of project information.

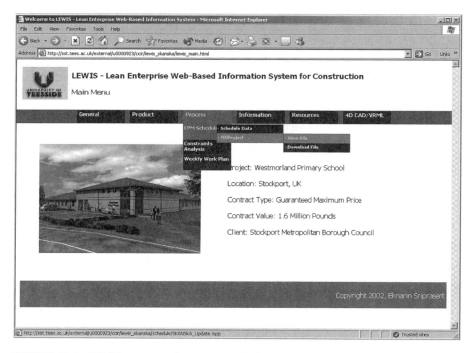

FIGURE 19.6   LEWIS main interfaces using real life case study.

These categories are ranging from general project information, product data, process data, information/documents, and resources, to visualization models presenting work progress and constraints in each period of construction.

19.5.4.2   General Project Information

This category of general project information covers project profile information, details and contacts of project participants, data dictionary (master tables), and calendar. The details of the menu and a description for general project information are listed in Table 19.1.

The menu listed in Table 19.1 allows users to access forms that contain data stored in the database tables. In this chapter, Microsoft Data Access Pages were used to design Web-based forms.

19.5.4.3   Product Information

This category of product information covers geometrical data of products, product classification (master table), and product models. The details of menu and description of product information are listed in Table 19.2.

Similar to the general project information, the menu listed in Table 19.2 allows users to access Web-based forms that contain data stored in the database tables. Figure 19.7 shows a view of the overall case study model.

TABLE 19.1 Menu and Description for General Project Information

Level	Menu	Description
0	General	
1	Project information	Stores project profile information, such as project name, type, location, size, duration, and cost.
1	Project participants	Keeps records of companies and staff that are participating in the project.
1	Data dictionary	
2	Area	Classifies project areas or zones.
2	Contract type	Used for procurement to identify types of contract such as direct purchasing or subcontracting.
2	Discipline	Used for documentation to categorize documents based on specific discipline (i.e., architecture, structure, civil, HVAC, landscape, electrical, mechanical, etc.).
2	Drawing phase	Used for drawing management to classify drawing phase (i.e., schematic document, issued for bid, issued for construction, approved for construction, as built, etc.).
2	Role	Classifies role of project participants (i.e., owner, designer, consultant, general contractor, construction manager, etc.).
2	Specification code	Provides references to an appropriate specifications section (i.e., 02220 for earthwork, 03100 for concrete formwork, 15400 for plumbing systems, etc.).
2	Document status	Identifies status of each document (i.e., open, closed, rejected, outstanding, all correct, etc.).
2	Material status	Identifies status of each material (i.e., purchase order sent, being manufactured, being transported, on site, used up, etc.).
1	Calendar	Simply presents calendar for general references.

TABLE 19.2 Menu and Description of Product Information

Level	Menu	Description
0	Product	
1	Architectural components	Provide a submenu that allows users to access data of a specific architectural component.
1	Building services	Provide a submenu that allows users to access data of a specific building services component.
1	Product Uniclass	List the Uniclass product code of all products in the database.
1	Model	
2	View file	
3	AutoCAD	To view the 3D model of the project in DWF (Drawing Web Format) using WHIP! plug-in.
3	IFC	To view the IFC STEP-Part 21 file of the project model on the Web (available only if the model is developed in IFC format).
2	Download file	
3	AutoCAD	To download the DWG file of project model.
3	IFC	To download the IFC file of project model.

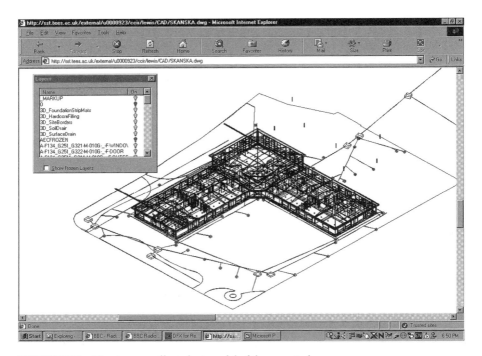

FIGURE 19.7   Viewing overall product model of the case study.

19.5.4.4   Process Information

The process information category covers scheduling information, analysis of physical, contract, information, and resources constraints, and weekly work plan. The details of menu and description for process information are listed in Table 19.3.

Similar to other types of information, users can access required process information directly from the menu listed in Table 19.3. Figure 19.8 and Figure 19.9 show the data access page for project schedule and viewing MS Project on the Web.

19.5.4.5   Information and Document

This category of information and document covers management of technical and administrative information. Technical information includes drawing, method statement, and safety assessment. Administrative information includes meeting minutes, change orders, request for information, memos, telephone records, and so on. It should be noted that features for managing administrative information are excluded in the development of LEWIS prototype since these features are generally available in most commercial packages. The details of menu and description for information and document are listed in Table 19.4.

19.5.4.6   Resources

This category of resources covers information for materials, labor, spaces, equipment, and temporary supports like scaffolding and temporary access. The details of menu and description for resources are listed in Table 19.5.

TABLE 19.3   Menu and Description for Process Information

Level	Menu	Description
0	Process	
1	CPM schedule	
2	Schedule data	Provides a form for accessing schedule data, which include list of activities, duration, start and finish dates, predecessors, successors, criticality, associated product groups, etc.
2	MS project	
3	View file	To view and edit MS Project schedule on the Web browser.
3	Download file	To download current version of MS Project schedule file.
1	Constraint analysis	(See details in subsection 5.4.6)
2	Physical constraints	
3	Prerequisite work	Provides completion status of prerequisite works of each activity.
3	Execution space	Calculates percent loading of execution spaces at each period.
2	Contract constraints	
3	Special agreement	Lists special agreements, such as milestones, restricted working time/areas, extension of time and cost, etc.
2	Information constraints	
3	Drawing	Identifies readiness status of drawings at each period.
3	Method statement	Identifies readiness status of method statements at each period.
3	Safety assessment	Identifies readiness status of safety assessment at each period.
2	Resource constraints	
3	Material	Identifies readiness status of materials at each period.
3	Equipment	Identifies readiness status of equipment at each period.
3	Temporary facility	Identifies readiness status of temporary facilities at each period.
2	All constraints	Presents readiness status of all constraints at each period.
2	Workable backlog	Lists scheduled activities that are free of constraints.
1	Weekly work plan	
2	Work plan data	Provides a form for weekly work planning. This form automatically picks up and allows only activities from the workable backlog to be further broken down into more details.
2	PPC	Presents a chart for weekly percent plan completion.
2	Reasons for variances	Presents a chart for reasons for variances.

19.5.4.7   Visualization Models

This category of visualization models covers video, VRML models, and presentation files for project visualization at different stages. The visualization models are prepared based on different schedules including baseline, actual, forecasted, and updated schedules. The details of menu and description for visualization models are listed in Table 19.6.

The menu listed in Table 19.6 provides users with direct access to various formats of visualization models including AVI (video format), VRML model, and Microsoft PowerPoint presentation file. Submenus are available for users to select a visualization model at a specific monitor date of construction.

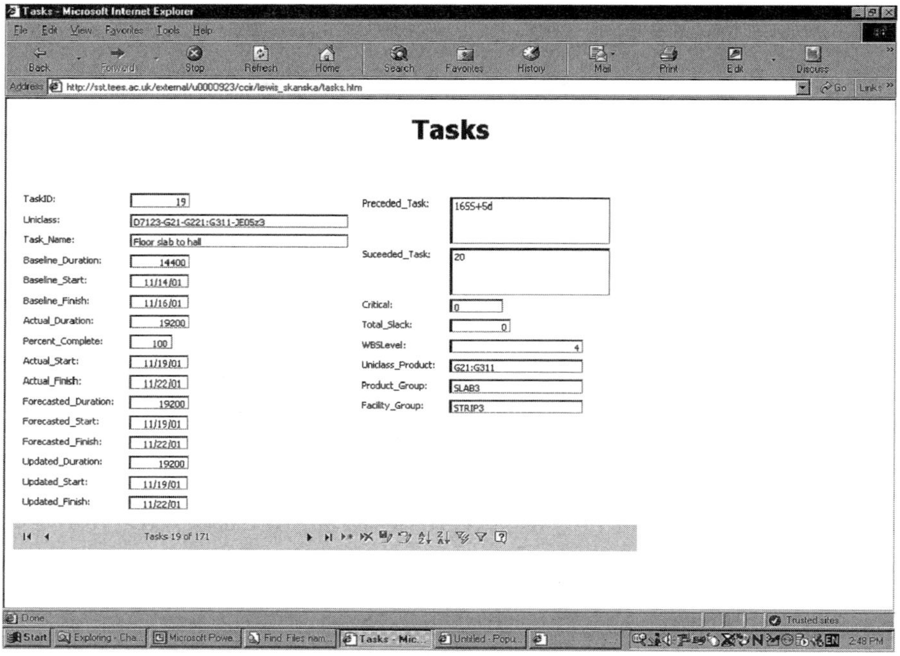

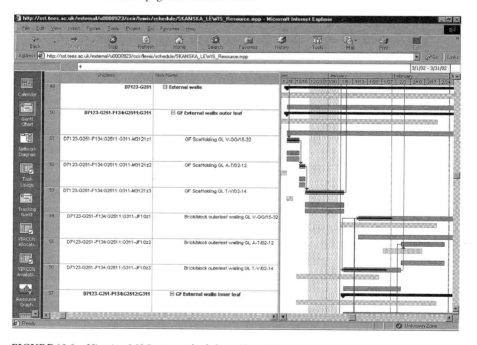

FIGURE 19.8   Data access pages for schedule data.

FIGURE 19.9   Viewing MS Project schedule on the Web.

TABLE 19.4 Menu and Description of Information and Document

Level	Menu	Description
0	Information	
1	Technical	
2	Drawing	
3	Revision	Lists details of all registered drawings and all of their revisions. Links are also provided to DWF drawings on the Web.
3	Allocation	Provides information about allocations of drawings to scheduled activities. Readiness status and problems of each allocation are also presented.
3	Distribution	Provides drawing distribution records, i.e., how many copies of which drawing has been sent when and to whom.
3	Status	Provides interactive query capability for tracking status of each drawing.
2	Method statement	
3	Allocation	Lists details of all registered method statements, their allocations to scheduled activities, as well as readiness status and problems of each allocation.
3	Status	Provides interactive query capability for tracking status of each method statement.
2	Safety assessment	
3	Allocation	Lists details of all registered safety assessments, their allocations to scheduled activities, as well as readiness status and problems of each allocation.
3	Status	Provides interactive query capability for tracking status of each safety assessment.
1	Administrative	Includes meeting minutes, change orders, request for information, memos, telephone records, etc. However, features for managing these types of information are excluded in the prototype development.

19.5.5   Multiconstraint Data Analysis

Once project participants update data in LEWIS as the construction progresses, look-ahead analysis of multiple constraints in each scheduled activity can be performed through the designed input forms on the Web. These multiple constraints may include completion status of prerequisite works, availability of labor and spaces, and readiness of information and materials. In this development, two approaches including: (1) a standard query using SQL and ASP; and (2) an interactive multidimensional query using OLAP were implemented.

19.5.5.1   Standard Query

Query is one of the most important features in every database management system. It allows users to generate a comprehensive view of data by combining data from various related tables and filtering them based on certain criteria. In this chapter, a number of queries were developed using "view design" functionality in the Microsoft SQL Server. Based on the multiconstraint planning and control concept, examples of these queries

TABLE 19.5 Menu and Description for Resources

Level	Menu	Description
0	Resources	
1	Material	
2	Delivery	Lists details of all registered materials and their delivery details, such as when and to what amount the materials have been delivered to site.
2	Allocation	Lists details of all registered materials, their allocations to scheduled activities, as well as readiness status and problems of each allocation.
2	Cost	Provides quantities and costs of each material.
2	Supplier	Provides supplier details of each material.
1	Labor	
2	Type and availability	Lists labor types and maximum units available at each period.
2	Utilization	Identifies a number of required resources against their availability at each period.
1	Space	
2	Zone and availability	Lists execution spaces/zones and their area sizes.
2	Utilization	Identifies a size of occupied spaces against their availability at each period (% loading).
1	Equipment	Includes equipment types, their availability and utilization at each period. Features for managing these types of information are, however, excluded in the prototype.
1	Temporary support	Includes types of temporary supports, their availability and utilization at each period. Features for managing these types of information are, however, excluded in the prototype.

are tracking status of individual or multiple constraints in each scheduled activity at a specific monitoring date.

Once the query has been generated, another feature called "Microsoft SQL Server Web Assistant" can be employed to quickly publish the query results on the Web. An advantage of using this feature is that a regular update schedule can be set for the result pages. Example of the query result is illustrated in Figure 19.10.

As seen in Figure 19.10, the example lists only activities that associated drawings, materials, prerequisite works, labours, or execution spaces are not ready for the activity to be executed. The bit data number 1 means "True" and number 0 means "False." For instance, drawing STEEL01 for Column GL12-32 is ready, while the Junior Wing, which is the execution space for this activity is overallocated. The <n/a> data indicates that documents or resources have not been allocated to the activities and their statuses have not been confirmed. It should be noted that advanced search pages can also be developed using ASP. These pages enables users to have full control over the set of constraints, period, and data fields that they want to see in the query results.

19.5.5.2 Interactive Multidimensional Query

Unlike the data warehouse, which is a huge collection of data used to support strategic decision-making process for the enterprise, the data mart is defined as "customized,

TABLE 19.6 Menu and Description of Visualization Models

Level	Menu	Description
0	Visualization	
1	Baseline	
2	4D CAD AVI	To download a prerecorded AVI file of a baseline 4D CAD model.
2	VRML model	To view a selected baseline VRML model at a specific monitor date using a Web-based VRML browser.
2	PowerPoint presentation	To download Microsoft PowerPoint file that contains precaptured screen shots of a baseline 4D CAD model.
1	Actual	
2	4D CAD AVI	To download a prerecorded AVI file of an actual 4D CAD model.
2	VRML model	To view a selected actual VRML model at a specific monitor date using a Web-based VRML browser.
2	PowerPoint presentation	To download Microsoft PowerPoint file that contains precaptured screen shots of an actual 4D CAD model.
1	Forecasted	
2	4D CAD AVI	To download a prerecorded AVI file of a forecasted 4D CAD model.
2	VRML Model	To view a selected forecasted VRML model at a specific monitor date using a Web-based VRML browser.
2	PowerPoint presentation	To download Microsoft PowerPoint file that contains precaptured screen shots of a forecasted 4D CAD model.
1	Updated	
2	4D CAD AVI	To download a prerecorded AVI file of an actual 4D CAD model.
2	VRML model	To view a selected updated VRML model at a specific monitor date using a Web-based VRML browser.
2	PowerPoint presentation	To download Microsoft PowerPoint file that contains precaptured screen shots of an updated 4D CAD model.

summarized data from the data warehouse tailored to support the specific analytical requirements of a given business unit" [Watterson, Shadish, and Wells 2000]. Apart from the standard query, the concept of data mart coupled with Microsoft OLAP tools was implemented. In the context of this chapter, this feature can provide decision makers with multidimensional and aggregated view of data; that is, readiness status of each constraint. Figure 19.11 illustrates the design of multidimensional query of drawing status using a cube editor view in the OLAP tools.

From Figure 19.11, the cube is designed based on a star schema. A star schema contains two types of tables—fact tables and dimension tables [Shumate 2000]. Fact tables contain the quantitative or factual data about a construction management entity. Dimension tables are smaller and hold descriptive data that reflect the dimensions of an entity. In this example, the "Drawing Allocation" table that contains the readiness status of drawing is designed as a fact table. This readiness status of drawing can be queried and filtered by a number of dimension tables, which are responsible person, area, discipline, allocated activity, and required date.

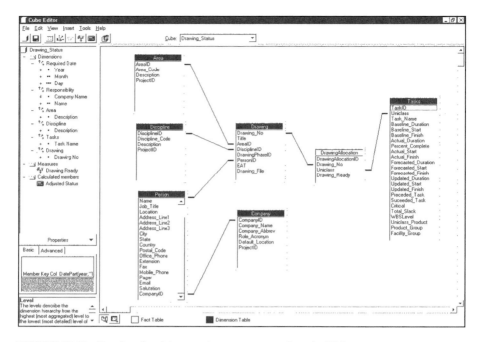

FIGURE 19.10   Results of multiconstraint query reported on the Web.

29	Column GL 12-32	STEEL01	1	P41512	1	20	1	Junior Wing	1
29	Column GL 12-32	STEEL02	1	P41512	1	20	1	Junior Wing	1
29	Column GL 12-32	STEEL03	1	P41512	1	20	1	Junior Wing	1
30	Column GL A-U	STEEL04	1	P41512	1	20	1	Infant Wing	1
30	Column GL A-U	STEEL01	1	P41512	1	20	1	Infant Wing	1
30	Column GL A-U	STEEL02	1	P41512	1	20	1	Infant Wing	1
31	Column to hall	STEEL01	1	P41512	1	20	1	Hall	1
31	Column to hall	STEEL02	1	P41512	1	20	1	Hall	1
31	Column to hall	STEEL05	1	P41512	1	20	1	Hall	1
33	Beam GL 12-32	STEEL01	1	P41513	1	29	1	Junior Wing	1
33	Beam GL 12-32	STEEL02	1	P41513	1	29	1	Junior Wing	1
33	Beam GL 12-32	STEEL06	1	P41513	1	29	1	Junior Wing	1
35	Beam to hall	STEEL01	1	P41513	0	31	1	Hall	1
35	Beam to hall	STEEL02	1	P41513	0	31	1	Hall	1
35	Beam to hall	STEEL00	0	P41513	0	31	1	Hall	1
35	Beam to hall	STEEL01	1	P41513	0	31	1	Steel Erector	0
35	Beam to hall	STEEL02	1	P41513	0	31	1	Steel Erector	0
35	Beam to hall	STEEL08	0	P41513	0	31	1	Steel Erector	0
39	Roof trusses to hall	STEEL11	0	<n/a>	<n/a>	35	0	Hall	1
39	Roof trusses to hall	STEEL01	1	<n/a>	<n/a>	35	0	Hall	1
39	Roof trusses to hall	STEEL02	1	<n/a>	<n/a>	35	0	Hall	1
39	Roof trusses to hall	STEEL11	0	<n/a>	<n/a>	35	0	Steel Erector	0

FIGURE 19.11   Cube editor view for multidimensional query of drawing status.

19.5.6 Work-Face Information

Three categories of work-face information including commitment planning, work-face instruction, and feedback are demonstrated in the following sections.

19.5.6.1 Commitment Planning

Commitment planning is a key to shield operation from constraints and improve plan reliability. To generate a work plan, the last planner (foreman or job superintendent) can simply add subactivities under a constraint-free activity (workable backlog). Other details, such as week number, day start, and day finish can then be added using drop-down lists.

19.5.6.2 Work-Face Instruction

Information such as drawings, specifications, method statements, work instructions, and testing regulations are needed for the work-face operation [De la Garza and Howitt 1998]. Based on a recent trial, site staff are more than ready to adopt mobile technology to assist their works. The development of wireless network cells (WNCs) has been shown to be viable, with the potential to provide a portable and scalable solution to the implementation of mobile networks for real-time data access and capture in the site environment.

19.5.6.3 Feedback

Based on the lean construction concept and the last planner method, weekly PPC, which is the number of finished activity over the number of planned activity, is regarded as a new key performance indicator for project control. The more the PPC obtained, the less the inflow variability is on construction site. Figure 19.12 shows the percent plan completion for the first 10 weeks of the case study project.

19.5.7 Conclusions from the Case Study

The integrated decision support system for multiconstraint planning and control introduced in this chapter was evaluated using real case study data of £1.6 million. The evaluation was categorized into six main phases: (1) preparation of preliminary data input, (2) generating initial multiconstraint model, (3) evaluating the original plan, (4) reestablishing baseline plan, (5) monitoring and forecasting plan, and (6) updating plan.

Three dimensions were considered for the system evaluation in each phase. Firstly, the system was verified whether it is functioning correctly as intended. Secondly, the system was validated through diagnosis of its benefits over other available systems. Thirdly, the system was evaluated in terms of its usability and barriers for implementation.

The system can greatly support planners to establish and maintain constraint-free optimum plans throughout the course of construction. To be more specific, three major advantages of the developed system over other currently available systems can be summarized as follows:

1. Multiconstraint information management feature enables capturing and incorporating constraint information regarding logistics of information and resources into the main contractor's schedule. Once the constraints are analyzed, the system

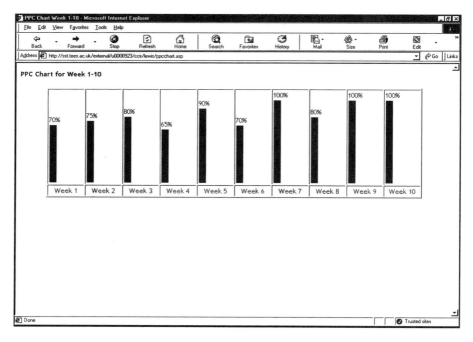

FIGURE 19.12   Percent plan completion chart for week 1–10 of the project.

manages the workable backlog and facilitates weekly work planning at work face operation level. Furthermore, the system acts as a central project repository where project participants can query up-to-date information and statuses of the project

2. Multiconstraint visualization feature enhances planner's ability to detect hidden problems in the CPM schedules. Based on the case study, invited evaluators could detect a large number of problems regarding illogical dependencies, site space congestion, and nonreadiness of information and resources. Furthermore, the system can be utilized to confirm the constraint-free schedule to project participants

3. Multiconstraint optimization feature enables generation of optimum constraint-free schedules by changing priority and construction method options of each activity. Based on the case study, the system could reduce the project duration from 294 days to 247 days or approximately 16%, while maintaining technological dependencies and constraint-free condition

Although the benefits of the system have been realized, a few issues about system implementation were raised by the invited evaluators. The most important one is relating to preparation of preliminary data input. It is widely acknowledged that many data, such as 3D CAD model and space allocations are not normally available in typical projects. Therefore, considerable efforts must be spent to generate and organize the data in compliance with certain standards, such as Uniclass and BS1192-5. It is argued that the use of 2.5D modelling approach, in which the 2D drawings are simply extruded up, would require much less efforts.

Another major barrier lies in the process of monitoring information and resource readiness due to the difficulty to make all upstream supportive organizations including designers, suppliers, and subcontractors subscribe to the LEWIS system. Fortunately, substantial research projects considering data standardization have been initiated, and could have potential to minimize these barriers for implementation. So far, two different schools in this area have emerged. One is supporting a view of implementing a universal data standard, such as industry foundation class (IFC), while another one is supporting a view of having a mediator to negotiate discrepancy among heterogeneous data sources. However, since the work on IFC has focussed on product models while limited process extensions have been developed, it seems more likely that firms will maintain the use of legacy applications/databases for process data. As a result, the importance of a research on integrating the LEWIS system with a mediator system (SEEK) has been realized and funded by the US government.

If all the major barriers can be removed in the future, it is envisaged that successful implementation of the developed system will enable generation of reliable plans and constraint-free execution assignments, in turn, reducing production risks and improving on-site productivity.

19.6   Summary

The emergence of the Internet has offered a new standard and cost-effective way for communication in construction project management. To enhance the vision for applications of the Web-based systems, this research has extended the basic capability of existing systems from being documentation management and communication tools, to become a DSS for the multiconstraint planning and control. The implementation of the proposed vision has resulted in a prototype called LEWIS. LEWIS acts as a backbone information infrastructure that enables project participants to gather various constraint statuses, perform look-ahead analysis, and generate constraint-free workable backlog for weekly work planning at the work face. In addition, it is designed as a platform for integration among planning system, multiconstraint visualization system, and multiconstraint optimization system.

Regarding LEWIS development, the Uniclass was employed as a standard for structuring product and process data in LEWIS database. The main reason for choosing the Uniclass is due to the insufficiency and immaturity of other standards, such as IFC and OCCS, in dealing with multiple constraints in construction processes. Given the appropriate data structure, a conceptual data model that presents relationships between different data types was designed using EXPRESS-G. This conceptual data model was then translated into a database physical diagram using Microsoft SQL Server. Product data from AutoCAD and process data from Microsoft Project were populated into LEWIS database using the customized macros and preidentified import/export maps. User-friendly, Web-based interfaces were also developed to facilitate capturing readiness status of information and resources, multiconstraint analysis, and weekly work planning. A real life case study has been used to validate LEWIS, and it was concluded that the system could reduce the project duration from 294 days to 247 days or approximately 16%, while maintaining technological dependencies and constraint-free condition. Limitations and issues with LEWIS were identified and documented.

References

Abdelsayed, M., and Navon, R. 1999. An information sharing, internet-based system for project control. *Civil Engineering and Environmental Systems*, 16 (3), 211–33.

Alavi, M. 1984. An assessment of the prototyping approach to information systems development. *Communications of the ACM*, 27 (6), 556–63.

Alshawi, M., and Ingirige, B. 2003. Web-enabled project management: An emerging paradigm in construction. *Automation in Construction*, 12, 349–64.

Amor, R., and Faraj, I. 2001. Misconceptions about integrated project database. *Electronic Journal of Information Technology in Construction*, 6, 57–68.

Anumba, C. J., Bouchlaghem, N. M., and Whyte, J. 2000. Perspectives on an integrated construction project model. *International Journal of Cooperative Information Systems*, 9 (3), 283–313.

Aouad, G., Sun, M., and Faraj, I. 2002. Automatic generation of data representations for construction applications. *Construction Innovation*, 2, 151–65.

Ballard, G. 1997. Lookahead planning: The missing link in production control. *Proceedings of the Fifth Annual Conference of the International Group for Lean Construction*, Berkeley, CA.

Boar, B. 1984. *Application prototyping: A requirements definition strategy for the 80s.* New York: John Wiley.

BS 1192-5. 1998. *Construction drawing practice: Guide for the structuring and exchange of CAD data.* London: British Standards Institute.

Chan, W.-T., Chua, D. K. H., and Lang, X. 1999. Collaborative scheduling over the internet. *Computer-Aided Civil and Infrastructure Engineering*, 14 (1), 15–24.

Chau, K. W., Cao, Y., Anson, M., and Zhang, J. 2002. Application of data warehouse and decision support system in construction management. *Automation in Construction*, 12, 213–24.

Choo, H. J., Tommelein, I. D., Ballard, G., and Zabelle, T. R. 1999. WorkPlan: Constraint-based database for work package scheduling. *Journal of Construction Engineering and Management*, 125 (3), 151–60.

Connell, J., and Shafer, L. B. 1989. *Structured rapid prototyping.* Englewood Cliffs, NJ: Prentice-Hall.

Crawford, M., Cann, J., and O'Leary, R. 1997. *Uniclass (unified classification for the construction industry).* London: Royal Institute of British Architects Publications (RIBA).

Davis, G. B., and Olson, M. H. 1985. *Management information systems conceptual foundations, structure and development*, 2nd ed. New York: McGraw-Hill Inc.

Dawood, N., Sriprasert, E., Mallasi, Z., and Hobbs, B. 2002. Product and process integration for 4D visualisation at construction site level: A Uniclass-driven approach. *Proceedings of the Fourth European Conference on Product and Process Modelling in the Building and Related Industries (ECPPM)*, Portorož, Slovenia, 409–16.

Dawood, N., Sriprasert, E., Mallasi, Z., and Scott, D. 2003. An industrial evaluation of the virtual construction site (VIRCON) tools. *Proceedings of CIB W78 Conference*, Auckland, NZ.

Deng, Z. M., Li, H., Tam, C. M., Shen, Q. P., and Love, P. E. D. 2001. An application of the Internet-based project management system. *Automation in Construction*, 10, 239–46.

Ekholm, A. 1996. A conceptual framework for classification of construction works. *Electronic Journal of Information Technology in Construction*, 1, 1–25.

Gupta, A. 1989. *Integration of information systems: Bridging heterogeneous databases.* New York: IEEE Press.

Hajjar, D., and AbouRizk, S. M. 2000. Integrating document management with project and company data. *Journal of Computing in Civil Engineering*, 14 (1), 70–77.

Heesom, D., and Mahdjoubi, L. 2003. Visualisation development. *VIRCON project report.* University of Wolverhampton, UK.

Heindel, L. E., and Kasten, V. A. 1997. P++: A prototype PC-based enterprise management system. *International Journal of Project Management*, 15 (1), 1–4.

Houghton Mifflin Company. 2000. *The American heritage dictionary of the English language*, 4th ed. Boston, MA: Houghton Mifflin Company.

Howard, H. C., Levitt, R. E., Paulson, B. C., Pohl, J. G., and Tatum, C. B. 1989. Computer integration: Reducing fragmentation in AEC industry. *Journal of Computing in Civil Engineering*, 3 (1), 18–32.

Isensee, S., and Rudd, J. 1996. *The art of rapid prototyping: User interface design for Windows™ and OS/2®.* London: International Thomson Computer Press.

ISO/DIS 10303-11. 1994. *Industrial automation systems and integration—product data representation and exchange—part 11: Description methods: The EXPRESS language reference manual.* Geneva, Switzerland: International Organization for Standardization.

Johnson, R. E., and Clayton, M. J. 1998. The impact of information technology in design and construction: The owner's perspective. *Automation in Construction*, 8, 3–14.

Kang, L. S., and Paulson, B. C. 2000. Information classification for civil engineering projects by Uniclass. *Journal of Construction Engineering and Management*, 126 (2), 158–67.

Lantz, K. E. 1986. *The prototyping methodology.* Englewood Cliffs, NJ: Prentice-Hall.

Mallasi, Z., and Dawood, N. 2003. Development of VRML visualization with AutoCAD. *Proceedings of Construction Applications of Virtual Reality Conference (CONVR)*, Blacksburg, VA, USA, 24–26 September, 122–33.

McMonnies, A. 2001. *Visual basic: An object oriented approach.* Harlow, UK: Addison-Wesley.

Mohamed, S., and Stewart, R. A. 2003. An empirical investigation of user's perceptions of web-based communication on a construction project. *Automation in Construction*, 12, 43–53.

O'Brien, M., and Al-Soufi, A. 1993. Electronic data interchange and the structure of the UK construction industry. *Construction Management and Economics*, 11 (6), 443–53.

O'Brien, W. J., and Hammer, J. 2002. Robust mediation of supply chain information. *ASCE Specialty Conference on Fully Integrated and Automated Project Processes (FIAPP) in Civil Engineering*, Blacksburg, VA, 415–25.

O'Brien, W. J., Issa, R. R. A., Hammer, J., Schamalz, M. S., Geunes, J., and Bai, S. X. 2002. SEEK: Accomplishing enterprise information integration across heterogeneous sources. *Electronic Journal of Information Technology in Construction*, (Special Issue: ICT for Knowledge Management in Construction), 101–24.

O'Connor, J. T., and Dodd, S. C. 2000. Achieving integration on capital projects with enterprise resource planning systems. *Automation in Construction*, 9, 515–24.

OCCS Development Committee. 2001. *The overall construction classification system: A strategy for classifying the built environment*. Preliminary draft for review and comment. http://www.occsnet.org (accessed November 2003).

Pakanen, J. E., Möttönen, V. J., Hyytinen, M. J., Ruonansuu, H. A., and Törmäkangas, K. K. 2001. A web-based information system for diagnosing, servicing and operating heating systems. *Electronic Journal of Information Technology in Construction*, 6, 45–56.

Penã-Mora, F., and Dwivedi, G. H. 2002. Multiple device collaborative and real time analysis system for project management in civil engineering. *Journal of Computing in Civil Engineering*, 16 (1), 23–38.

Rojas, E. M., and Songer, A. D. 1999. Web-centric systems: A new paradigm for collaborative engineering. *Journal of Management in Engineering*, 15 (1), 39–45.

Schenck, D., and Wilson, P. 1994. *Information modelling the EXPRESS Way*. Oxford: Oxford University Press.

Shumate, J. 2000. *A practical guide to Microsoft OLAP server*. Upper Saddle River, NJ: Addison-Wesley.

Song, Y., Clayton, M. J., and Johnson, R. E. 2002. Anticipating reuse: Documenting buildings for operations using web technology. *Automation in Construction*, 11, 185–97.

Sriprasert, E., and Dawood, N. 2001. Potential of integrated digital technologies for construction work-face instruction. *Proceedings of AVRII and CONVR 2001*, Chalmers, Gothenburg, Sweden, 136–45.

Tam, C. M. 1999. Use of the Internet to enhance construction communication: Total information transfer system. *International Journal of Project Management*, 17 (2), 107–11.

Turban, E. 1995. *Decision support and expert systems: Management support systems*, 4th ed. Englewood Cliffs, NJ: Prentice-Hall.

Turk, Z. 2001. Phenomenological foundations of conceptual product modelling in architecture, engineering, and construction. *Artificial Intelligence in Engineering*, 15 (2), 83–92.

Ward, M. J., Thorpe, A., and Price, A. D. F. 2003. SHERPA: Mobile wireless data capture for piling works. *Computer-Aided Civil and Infrastructure Engineering*, 18, 299–312.

Watson, A., and Underwood, J. 2002. ProCure – a review of construction-related information technology. *Proceedings of Institute of Civil Engineering*, 150 (6), 18–23.

Watterson, K., Shadish, B., and Wells, G. 2000. *10 projects you can do with Microsoft®SQL Server™ 7*. New York: John Wiley & Sons, Inc.

Yamazaki, Y., and Maeda, J. 1998. The SMART system: An integrated application of automation and information technology in production process. *Computers in Industry*, 35, 87–99.

Zamanian, M. K., and Pittman, J. H. 1999. A software industry perspective on AEC information models for distributed collaboration. *Automation in Construction*, 8 (3), 237–48.

Zhu, Y., Issa, R. A., and Cox, R. F. 2001. Web-based construction document processing via malleable frame. *Journal of Computing in Civil Engineering*, 15 (3), 157–69.

III

Commentary

Richard H.F. Jackson
FIATECH

R EADING THESE CHAPTERS WAS BOTH EDUCATIONAL and heartening
for me. I learned much more about current research interests of some of the
industry's leading academics, and I discovered that their interests and future
vision match closely to the needs and the key challenges that the members of FIATECH
have expressed. Below, I discuss the efforts of FIATECH in the context of the chapters
in this section.

FIATECH is an industry-based consortium whose mission is to provide global
leadership in identifying and accelerating the development, demonstration, and
deployment of fully integrated and automated technologies to deliver the highest
business value throughout the life cycle of all types of capital projects. In achieving
this mission, FIATECH has developed a "Capital Projects Technology Roadmap"
(CPTR) to ensure right technologies are developed and in the order that delivers
the highest business value across all phases and processes of the capital project life
cycle. The CPTR is available at www.fiatech.org and is encompassed in the graphic
(Figure 1).

The central role of automation and integration technologies envisioned in the CPTR
and illustrated in Figure 1 should come as no surprise to readers of this book. Technology is the engine of economic growth: it transforms economies and the nature of
business competition. It is an increasingly important source of leverage. Being first
to discover or invent, and first to apply effectively, leads to crucial advantages. These
advantages fundamentally change the ecology of business, redefine the nature of
competition, and place a high premium on cooperation. Advances in technology and
improvements in performance—in other words, innovation and execution—are integral to what the capital projects (construction and building) industry can accomplish
in the future.

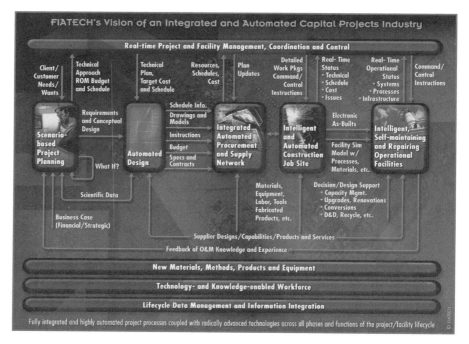

FIGURE 1 The FIATECH CPTR is composed of nine elements addressing all aspects of the automation and integration supporting capital projects across their life cycle.

In looking to the future of this industry, there are challenges on many fronts:

Competition due to globalization and offshoring
Demand for better quality, faster, and less costly construction
Climate change: sustainability and environmental security
Renewal of aging physical infrastructure
Homeland security and disaster loss reduction

Even for an industry whose global investment tops $3 trillion, these challenges are too complex for any one company, and in some cases, any one country to overcome. They have taken on global dimensions. Moreover, the basis of competition has broadened from price, to quality, to time-to-market, to innovation, to design, all the way to the dexterity and capabilities of supply chains.

Early in 2007, FIATECH convened a group of industry leaders to consider the challenges of the next few years, and to lay out a new focus for FIATECH efforts. The result was a strategic plan for FIATECH, entitled FIATECH Outlook 2007–2010, published in May 2007, and available also at www.fiatech.org. In that document, we acknowledged the increasing importance of the supply chain and managing all things related to it. This increased focus on the supply chain in FIATECH projects includes activities addressing improved information flow as well as increased use of wireless sensors to gather better information. It also includes activities to accelerate the development and deployment of standards for streamlining access to all that information. These initiatives are closely linked to several chapters in this section.

For example, in our Smart Chips project, we investigate commercial-ready and near-ready radio-frequency identification (RFID) and sensor technologies that can be easily adapted for field construction, and operations and maintenance applications, in areas ranging from material control to structural strength determination. We have planned, resourced, and completed eight in-field pilot tests. The chapter by Kiziltas et al. provides an excellent overview of the technologies used in the field to improve performance in the areas of tracking and locating resources, compliance, and progress monitoring. FIATECH fully expects these technologies for improving field productivity will rapidly mature and be commonly deployed. Although not expressly linked to common metrics of supply chain performance such as productivity, it is important to note that these field technologies have applications to safety.

Another project, Automating Equipment Information Exchange (AEX), is developing, demonstrating, and deploying eXtensible Markup Language (XML) specifications to automate information exchange for the design, procurement, delivery, operation, and maintenance of engineered equipment of all types. It has delivered XML specifications, detailed object information models, and example files for a wide range of equipment, from pumps, to valves, to heat exchangers, and will deliver many more in the future.

In our Accelerating the Deployment of ISO 15926 (ADI) project, we delivered a standard for deploying a building information model (BIM) for the process industry. We worked with other industry associations and the International Standards Organization to accelerate the development and deployment of the ISO 15926 standard for the integration of life-cycle data for process plants, and developed a tool to deploy it rapidly. The result has been the creation of a functional Work-in-Progress (WIP) database that companies can access at no cost. This project represents the first really new idea in standards development since standards bodies began working on software interoperability about 30 years ago. No one else has ever been able to work so closely and so quickly with these bodies to develop and deploy a useable WIP version of a standard in this way.

The AEX and ADI projects are specific initiatives that complement descriptions of the deployment of standards in the chapters by Cutting-Decelle et al. and Dawood. Cutting-Decelle et al. describe the role of data standards and a methodology for evaluating them that provides a useful adjunct to the bottom-up approach behind the AEX project and support the rapid development of standards in the ADI project. The challenges of data integration are not limited to development of a common language; as in any communication problem, context matters. The chapter by Dawood clearly underscores this by showing through a case study the close connection between data standards and application development and deployment. The complexity and scale of capital projects makes it difficult to generate a unifying data standard. Approaches in the past, as Dawood shows, have been most fruitful around specific problems and limited groups of firms that can band together around a specific but common issue. The present and future, however, are and will be vastly different. There are several key emerging standards that affect the construction and buildings industry, and FIATECH and several other organizations have reached agreement on how to tie these together, and create an interoperability platform for interoperability standards using the ISO 15926 WIP as the mass collaboration tool to get the job done. Much of this structure is already in place; much is yet to be done. Opportunities and challenges in this area still remain.

The chapter by Alarcón et al. presents a case study of the application of an e-marketplace for bidding and procurement in Chile. This chapter presents a hopeful message in several aspects. First, it shows the range of application of information technologies to support early stages of the project life cycle. Second, it shows that the benefits obtained—and implementation challenges—cross cultures. Third, it is a project that sits outside the scope of FIATECH projects to-date. This is not a bad thing; the industry needs and opportunities are too big for any single organization to address. That we can share and learn from global successes is important; the challenge will be to learn how to translate these successes. Here, the link between information technology development and organization research can be seen as an explicit need.

Finally, the chapter by Vaidyanathan provides both an excellent overview of the possibilities for information technology to support the supply chain (in a manner that complements the FIATECH CPTR), and a comparison of manufacturing and construction supply chain applications. This comparison is perhaps most instructive as it provides concrete examples of what has been accomplished in other industries and, by comparison, how much the capital projects industry can benefit from advanced supply chain technologies. The role of software for strategic and tactical decision support that is now common in manufacturing has not yet been realized in construction. The developments in data standards, field technologies, supporting CAD analysis provides a solid foundation for such advanced applications.

Author Index

Subject Index